Selected Titles in This Series

Volume

11 Colleen D. Cutler and Daniel T. Kaplan, Editors
Nonlinear dynamics and time series: Building a bridge between the natural and statistical sciences
1997

10 Jerrold E. Marsden, George W. Patrick, and William F. Shadwick, Editors
Integration algorithms and classical mechanics
1996

9 W. H. Kliemann, W. F. Langford, and N. S. Namachchivaya, Editors
Nonlinear dynamics and stochastic mechanics
1996

8 Larry M. Bates and David L. Rod, Editors
Conservative systems and quantum chaos
1996

7 William F. Shadwick, Perinkulam Sambamurthy Krishnaprasad, and Tudor Stefan Ratiu, Editors
Mechanics day
1996

6 Anna T. Lawniczak and Raymond Kapral, Editors
Pattern formation and lattice gas automata
1996

5 John Chadam, Martin Golubitsky, William Langford, and Brian Wetton, Editors
Pattern formation: Symmetry methods and applications
1996

4 William F. Langford and Wayne Nagata, Editors
Normal forms and homoclinic chaos
1995

3 Anthony Bloch, Editor
Hamiltonian and gradient flows, algorithms and control
1994

2 K. A. Morris, Editor
Control of flexible structures
1993

1 Michael J. Enos, Editor
Dynamics and control of mechanical systems: The falling cat and related problems
1993

Nonlinear Dynamics and Time Series

Fields Institute Communications

The Fields Institute for Research in Mathematical Sciences

Nonlinear Dynamics and Time Series

Building a Bridge between the Natural and Statistical Sciences

Colleen D. Cutler
Daniel T. Kaplan
Editors

American Mathematical Society
Providence, Rhode Island

The Fields Institute for Research in Mathematical Sciences

The Fields Institute is named in honour of the Canadian mathematician John Charles Fields (1863–1932). Fields was a visionary who received many honours for his scientific work, including election to the Royal Society of Canada in 1909 and to the Royal Society of London in 1913. Among other accomplishments in the service of the international mathematics community, Fields was responsible for establishing the world's most prestigious prize for mathematics research—the Fields Medal.

The Fields Institute for Research in Mathematical Sciences is supported by grants from the Ontario Ministry of Education and Training and the Natural Sciences and Engineering Research Council of Canada. The Institute is sponsored by McMaster University, the University of Toronto, the University of Waterloo, and York University and has affiliated universities in Ontario and across Canada.

2000 *Mathematics Subject Classification.* Primary 62M10, 58-XX; Secondary 62M20.

Library of Congress Cataloging-in-Publication Data

Nonlinear dynamics and time series : building a bridge between the natural and statistical sciences / Colleen D. Cutler, Daniel T. Kaplan, editors.

p. cm.—(Fields Institute communications, ISSN 1069-5265 ; 11)

Includes bibliographical references.

ISBN 0-8218-0521-5 (hard cover : alk. paper); ISBN 0-8218-4185-8 (soft cover : alk. paper)

1. Time-series analysis—Congresses. 2. Nonlinear theories—Congresses. 3. Dynamics—Congresses. I. Cutler, Colleen D. (Colleen Diane), 1957– . II. Kaplan, Daniel. III. Series: Fields Institute communications ; v. 11.

QA280.N64 1997

519.5′5—dc21

96-44411

CIP

∞ The paper used in this book is acid-free and falls within the guidelines
established to ensure permanence and durability.
This publication was prepared by the Fields Institute.
`http://www.fields.utoronto.ca`
Visit the AMS home page at `http://www.ams.org/`

10 9 8 7 6 5 4 3 2   11 10 09 08 07 06

Contents

Preface vii

Opening Lectures

Tools for the Analysis of Chaotic Data 1
HENRY D. I. ABARBANEL

Some Comments on Nonlinear Time Series Analysis 17
HOWELL TONG

Embeddings, Dimension, and System Reconstruction

A General Approach to Predictive and Fractal Scaling Dimensions in Discrete-Index Time Series 29
COLLEEN D. CUTLER

Statistics for Continuity and Differentiability: An Application to Attractor Reconstruction from Time Series 49
LOUIS M. PECORA, THOMAS L. CARROLL and JAMES F. HEAGY

Reconstruction of Integrate-and-Fire Dynamics 63
TIM SAUER

Surrogate Data Methodology

On the Validity of the Method of Surrogate Data 77
KUNG-SIK CHAN

Using "Surrogate Surrogate Data" to Calibrate the Actual Rate of False Positives in Tests for Nonlinearity in Time Series 99
JAMES THEILER and DEAN PRICHARD

Local Lyapunov Exponents

Chaos with Confidence: Asymptotics and Applications of Local Lyapunov Exponents 115
BARBARA A. BAILEY, STEPHEN ELLNER and DOUGLAS W. NYCHKA

Estimating Local Lyapunov Exponents 135
ZHAN-QIAN LU and RICHARD L. SMITH

Long-Range Dependence

Defining and Measuring Long-Range Dependence 153
PETER HALL

Modelling Nonlinearity and Long Memory in Time Series 161
PETER M. ROBINSON and PAOLO ZAFFARONI

Data Analysis and Applications

Ergodic Distributions of Random Dynamical Systems 171
L. MARK BERLINER, STEVEN N. MACEACHERN and CATHERINE SCIPIONE FORBES

Detecting Structure in Noise 187
LISA BORLAND

Characterizing Nonlinearity in Weather and Epilepsy Data: A Personal View 201
MARTIN C. CASDAGLI

Assessment of Linear and Nonlinear Correlations Between Neural Firing Events 223
ANDRÉ LONGTIN and DANIEL M. RACICOT

Markov Chain Methods in the Analysis of Heart Rate Variability 241
STEPHEN J. MERRILL and JOHN R. COCHRAN

Preface

The papers collected in this volume reflect an attempt to bring together researchers in two fields: Nonlinear Dynamics, and Statistics, with the goal of providing an exchange and creating a link between two diverse groups with a common interest in the analysis of nonlinear time series data.

The workshop in which these papers originated was entitled "Nonlinear Dynamics and Time Series: Building a Bridge Between the Natural and Statistical Sciences," and was held at the Centre de Recherches Mathématiques (CRM) in Montréal, Canada in July 1995. The workshop was a joint venture funded partly by the CRM and partly by The Fields Institute for Research in the Mathematical Sciences. It followed in the footsteps of recent similar interdisciplinary meetings, such as the one-day meeting on chaos organized by the Royal Statistical Society in 1991, and the two-day meeting on chaos and forecasting sponsored by the Royal Society of London in 1994.

With ideas from nonlinear dynamics being applied more and more to real-world problems, it has become of increasing importance to develop and understand statistical methodology designed specifically for nonlinear and dynamical approaches to time series. And as nonlinear dynamicists attempt to cope with the limitations of noisy real-world data, the strategies and expertise of statisticians become of growing relevance and significance to them.

For statisticians, nonlinear dynamics offers new perspectives on the modelling and analyses of time series data, and has provided an influx of unfamiliar terminology, such as chaos, sensitivity to initial conditions, attractors, fractal dimension, embedding dimension, $1/f$ noise, Lyapunov exponents, and bifurcations. Ideas from nonlinear dynamics stand to make an impact on the way in which statisticians view data, while at the same time presenting a wealth of open statistical problems to which statisticians can contribute their expertise, thereby providing a substantial return to the dynamics community.

Dynamicists and statisticians often need access to the concepts and tools developed in each discipline, but it can be difficult to cross the terminological and conceptual gulf that separates the two fields. We felt that by bringing together workers in both fields, charged with the task of making their own work accessible to those from the other side, we could lay the foundations for a bridge. For those who participated in the workshop, we feel that this ambition was fulfilled.

The normal mode of scientific communication, that of publication in peer-reviewed journals, is not always well suited to maintaining communication between fields. It can be daunting to pick up a journal from another field; the vocabulary and methods are often unfamiliar, and the motivation behind a particular approach (or even the significance and meaning of the problem being investigated) may be mysterious and obscure to the uninitiated. This problem is exacerbated by the fact that introductory material, essential to outsiders, is often suppressed in the name of conciseness and the desire to emphasize new results. It is even more difficult to write a paper for a journal in a different field; the reviewers may not be sympathetic to or cognizant of the methods and motivation behind the paper, or the style may simply jar them.

In reviewing the papers presented here, the editors and the anonymous peer-reviewers have attempted to minimize these problems. The result, we believe, is a collection of interrelated papers that highlight current areas of research in statistics that might have particular applicability to nonlinear dynamics, and new methodology and open data analysis problems in nonlinear dynamics that might find their way into the toolkits and research interests of statisticians.

We are of course very grateful for the funding, from both The Fields Institute and CRM, that made this workshop and volume possible. We thank John Chadam, then director of The Fields Institute, for providing the impetus to organize the workshop, and also the support staff, particularly Judy Motts of Fields and Louis Pelletier of CRM, for their excellent organizational work. A special note of thanks must go to the publications assistant Erna Unrau of The Fields Institute, who was responsible for getting this volume into the final form you see before you now, and who offered much sage advice throughout the publication process, not the least of which was "Relax, calm down, have a doughnut."

Finally, and most importantly, we thank the participants and authors themselves, for contributing their time, talents, and enthusiasm to making the workshop a great success.

Colleen Cutler
Danny Kaplan

August 1996

Fields Institute Communications
Volume **11**, 1997

Tools for the Analysis of Chaotic Data

Henry D. I. Abarbanel*
Department of Physics
and
Marine Physical Laboratory
Scripps Institution of Oceanography
University of California, San Diego
Mail Code 0402
La Jolla, CA, USA 92093-0402

Abstract. As a class of time series, chaos lies between regular periodic or quasiperiodic signals and the unpredictable, irregular signals we call "noise". Chaos has structure in state space and is predictable in the short run. We have developed a set of useful, computationally efficient tools for the analysis of temporal signals, chaotic as well as regular, in the appropriate multivariate time domain. These tools are geometric as well as information theoretic. The tools allow one to determine the global dimension in which to "unfold" the observed time series, which is viewed as a projection from the underlying attractor associated with the signal source. Using the tools we also determine the local or dynamical dimension of the signal source; this adds predictability or a dynamical criterion to the geometric ideas. In the space where the attractor is unfolded, we can use the tools to determine the Lyapunov exponent spectrum for the signal source as well as the fractal dimensions of the attractor, if for some reason those are desired. Predictive models on the attractor can be made using the observed information in state space.

These tools have been used for the analysis of real world data, from machining chatter, nonlinear circuits, and laser operation, to various geophysical data sets. Examples will be given from these data sets.

This paper will also try to identify those parts of the tool kit which are amenable to serious statistical analyses—on the whole, that analysis has not been done.

The tools described here are available as **TOOLS FOR DYNAMICS** for Windows 95/NT as **cspW** and for UNIX/Motif/X as **cspX**. To see more details of the packages of algorithms which come organized under a useful graphical user interface, visit

http://www.zweb.com/apnonlin/.

1 Reconstructing Phase Space

The analysis of time series has long been focused on regular, periodic, or quasiperiodic time dependence—with all other time courses considered as "noise".

1991 *Mathematics Subject Classification.* Primary: 58F13, 62M20; Secondary: 62M10.
*Institute for Nonlinear Science.

Chaotic motions lie between these extremes as they possess structure in state space and have limited predictability while they possess continuous broad Fourier spectra, are nonperiodic, and are unstable everywhere in state space. These characteristics mean chaotic signals are not amenable to the enormous body of signal analysis tools built up over the past sixty or more years which are tuned to globally linear signal sources and rest heavily on Fourier analysis to simplify the tasks at hand.

We have developed a set of analysis tools for the study of time series (Abarbanel [1996], Abarbanel et al. [1993]) which rest on geometric and dynamical lessons from the study of nonlinear systems—the only systems in which chaos can appear. These are time domain tools. They are also useful for regular signals from linear sources, but probably they do not compete well with the vast selection of methods developed in the frequency domain or based on newer linear ideas such as the use of wavelets. They are indispensable for the analysis of chaotic signals.

The analysis tools rest on the idea of phase space reconstruction (Abarbanel [1996], Fraser and Swinney [1986], Fraser [1989]) which takes scalar observations $s(n) = s(t_0 + n\tau_s)$ made on some system at intervals τ_s and creates a d-dimensional multivariate space from $s(n)$ and its time lags at intervals $T\tau_s$. The observed scalar data is replaced by the d-dimensional vectors

$$\mathbf{y}(n) = [s(n), s(n+T), s(n+2T), \dots, s(n+(d-1)T)], \tag{1.1}$$

which are points in a phase space or state space which is dynamically identical to the original (but unknown) space in which the signal source operates, as long as d is large enough. The geometric theorem underlying this construction tells us that if we have an infinite amount of infinitely accurate data the value of the lag T is not important, and when the dimension of the system attractor is d_A (which could be fractional) then, choosing the integer d such that $d > 2d_A$, we have a sufficient condition to capture the properties of the underlying dynamics. The condition $d > 2d_A$ is not always necessary. We never have an infinite amount of data—if we did, then accounting for the time to collect it, we would have little time left for its analysis. We never have infinitely accurate data. Clearly if we wish to move beyond entertaining, but useless, mathematics, we need other assistance.

To use the vectors $\mathbf{y}(n)$ we need to know "good" values for the lag T and the necessary dimension $d = d_E$—the embedding dimension. If we have selected a value for T on the basis of some prescription, then we need to choose d_E so that strands of the system orbit in d_E-dimensional space do not overlap. This overlap, due to projection from the correct higher dimensional space, decreases as we increase the utilized embedding dimension d until the last overlaps occur as points. When the utilized embedding dimension of the vectors $\mathbf{y}(n)$ is large enough to remove overlaps in points, then this or any larger dimension will be perfectly acceptable for the analysis of the observed data, that is, acceptable as d_E. Additional coordinates will not serve to remove any further overlaps—there are none—so the dimension where overlaps cease is the necessary dimension for the unfolding of the attractor.

Now the criterion of necessity is dictated by the goals one has in mind. We have prediction in mind, and if two strands of the attractor overlap there are regions or points in state space where it is ambiguous as to where the orbit should go, and this can only be resolved by viewing the orbits in higher dimensional space. If one wishes to compute only the fractal dimension of the attractor, $d > d_A$ will do (Abarbanel [1996]).

The analysis of chaotic signals in the space in which they are observed is sure to fail. Chaos only occurs when the dynamics takes place in three or more dimensions if the system is governed by differential equations; that is, continuous time, or in two or more dimensions if the time label is discrete. Using $s(n)$ directly can only lead to erroneous conclusions about prediction or control or classification or system identification; that is, it is generically the wrong way to proceed to answer the usual questions of signal processing for nonlinear systems. The methods we present here and elaborate on in our papers (Abarbanel [1996], Abarbanel et al. [1993]) may be replaced by newer, more direct and computationally more efficient methods over the years, but they address directly the multivariate nature of the state space for nonlinear dynamical systems. If one wishes to know the answer to the usual class of questions about the source of observed signals and that source is nonlinear—that is, it is remotely connected to properties of the real physical or biological world—methods such as the ones we discuss are required.

Very little serious analysis of the methods we will discuss have been performed by statisticians who wish to answer questions bearing on the number of data one requires for this or that accuracy in each of the analyses we shall present. Also the answers to questions bearing on the distribution of errors expected in making this or that kind of analysis discussed below are absent. In other words, the field is rather wide open and cries out for the application of quantitative statistical analyses of the methods one utilizes to understand nonlinear dynamics when presented with observed data. I have no doubt that the road is rich with remarkable results, and I compliment the organizers of this meeting in bringing together some persons who have the statistical skills and interests to address these unaddressed and unanswered matters and those of us who have struggled to make do without theorems to guide us in applying these ideas to real world data of direct physical or biological interest.

2 False Nearest Neighbors

To determine d_E for the data vectors $\mathbf{y}(n)$ we employ a statistic about these data known as *false nearest neighbors* (Kennel et al. [1992]). The construction of the d_E-dimensional data vectors is purely geometric once a time delay T has been chosen. One is seeking a multivariate space in which the orbits $\mathbf{y}(n)$ do not overlap because of projection from a higher dimension. The main idea is that if the data comes from a system governed by autonomous differential equations or discrete time maps with time independent parameters, then orbits cannot cross each other. If they appear to cross, it is because we are viewing them in a dimension which is too small or which is smaller than the number of active degrees of freedom in the source of our observations.

To establish whether the space has become big enough we look at each point $\mathbf{y}(n)$ and its nearest neighbor $\mathbf{y}^{\mathrm{NN}}(n)$ in dimension d. The time index k associated with $\mathbf{y}^{\mathrm{NN}}(n) = \mathbf{y}(k)$ bears little resemblance to the time index n of the data point we are examining. As data folds back on itself in dissipative systems, the neighborhood of $\mathbf{y}(n)$ may be populated by points such as $\mathbf{y}^{\mathrm{NN}}(n)$ of quite different time index. Now we ask what happens to the distance between $\mathbf{y}(n)$ as seen in dimension $d+1$ where it is the vector

$$\mathbf{y}(n) = [s(n), s(n+T), \dots s(n+Td)], \tag{2.1}$$

and the vector $\mathbf{y}^{\mathrm{NN}}(n)$ in dimension $d+1$, where it becomes

$$\mathbf{y}^{\mathrm{NN}}(n) = \mathbf{y}(k) = [s(k), s(k+T), \dots, s(k+Td)]. \tag{2.2}$$

If the distance in dimension $d+1$ is large, we label $\mathbf{y}^{\mathrm{NN}}(n)$ a false nearest neighbor of $\mathbf{y}(n)$. When d becomes large enough, the number of false neighbors will

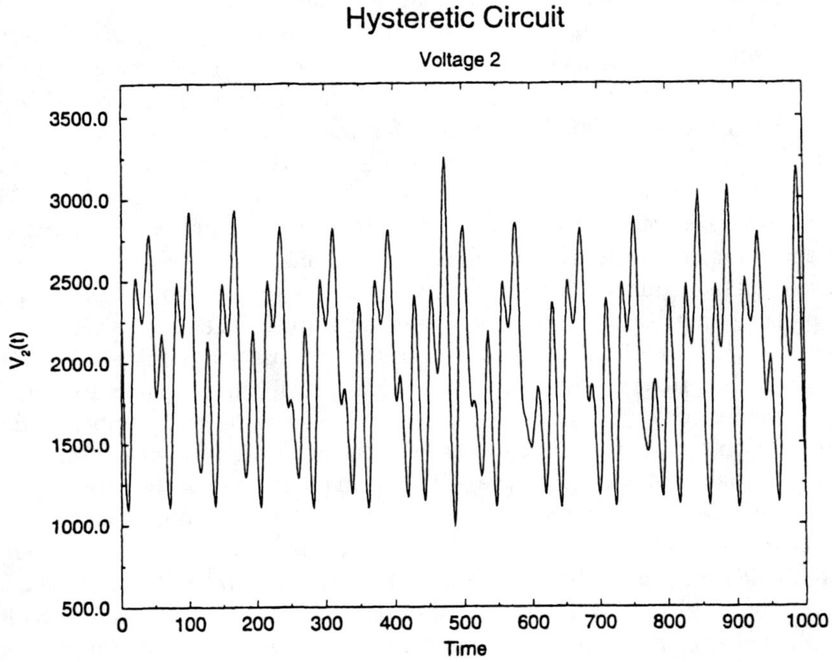

Figure 1

go to zero, for the overlaps due to projections will be removed. At that dimension the attractor is unfolded in the coordinate system defined by our choice of time delay T. Now the idea of "large" distances requires some threshold for deciding that a nearest neighbor $\mathbf{y}^{\text{NN}}$ is false. The choice of false neighbor turns out to be independent of this threshold over a wide range of choices for its value (Kennel et al. [1992]).

An example of the false nearest neighbors method is seen in Figures 1, 2, and 3. In Figure 1 we see a sample of the original time series from a nonlinear hysteretic circuit operated at the US Naval Research Laboratory by Tom Carroll and Lou Pecora; $\tau_s = 0.1\text{ms}$. Figure 2 is the percentage of false nearest neighbors when the vectors $\mathbf{y}(n)$ are constructed with a time lag $T = 6$; clearly $d_E = 3$ for these data. In Figure 3 we show the attractor reconstructed in the space $\mathbf{y}(n) = [s(n), s(n+6), s(n+12)]$ for the hysteretic circuit. The structure in state space is clearly revealed only in the last view of the data. The time series in Figure 1 and the Fourier spectrum of these data are of little value in characterizing the source of these signals.

For a statistician (if the members of this audience will permit my presuming to suggest problems of interest—which I do with humility as I make no (false) claims to have expertise or experience in statistics!) it may be of interest to determine the distribution of the false nearest neighbor statistic as a function of the number of data, the amount of the attractor visited by the orbit (similar to the number of data issue), the signal to noise ratio of the observations, the time lag T, and other questions of interest in the practical use of this statistic for the determination of d_E.

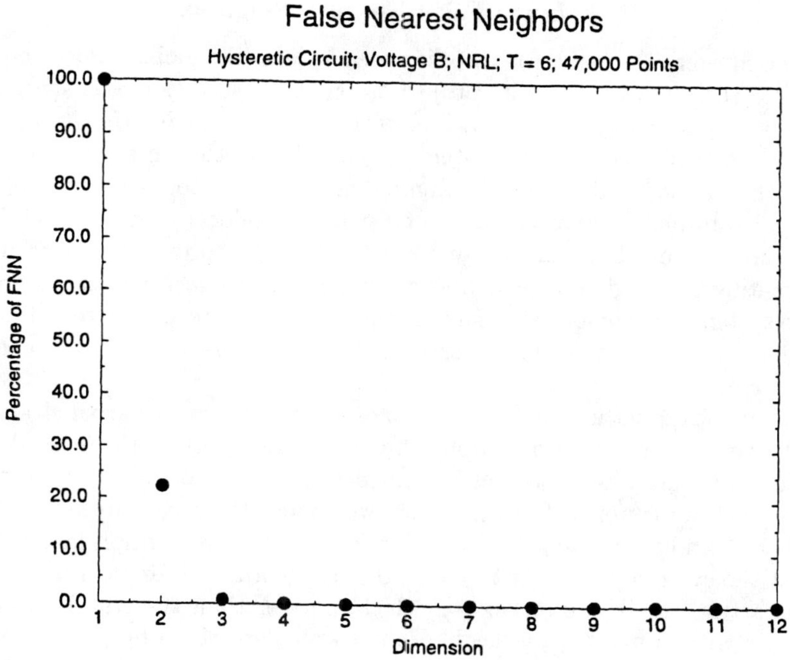

Figure 2

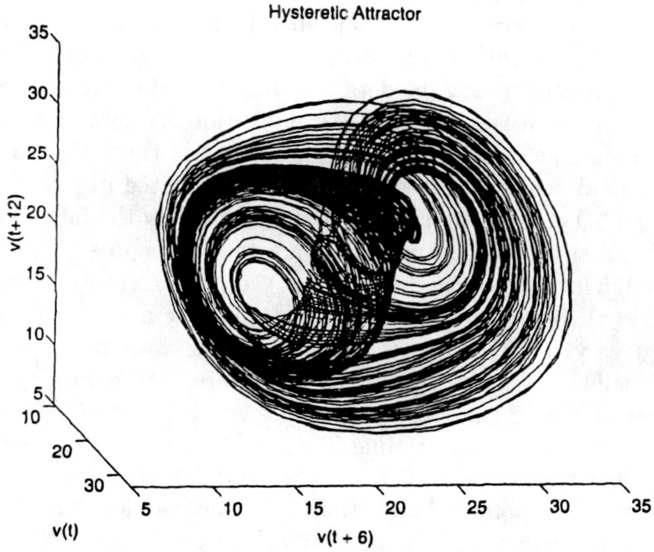

Figure 3

3 Local False Nearest Neighbors

The integer dimension d_E just discussed tells us the global dimension required to unfold the attractor from its projection on the observation axis $s(n)$. The *local dimension* d_L of the dynamics may be less than or equal to d_E. If $d_E > d_L$, it means that the particular coordinate system, namely that of the time delay vectors $\mathbf{y}(n)$, twists the attractor about so that unfolding the attractor from the observations requires additional dimensions to "untwist it". The local or dynamical dimension is the same in any coordinate system of $\mathbf{y}(n)$ or other "proxy" multivariate vectors representing the original coordinates as long as the connections among these coordinate systems are smooth enough. d_L represents the integer number of differential equations (or maps, if time is discrete) required to describe the local evolution of the system.

We use the method of local false nearest neighbors (Abarbanel and Kennel [1993]) which starts with a working dimension $d_W \geq d_E$ with d_E as determined by the global false neighbors algorithm described above. All distances between points $\mathbf{y}(n)$ are computed in d_W. Now we choose the local dimension d starting with $d = 1$ and proceeding to $d = d_W$ asking at each dimension how well we can predict the evolution of a cluster of neighbors about $\mathbf{y}(n)$ for each $n = 1, 2, \ldots, N$ on the attractor. In a dimension d which is lower than the true local dimension d_L, there will be points in a neighborhood which got there by projection and not by a dynamical rule. The ability to predict where these points go will be impaired for $d < d_L$. As we increase d up to d_L our ability to predict will improve until it levels off and becomes independent of the estimated local dimension d as well as the number of neighbors we use to define a neighborhood whose evolution we predict. In dimensions $d_L \leq d \leq d_W$ predictability should become independent of d as we have enough dynamical degrees of freedom already to capture the evolution of the data. The quality of prediction is determined by how far ahead we wish to predict, and by how large an error we tolerate until we say the prediction has failed.

To be more precise about the details of the method of local false nearest neighbors we note that what is involved in the prediction is a local map from one neighborhood of the attractor to another neighborhood of the attractor. The members of a neighborhood are determined by distances evaluated in a dimension $d_W \geq d_E$ with d_E the global embedding dimension determined by the false nearest neighbors method. This assures that all neighbors are true neighbors. Then in a neighborhood, we ask if a local model in dimension $d \leq d_W$ accurately relating the neighbors of the point $\mathbf{y}(n)$ to the neighbors of $\mathbf{y}(n+1)$ can be made. This local model is a rule which gives $\mathbf{y}(n+1)$ in terms of a polynomial constructed out of the vectors $\mathbf{y}(n)$. The coefficients in this polynomial rule are determined by a least squares fit minimizing the residual errors in the rule which takes $\mathbf{y}(n) \to \mathbf{y}(n+1)$. This rule is used to predict ahead a time T equal to the time lag used before. If, in predicting ahead this amount of time, the error between the prediction and the known observed data point is larger than a certain fraction β of the attractor size, we deem it a bad prediction. The fraction of bad predictions P_K is collected as we move over the whole attractor, and this is displayed against the dimension d of the local model for various choices of the number of neighbors N_B of the points $\mathbf{y}(n)$ on the orbit. We seek a dimension where the model, represented by its fraction of bad predictions, becomes independent of the local dimension and the number of neighbors used to establish the model. The d at which this occurs determines the dimension d_L of a good local model. As one changes the fraction β of the attractor

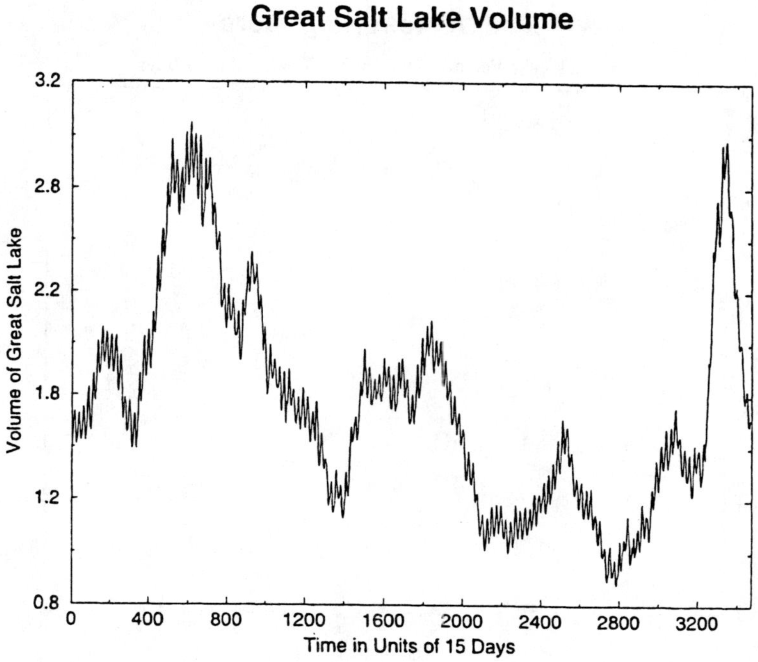

Figure 4

size which defines the allowed error sphere or changes the time which one predicts ahead, the fraction of bad predictions P_K will move up and down in absolute value. The dimension d_L at which P_K becomes independent of d and N_B will remain the same. As it is d_L which one wishes to extract from the local false nearest neighbors statistic, the absolute size of P_K is of no special significance.

We illustrate the method of local false nearest neighbors with data from the volume of the Great Salt Lake (GSL) in Utah. The GSL, located at approximately 40° to 42° N and 110° to 112° W, is the fourth largest, perennial, closed basin, saline lake in the world. It drains an area of $90,000\,\text{km}^2$. The GSL' s immediate atmospheric catchment is the Pacific Ocean, and its fluctuations exhibit high spectral coherence with the Pacific North American Circulation Index, and the Southern Oscillation index at interannual time scales in the Northern Pacific atmospheric flow and tropical Pacific atmospheric variability related to the El Niño Southern Oscillation, respectively. Its fluctuations have been related (Mann et al. [1994]) to those in the hemispheric surface temperature and sea level pressure records.

The levels of the GSL have been recorded by various methods since 1847. From these data a biweekly time series of GSL volumes has been compiled by Sangoyomi [1993] which shows substantial variation both on an annual time scale as well as on interannual and interdecadal time scales. Lall and Mann [1994] speculate that the low frequency oscillations found on spectral analysis of this and related hydro-climatic data sets may represent unstable oscillations that arise from the nonlinearity of the climate dynamics, rather than stable periodic motion.

Figure 4 shows the time series for the GSL data, and the false nearest neighbors for $T = 12$ is seen in Figure 5. $d_E = 4$ is clearly selected. Figure 6 shows the local

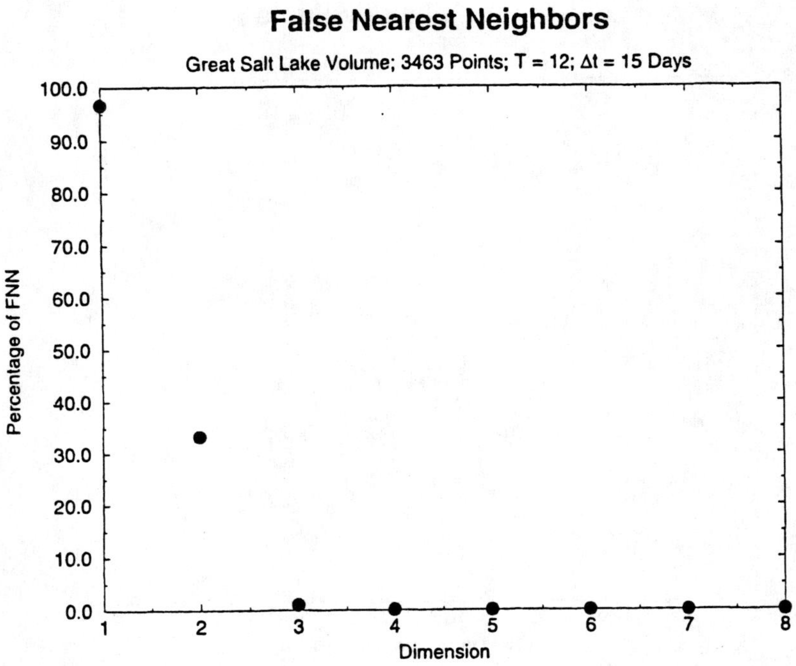

Figure 5

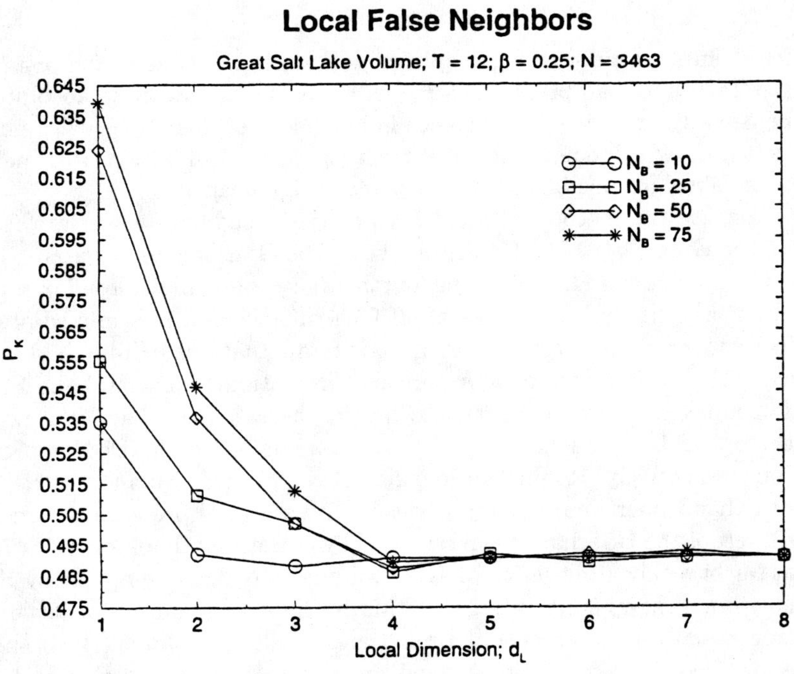

Figure 6

false nearest neighbors, and we also see quite clearly that $d_L = 4$ is required by these data.

In the statistic we call local false nearest neighbors we have added dynamics to the geometrical considerations underlying global false nearest neighbors. The dynamics is added in the form of local predictions from neighborhood to neighborhood of the attractor in multidimensional space. This prediction assumes that underlying the data is a discrete time map (for finite τ_s all dynamics is discrete or a discrete approximation to a flow)

$$\mathbf{y}(n) \to \mathbf{F}(\mathbf{y}(n)) = \mathbf{y}(n+1), \tag{3.1}$$

and the main job is to estimate $\mathbf{F}(\mathbf{x})$ accurately. The estimation is made in terms of some local basis function set $\phi_k(\mathbf{x})$ as

$$\mathbf{F}(\mathbf{x}) = \sum_{k=1}^{M} \mathbf{c}(k)\phi_k(\mathbf{x}). \tag{3.2}$$

We want to evaluate the coefficients $\mathbf{c}(k)$ locally in phase space by selecting N_B neighbors of the point $\mathbf{y}(n)$ and observing how they evolve into the points near $\mathbf{y}(n+1)$. So we use phase spatial information to capture temporal evolution. The determination of the coefficients is made using a least squares fit to the points $\mathbf{y}(n+1)$ compared to

$$\sum_{k=1}^{M} \mathbf{c}(k)\phi_k(\mathbf{y}(n)). \tag{3.3}$$

There are a number of clear statistical issues:

- what is the "best" choice of basis functions $\phi_k(\mathbf{x})$ for each problem?
- how many neighbors N_B do we need in dimension d_E with local dimension d_L for N data points?
- what is the convergence of the approximation as a function of these parameters ?
- what order do we require for what quality of approximation to $\mathbf{F}(\mathbf{x})$?
- what is the distribution of errors as a function of the properties of the attractor $(d_E, d_L, d_A, \lambda_a(\mathbf{x}, L), \ldots)$ and the given data set $(N, N_B, \ldots)$?
- etc.

It seems fair to assume that, as these questions are addressed, we shall see new algorithms for each of these questions. Their implementation and analysis will be most valuable to the full development of the kind of nonlinear dynamics analysis toolkit discussed here.

4 Local and Global Lyapunov Exponents

The evolution of small perturbations to an observed orbit $\mathbf{y}(n)$ is governed by the linearized equations of motion, whatever they may be. As we do not know them from looking at the data alone, we assume there is an underlying nonlinear process $\mathbf{y}(n) \to \mathbf{y}(n+1) = \mathbf{F}(\mathbf{y}(n))$ which moves the system ahead one sampling time τ_s. A small perturbation $\mathbf{w}(n)$ to the orbit $\mathbf{y}(n)$ satisfies

$$\begin{aligned} \mathbf{y}(n+1) + \mathbf{w}(n+1) &= \mathbf{F}(\mathbf{y}(n) + \mathbf{w}(n)) \\ \mathbf{w}(n+1) &= \mathbf{DF}(\mathbf{y}(n)) \cdot \mathbf{w}(n) + O(\mathbf{w}(n)^2) \end{aligned} \tag{4.1}$$

where the Jacobian matrix is

$$\mathbf{DF}(\mathbf{x})_{ab} = \frac{\partial F_a(\mathbf{x})}{\partial x_b}, \tag{4.2}$$

and $a, b = 1, 2, \ldots d_L$.

The eventual growth or shrinkage of the perturbation under this linear evolution rule is determined by the eigenvalues $e^{\lambda_a(\mathbf{x},L)}$ of the Oseledec matrix (Abarbanel [1996], Abarbanel et al. [1993], Oseledec [1968]).

$$\mathbf{OSL}(\mathbf{x}, L) = \left[(\mathbf{DF}^L(\mathbf{x}))^T \cdot \mathbf{DF}^L(\mathbf{x})\right]^{1/2L}. \tag{4.3}$$

The matrix $\mathbf{DF}^L(\mathbf{x})$ is the composition of L Jacobian matrices along the observed orbit $\mathbf{y}(n)$ starting at location $\mathbf{x}$. As $L \to \infty$, the $\lambda_a(\mathbf{x}, L) \to \lambda_a$, which are the usual global Lyapunov exponents. The λ_a are independent of $\mathbf{x}$ in the basin of attraction of the attractor. They are invariants of the dynamics and characterize it. They are also independent of the coordinate system in which they are evaluated. To determine the λ_a reliably we need to know the value of d_L as this is the dimension of the dynamics. If we work in a space with $d > d_L$, then $d - d_L$ of the eigenvalues of the Oseledec matrix will be false, and we need a reliable rule to establish which are true and which, not. Similarly if we work in $d < d_L$, we will not have unfolded the local dynamics in such a way which would allow the correct evaluation of $\mathbf{DF}(\mathbf{x})$, and thus the λ_a would be in error.

Using $d_E = d_L = 4$, we computed the $\lambda_a(\mathbf{x}, L)$ for a large number of starting locations $\mathbf{x}$ on the GSL attractor, then determined the value of these quantities as a function of the number of steps we look ahead of these starting points. This allows us to define an average local Lyapunov exponent

$$\bar{\lambda}_a(L) = \frac{1}{N_S} \sum_{k=1}^{N_S} \lambda_a(\mathbf{y}(k), L) \tag{4.4}$$

for N_S starting locations $\mathbf{y}(k)$. For the GSL data we estimated the local Jacobian matrices by making local cubic neighborhood-to-neighborhood maps in $d_E = d_L = 4$ state space and picking off the linear part of these local maps. These maps can be thought of as regressions of the GSL volume $V(n+1)$ on the vector defined by the embedding $[V(n), V(n+T), ..V(n+(d-1)T)]$. These estimated local Jacobians were then multiplied to give the Oseledec matrix and diagonalized as indicated. In Figure 7 we display the average local Lyapunov exponents as a function of number of steps $2^{(L-1)}$ along the orbit for $L = 0, 1, \ldots 12$. The average exponents have been averaged over 750 different starting locations on the attractor. These locations are spaced approximately equally in time from the center of the measured data set. Had one chosen them randomly from the data set or essentially any other way, when the number is large enough, the results will be the same as implied by the multiplicative ergodic theorem (Oseledec [1968]). The exponents converge to the approximate global values: $\lambda_1 = 0.17, \lambda_2 = 0, \lambda_3 = -0.13$, and $\lambda_4 = -0.68$. These estimates used a reconstructed state space with time lag $T = 17$. In principle, they are independent of the lag T.

We conclude two important things from these values for the λ_a. First, one of the exponents is zero within numerical accuracy. This means that the underlying dynamics comes from a differential equation as a zero global exponent is always associated with a flow in state space. Secondly, the value of the largest exponent is $\lambda_1 \approx 1/5.5$, in units of $1/\tau_s$. Errors along the orbit or in initial conditions grow as

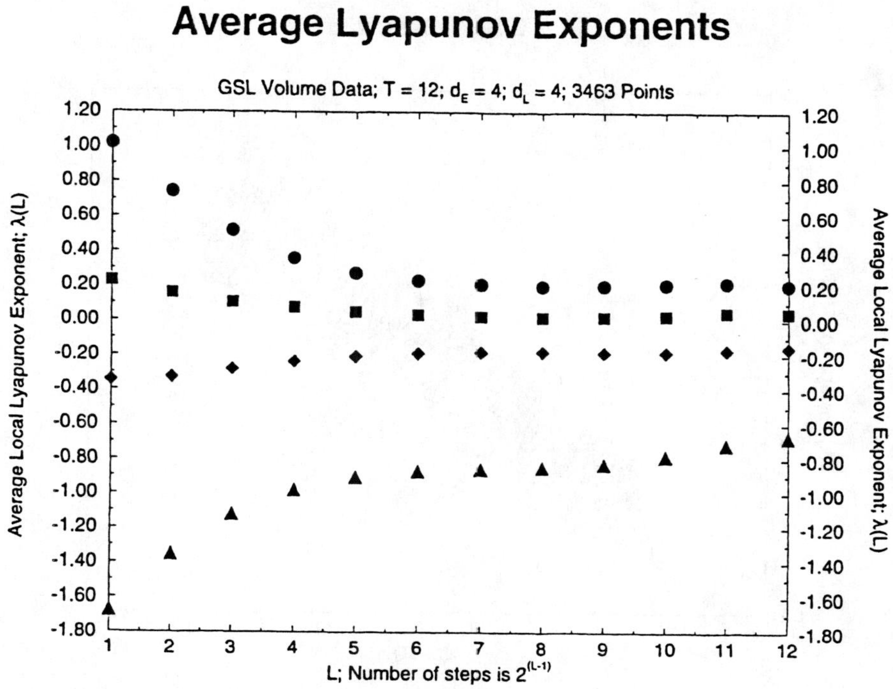

Figure 7

$\exp[t/5.5\tau_s]$, so when the time after some perturbation or initial state is of order a few times $5.5\tau_s$ or a few hundreds of days (or a year or so), we lose the ability to predict the behavior of the volume of the GSL.

These estimates for the Lyapunov exponents can be used to estimate the fractal dimension of the attractor using the idea of Lyapunov dimension (Kaplan and Yorke [1979]) which has been conjectured to be the same as the information dimension. This Lyapunov dimension is evaluated by finding at which K the sum

$$\sum_{a=1}^{K} \lambda_a, \tag{4.5}$$

is positive, but the sum

$$\sum_{a=1}^{K+1} \lambda_a, \tag{4.6}$$

is negative. The Lyapunov dimension is defined as

$$D_L = K + \frac{\sum_{a=1}^{K} \lambda_a}{|\lambda_{K+1}|}. \tag{4.7}$$

For the GSL data we find $D_L \approx 3.05$. This means that we should be able to get a good idea of the structure of the attractor by looking at it in three dimensions with $T = 12$. This is displayed in Figure 8. The three dimensional view of the attractor suggests that the annual cycle is approximately motion around the smaller radius of the "spool" on which the attractor lies while the longer term motion which has larger amplitude moves the orbits along the longer axis of this "spool". Such motions may correspond to the spectral power identified by Lall and Mann [1994]

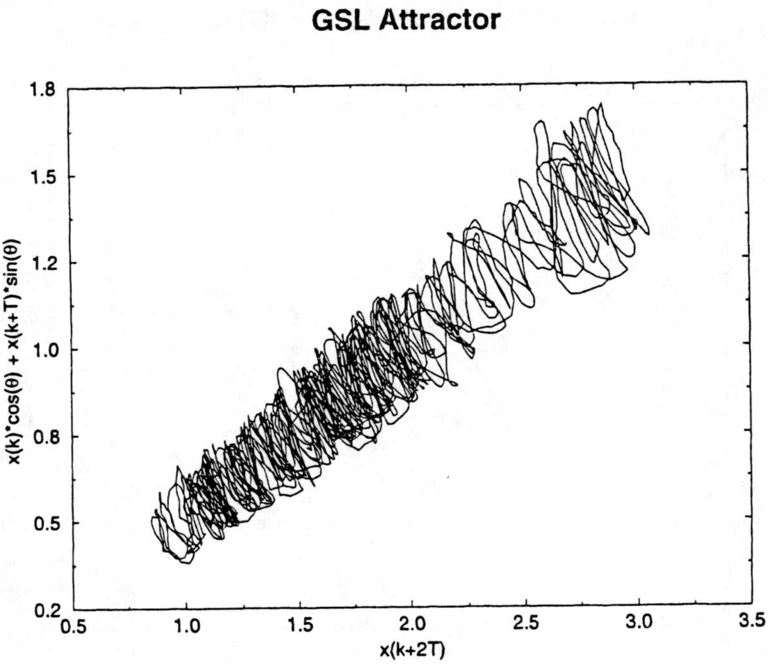

Figure 8

at certain interannual and interdecadal frequencies. Interestingly, such structure persists even if larger delays T that obscure the annual cycle are used.

The issues requiring statistical analysis here are initially those mentioned earlier with some fascinating twists. One clearly needs to make a good local map

$$\mathbf{y}(n) \rightarrow \sum_{k=1}^{M} \mathbf{c}(k)\phi_k(\mathbf{y}(n)), \tag{4.8}$$

as above, but then one must very accurately pick off from this map the local Jacobian with components

$$DF(\mathbf{y}(n))_{ab} = \sum_{k=1}^{M} c_a(k)\frac{\partial \phi_k(\mathbf{y}(n))}{\partial y_b(n)}, \tag{4.9}$$

and one must then combine many of these local Jacobians to construct the very ill conditioned Oseledec matrix whose eigenvalues we seek. How well this construction can be expected to work as a function of the usual dynamical quantities $d_E, d_L, \ldots$ and the number of data is quite important. Similarly, how the errors in these estimates of the eigenvalues of the Oseledec matrix are distributed as a function of the number of data, the size of neighborhoods, etc., is well worth investigating.

5 Predicting in Reconstructed Phase Space

Even without knowledge of the dynamical equations for the GSL system we can use the information we have acquired so far to make predictive models for the GSL volume evolution. The method (Abarbanel et al. [1993]) utilizes the compactness of the attractor in $\mathbf{y}(n)$ space by noting that we have knowledge of the evolution of whole phase space neighborhoods into whole neighborhoods later in time. We can use this to make local models of this evolution and then use these models as interpolating rules for the evolution of new phase space points near the attractor. We use only local polynomial models although basis sets other than polynomials are certainly quite useful. Indeed, in this work we use only local linear models as we have substantial amounts of data and thus good coverage of the attractor; namely, every neighborhood is rather well populated.

The idea is that locally on the attractor we find the N_B neighbors $\mathbf{y}^{(r)}(n)$; $r = 1, 2, \ldots N_B$ of each point $\mathbf{y}(n)$ and make the local linear model $\mathbf{y}(r; n+1) = \mathbf{A}_n + \mathbf{B}_n \cdot \mathbf{y}^{(r)}(n)$ where $\mathbf{y}(r; n+1)$ is the point to which $\mathbf{y}^{(r)}(n)$ goes in one time step. The coefficients $\mathbf{A}_n$ and $\mathbf{B}_n$ are determined by minimizing at each time location n

$$\sum_{r=0}^{N_B} |\mathbf{y}(r; n+1) - \mathbf{A}_n - \mathbf{B}_n \cdot \mathbf{y}^{(r)}(n)|^2, \tag{5.1}$$

where $r = 0$ means $\mathbf{y}(n)$ itself and $\mathbf{y}(n+1)$. When we have a new point $\mathbf{z}(k)$ on or near the attractor, we seek the nearest neighbor $\mathbf{y}(Q)$ among all the data in the set we used to determine the $\mathbf{A}_n$ and the $\mathbf{B}_n$. The predicted point $\mathbf{z}(k+1)$ is then

$$\mathbf{z}(k+1) \approx \mathbf{A}_Q + \mathbf{B}_Q \cdot \mathbf{z}(k). \tag{5.2}$$

This works remarkably accurately within the limits of prediction dictated by the largest Lyapunov exponent λ_1. When we try to predict beyond the instability horizon, that is for times much greater than

$$\frac{\tau_s}{\lambda_1}, \tag{5.3}$$

our predictions should rapidly lose accuracy.

5.1 Prediction on the GSL Attractor. These methods were used to predict the volume of the Great Salt lake in $d_E = d_L = 4$. In Figure 9 we display the result of predicting one step ahead for the 100 time points after time index 3301; April 24, 1985 to June 2, 1989. The data used to determine the local linear maps was taken from the beginning of the time series up to point 3263. It is clear that the ability to predict one step (15 days) ahead is very high. Next we determined the local linear maps from data starting at the beginning of the data set and using points up to index 3050. In Figure 10 we display the result of predicting 10 time steps ahead, that is 150 days or about a half year, again starting with time index 3248. In Figure 11 we trained local linear maps from the first 3200 points of the data set, and then predicted ahead ten steps from time index 3348. These predictions are actually rather good, and, as we can see, the prediction ability degrades with the number of steps ahead one wishes to predict, as is expected, given that the largest Lyapunov exponent is positive (implying that the system is chaotic). The main message in these predictions is that we can accurately follow the features of the data without elaborate computations by working in dimension four as our systematic analysis would suggest. Of course, it is possible to devise even better predictors using more elaborate methods.

Many of the potentially interesting statistical issues associated with prediction on the attractor are extensions of well studied matters in conventional statistics with the very interesting twist that the focus is now on the local predictability of a system which is everywhere unstable in the state space in which one is predicting. Many of the issues we have thought to be interesting were raised above in the consideration of local false nearest neighbors, and we shall not simply list them again.

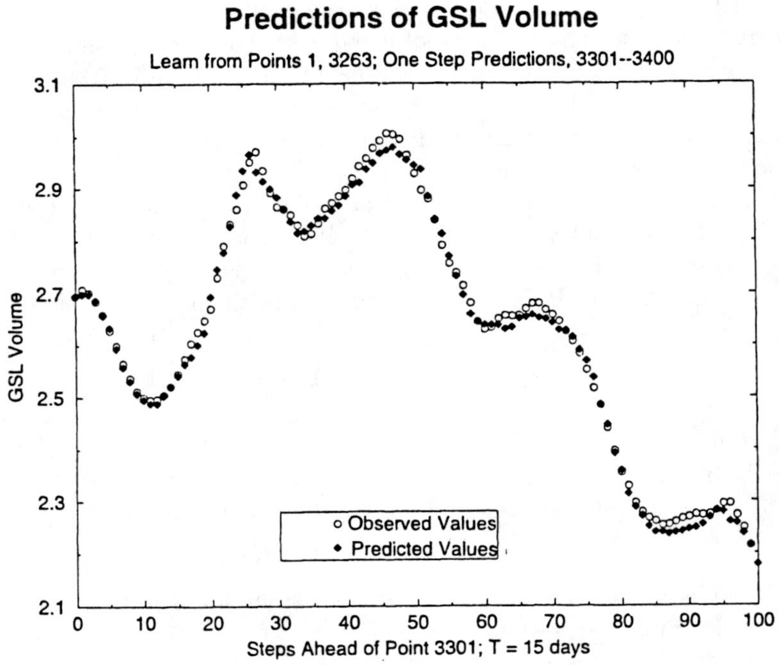

Figure 9

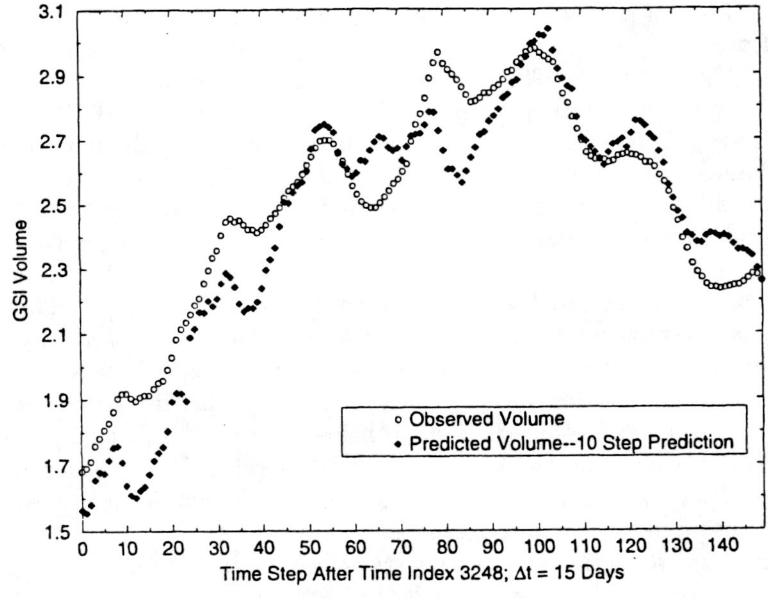

Figure 10

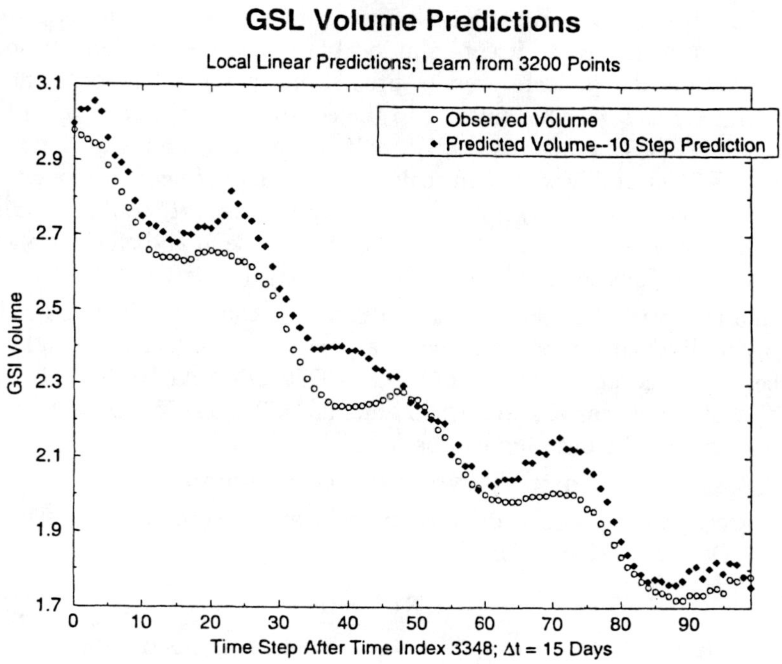

Figure 11

6 Conclusions

The tools we have developed for the analysis of observed data apply equally well for linear and nonlinear sources. If linear sources are encountered they would reveal themselves in the Fourier spectra and in zero or negative Lyapunov exponents, if not before from a close examination of the periodic structures in the time series. In that case, we can recommend without shame the use of the vast selection of Fourier based methods for the systematic analysis of such data. If the data comes from a nonlinear source operating in a chaotic regime, we can recommend without shame the use of our methods as presented here, or improved versions of these methods which we hope the readers will construct and share with us. Fourier methods tuned to linear data sources will prove misleading and without foundation in such a case.

As we have pointed out at various points in this paper, there is an enormous body of analysis to be done to determine the limits on the geometric methods we have discussed. My colleagues and I hope that the quite striking results of the analyses we have presented will generate interest in the statistical community in establishing firm foundations for the statistical reliability of these methods as a function of critical parameters such as number of data, dimension of reconstructed space, time delay used in reconstructed state space vectors, etc.

Two key ingredients recur in the analyses we have presented, and I personally hope that there can be some firm foundations and new ideas for them as well:

- **multi-dimensional searches** occur again and again. We have used kd-trees for these searches, and we have made no attempt to optimize these searches to reduce the operations count as much as possible. Doing this and delineating the conditions under which the optimized searches can be expected would be a quite useful contribution.

- **interpolations in multi-dimensional space** are required over and over again in this work. It enters in local false nearest neighbors, in determining Lyapunov exponents, and in prediction on the attractor. The literature, frankly, is somewhat unclear to the consumer (me) about when to use what kind of basis functions (radial, polynomial, etc.) and what kind of accuracy to expect and how this all scales with the size of neighborhoods, the dimension of the space d_E, etc. Knowledge of this sort would be most valuable both academically and for the planning of useful and efficient algorithms for carrying out the kinds of analyses discussed here.

Finally, let me reiterate that the algorithms considered here as well as others in our "toolkit" are now available as a package with a useful graphical user interface. The package is called **TOOLS FOR DYNAMICS**. It is available for UNIX/Motif/X as **cspX** and for Windows 95/NT as **cspW**. Details on its features and instructions for ordering can be found at

http://www.zweb.com/apnonlin/.

The package comes with a detailed manual and a copy of Analysis of Observed Chaotic Data (Abarbanel [1996]).

References

Abarbanel, H. D. I. [1996] *Analysis of Observed Chaotic Data*, Springer-Verlag.

Abarbanel, H. D. I., Brown, R., Sidorowich, J. J. ("Sid") and Tsimring, Lev Sh. [1993] *The analysis of observed chaotic data in physical systems*, Reviews of Modern Physics, **65**, 1331–1392.

Abarbanel, H. D. I. and Kennel, M. B. [1993] *Local false nearest neighbors and dynamical dimensions from observed chaotic data*, Phys. Rev. E., **47**, 3057–3068.

Fraser, A. M. [1989] *Information and entropy in strange attractors*, IEEE Trans. on Info. Theory, **35**, 245.

Fraser, A. M. and Swinney, H. L. [1986] *Independent coordinates for strange attractors*, Phys. Rev., **33A**, 1134.

Kaplan, J. L. and Yorke, J. A. [1979] *Chaotic behavior in multidimensional difference equations*, Lecture Notes in Mathematics, **730**, 228.

Kennel, M. B., Brown, R. and Abarbanel, H. D. I. [1992] *Determining minimum embedding dimension using a geometrical construction*, Phys. Rev. A. **45**, 3403-3411.

Lall, U. and Mann, M. [1994] *The Great Salt Lake: A barometer of interannual climatic variability*, Water Resources Research, submitted.

Mann, M., Lall, U. and Saltzman, B. [1994] *Decadal and secular climate variability: understanding the rise and fall of the Great Salt Lake*, Geophys. Res. Lett., submitted.

Moon, Y. I. and Lall, U. [1995] *Atmospheric flow indices and interannual Great Salt Lake variability*, J. of Hydrologic Eng., submitted.

Oseledec, V. I. [1968] *A multiplicative ergodic theorem. Lyapunov characteristic numbers for dynamical systems*, Trudy Mosk. Mat. Obsc, **19**, 197; Moscow Math. Soc., **19**, 197.

Sangoyomi, T. B. [1993] *Climatic Variability and Dynamics of Great Salt Lake Hydrology*, PhD Dissertation, Department of Civil and Environmental Engineering, Utah State University.

Fields Institute Communications
Volume 11, 1997

Some Comments on Nonlinear Time Series Analysis

Howell Tong
Institute of Mathematics and Statistics
University of Kent
Canterbury, Kent CT2 7NF UK

Abstract. Through two simple examples, we reveal some basic similarities as well as basic differences between two recent approaches to defining initial-value sensitivity for a stochastic dynamical system. We also make some comments relevant to the practice of prediction.

1 Introduction

Recent developments in nonlinear time series have substantially widened the scope of traditional time series analysis. They have deepened our understanding of prediction and the role of initial values. This paper makes a few comments and suggests a couple of issues which require further research.

2 Initial-Value Sensitivity for a Stochastic Dynamical System

There are at least two different approaches to defining initial-value sensitivity for a stochastic dynamical system. (For references see, e.g., Tong [1995]). The earlier approach starts with a deterministic dynamical system typified by

$$X_t = F(X_{t-1}), \tag{2.1}$$

where X_t denotes a state vector in R^d and $t \geq 1$. For simplicity of discussion, let $d = 1$, which will suffice for our purpose of raising the forthcoming issues. If the system is ergodic, then it is well known that the so-called *Lyapunov exponent* λ defined by $\lim_{t\to\infty} t^{-1} \sum_{m=0}^{t-1} \ln |\dot{F}(X_m)|$, or equivalently $E \ln |\dot{F}(X)|$ by ergodicity, plays a central role in quantifying the initial-value sensitivity. Here, the expectation is taken with respect to the invariant measure associated with the ergodic deterministic dynamical system. If the time series $\{X_t, -\infty < t < \infty\}$ is generated by an ergodic stochastic dynamical system of the form

$$X_t = F(X_{t-1}) + \epsilon_t, \tag{2.2}$$

where $\{\epsilon_t, t \geq 1\}$ is a (dynamic) noise sequence taking the form of a sequence of independent identically distributed random variables, then it would seem quite natural to adopt essentially the same Lyapunov exponent as above to quantify the initial-value sensitivity of the stochastic dynamical system (2.2), which is obtained

1991 *Mathematics Subject Classification.* Primary: 62M20, 62M10; Secondary: 58F13.

by "clothing" its obvious deterministic counterpart (2.1), also called the *skeleton*. The only modification required is to take the expectation with respect to the invariant measure of the stochastic dynamical system instead, assuming that the latter is ergodic. Specifically, for almost all $\omega \in \Omega$, where Ω is the underlying probability space, we get

$$\lambda(\omega) = \lim_{t\to\infty} t^{-1} \sum_{m=0}^{t-1} \ln |\dot{F}(X_m(\omega))|, \tag{2.3}$$

which, by ergodicity, yields a Lyapunov exponent independent of ω, namely

$$\lambda = E \ln |\dot{F}(X)|, \tag{2.4}$$

where the expectation is taken with respect to the invariant measure associated with the stochastic dynamical system (2.2). Note that in this approach, the same $\epsilon_t(\omega)$ is used when we study the effect of perturbing $X_0(\omega)$ to, say, $X_0(\omega) + \delta$. For future reference, we call this the *identical noise realization approach.* K. S. Chan in his contribution to the discussion of Tong [1995] has generalized (2.3) to

$$\lambda^*(\omega) = \lim_{t\to\infty} t^{-1} \ln |\partial X_t(\omega)/\partial X_0(\omega)|. \tag{2.5}$$

As pointed out by Chan (*op. cit.*), one advantage of this generalization is that it can be applied to the more general stochastic dynamical system of the form

$$X_t = F(X_{t-1}, \epsilon_t), \tag{2.6}$$

where the dynamic noise need not be additive. Another advantage is that the new definition is invariant under a 1-1 co-ordinate transformation of the state. We shall adopt the generalized version in this paper. Note that it is obviously still in keeping with the identical noise realization scenario.

A recent alternative is to take full account of the fact that the states are now random variables and study the effect of perturbing the initial value on the distributions (or some summary characteristics) of the later states. Now, let $\{X_t\}$ denote a *general* real-valued stochastic process indexed by $t \in N$ and with finite variance. Assume that the conditional density of X_m given $X_0 = x$ exists and let us denote it by $g_m(.|x)$. We suppose that $g_m(.|x)$ is sufficiently smooth in x. For two neighbouring initial values x, $x + \delta \in R$, after time $m \geq 1$, the *divergence* of the conditional distribution of X_m may be defined once the choice of a distance function over the space of distributions is made. The choice is clearly non-unique. For example, if the negative Kullback-Leibler mutual information is used, then we may define the divergence as

$$K_m(x;\delta) = \int \{g_m(z|x+\delta) - g_m(z|x)\} \ln\{g_m(z|x+\delta)/g_m(z|x)\} dz. \tag{2.7}$$

Note that stationarity is not required for this equation. Now, for small δ, we can expand the right hand side using Taylor's series about x. This gives the approximation

$$K_m(x;\delta) = \delta^2 I_{1,m}(x) + o(\delta^2), \tag{2.8}$$

where

$$I_{1,m}(x) = \int \{\partial g_m(z|x)/\partial x\}^2 / g_m(z|x) dz. \tag{2.9}$$

If the L_2 norm is used, then we may define another divergence as

$$D_m(x;\delta) = \int \{g_m(z|x+\delta) - g_m(z|x)\}^2 dz. \tag{2.10}$$

It follows from the Taylor's expansion that

$$D_m(x;\delta) = \delta^2 I_{2,m}(x) + o(\delta^2), \tag{2.11}$$

where

$$I_{2,m}(x) = \int \{\partial g_m(z|x)/\partial x\}^2 dz. \tag{2.12}$$

Sometimes it is convenient to specialize the above setup to just the conditional mean function when the latter exists. Consider model (2.2) but now we weaken the assumption of independence and identical distribution of the dynamic noise ϵ_t to $E(\epsilon_t|X_k, k < t) = 0$. Let us refer to this model as model (2.2*). Let $F_m(x) = E[X_m|X_0 = x]$, $x \in R$ and $m \geq 1$. For $\delta \in R$,

$$F_m(x+\delta) - F_m(x) = \dot{F}_m(x)\delta + o(|\delta|), \tag{2.13}$$

where $\dot{F}_m(x)$ denotes $dF_m(x)/dx$. For model (2.2*), $F_1(x) = F(x)$ and

$$\begin{aligned} F_m(x) &= E\{F(X_{m-1})|X_0 = x\} \\ &= E\{F(F(X_{m-2}) + \epsilon_{m-1})|X_0 = x\} \\ &= E\{F((\cdots(F(x) + \epsilon_1) + \cdots) + \epsilon_{m-1})|X_0 = x\}. \end{aligned} \tag{2.14}$$

By the chain rule, differentiation of the right hand side of the above equation gives

$$\dot{F}_m(x) = E\{\prod_{k=1}^{m} \dot{F}(X_{k-1})|X_0 = x\}, \tag{2.15}$$

where we have assumed that the differentiation under the integral sign is justified. We may interpret equations (2.13) and (2.15) as stochastic generalizations of their obvious deterministic counterparts.

Given the existence of different approaches, it is natural to enquire what different aspects of initial-value sensitivity they reflect. The following very simple examples are quite instructive.

Example 2.1 Consider a simple linear first order autoregressive process

$$X_m = \epsilon_m + \alpha X_{m-1}, \tag{2.16}$$

where $\{\epsilon_j\}$ is a sequence of independent random variables each distributed as $\mathcal{N}(0,1)$. Clearly,

$$X_m = \epsilon_m + \alpha\epsilon_{m-1} + \ldots + \alpha^{m-1}\epsilon_1 + \alpha^m X_0. \tag{2.17}$$

Therefore , given $X_0 = x$, we have $X_m \sim \mathcal{N}(\alpha^m x, \sigma_m^2)$, where

$$\sigma_m^2 = \left(\frac{1-\alpha^{2m}}{1-\alpha^2}\right), \tag{2.18}$$

for $\alpha \neq 1$, and

$$\sigma_m^2 = m, \tag{2.19}$$

for $\alpha = 1$. Let us consider the conditional distribution approach first since it does not require stationarity. Now, simple calculations yield that

$$I_{1,m}(x) = \alpha^{2m}/\sigma_m^2. \tag{2.20}$$

Equation (2.20) shows a sensitivity measure which differs from the classical Lyapunov exponent in that it incorporates directly the effect of the dynamic noise in the form of a diffusion term σ_m^2 in order to adjust the impact of the disturbance δ on the drift term. Substituting the value of σ_m^2 from (2.18) in (2.20), we get

$$I_{1,m}(x) = \alpha^{2m}(1-\alpha^2)/(1-\alpha^{2m}). \tag{2.21}$$

To investigate the asymptotic behaviour of $I_{1,m}$, we consider three cases separately as follows.

(i) $|\alpha| < 1$: In this case, $I_{1,m}(x) \to 0$ as $m \to \infty$. This mimics the behaviour of the globally stable skeleton, i.e., the case with the dynamic noise switched off. Thus, even after clothing the skeleton remains initial-value insensitive.

(ii) $|\alpha| > 1$: In this case, $I_{1,m}(x) \to (\alpha^2 - 1)$ as $m \to \infty$. It is interesting to note that the limit is positive but *finite*. Thus, the stochastic model is sensitive to initial values; the sensitivity is clearly induced by the instability of the skeleton. (Recall that stationarity is not required in the definition of $K_m(x;\delta)$.)

(iii) $|\alpha| = 1$: This has points of contact with the well-known *unit-root* model in econometrics. In this case, $I_{1,m}(x) = m^{-1} \to 0$ as $m \to \infty$.

Next, let us turn to the conditional mean specialization and the identical noise realization approach. For these, we need to impose the stationarity condition $|\alpha| < 1$. Plainly for the conditional mean specialization, $|\dot{F}_m(x)| = |\alpha|^m$ for each m. Therefore

$$|F_m(x+\delta) - F_m(x)| = |\alpha|^m \delta + o(|\delta|). \tag{2.22}$$

For the identical noise approach, we have

$$|\partial X_m(\omega)/\partial X_0(\omega)| \equiv |\alpha|^m, \tag{2.23}$$

in agreement with the conditional mean specialization. In fact, all the above results are broadly in agreement if we note that under stationarity $|I_{1,m}|^{\frac{1}{2m}} \to |\alpha|$.

Example 2.2 Consider the following simple model which concentrates on the diffusion term

$$X_t = \epsilon_t X_{t-1}, \tag{2.24}$$

where $\epsilon_t \sim \mathcal{N}(0,1)$ and ϵ_t is independent of $X_s, s < t$. Clearly $\{X_t\}$ is a sequence of uncorrelated random variables, i.e., a white noise sequence.

First, the conditional mean specialization has nothing to say on the initial-value sensitivity because $E[X_m|X_0 = x] \equiv 0$, all m.

For the identical noise realization approach, simple calculation gives

$$\lambda^* = E[\ln|\epsilon_1|] = (2/\pi)^{\frac{1}{2}} \int_0^\infty (\ln x)\exp(-x^2/2)dx = -(\gamma + \ln 2)/2, \tag{2.25}$$

where γ denotes Euler's constant. Thus,

$$\lambda^* \approx -0.635 < 0.$$

Apparently the identical noise realization approach attends to characteristics of the random system other than the conditional mean. Equations (2.4) and (2.5) seem to suggest that it is unlikely that the approach attends to any particular (conditional or unconditional) moments associated with the joint distributions of $\{X_t\}$. On noting that $X_m = X_0 \prod_{j=1}^m \epsilon_j$, clearly any perturbation of X_0 will be successively shrunk by $\epsilon_j, j = 1,..,m$. The identical noise realization approach quantifies the shrinkage factor (i.e., $\exp\{\lambda^*\}$) to be roughly 0.5 for the case with standard Gaussian noise. Clearly, λ^* depends on the noise distribution. For example, for $\epsilon_1 \sim U(-\sqrt{3}, \sqrt{3})$, we have $\exp\{\lambda^*\} \approx 0.2$. Obviously, the assumption of identical noise realization yields for each m, $\partial X_t(\omega)/\partial X_0(\omega) = \prod_{j=1}^m \epsilon_j(\omega)$, in which $X_0(\omega)$ is absent. Suppose we remove this assumption. Let $X_0 = x_0$. Then $X_m(\omega) = x_0 \prod_{j=1}^m \epsilon_j(\omega)$. After perturbing x_0 to $x_0 + \delta$ and employing a *different* noise realization, say $\{\epsilon_j(\omega^*), j = 1, .., m\}$, we get $X_m(\omega^*) = (x_0 + \delta) \prod_{j=1}^m \epsilon_j(\omega^*)$. We note that in contrast to the above, x_0 will now be present in $X_m(\omega^*) - X_m(\omega)$ at least for each finite $m > 0$.

For the conditional distribution approach, simple calculation yields

$$\sqrt{I_{1,1}(x_0)} = \sqrt{2}/|x_0|.$$

The result suggests that any perturbation to the initial value x_0 is amplified or reduced accordingly as $|x_0|$ is smaller than or larger than $\sqrt{2}$. Note that conditional on x_0, $X_1 \sim \mathcal{N}(0, x_0^2)$. It is intuitively obvious that the discrepancy between say $\mathcal{N}(0, 1000)$ and $\mathcal{N}(0, 1000.5)$ is insignificant when compared with that between say $\mathcal{N}(0, 0)$ (a Dirac delta function) and $\mathcal{N}(0, 0.5)$. For $I_{1,m}(x_0), m > 1$, it seems decidedly difficult to obtain a closed form expression. However, the following approximate result holds for large m (mostly due to Professor K. S. Chan–private communication). First, we note that $X_m/X_0 = M_m S_m$, where $M_m = |\epsilon_m||\epsilon_{m-1}|...|\epsilon_1|$ and $S_m = \text{sign}(\epsilon_m \epsilon_{m-1} ... \epsilon_1)$. Second, we note that for large m, we may appeal to the Central Limit Theorem and deduce that $\ln M_m$ is approximately $\mathcal{N}(m\mu, m\sigma^2)$, where $\mu = E(\ln|\epsilon_1|)$ and $\sigma^2 = \text{Var}(\ln|\epsilon_1|)$. Finally, we deduce that

$$\sqrt{I_{1,m}(x_0)} \approx \{\sqrt{m}\sigma|x_0|\}^{-1}.$$

It is interesting to note the continuing presence of $|x_0|$ in the denominator. Further, it is clear that $I_{1,m}(x_0)$ tends to 0 as m tends to ∞ for all $x_0 \neq 0$.

Clearly the identical noise realization approach and the conditional distribution approach have overlapping as well as non-overlapping targets. The above simple examples have revealed some of their basic similarities as well as basic differences but a more systematic study is clearly necessary.

3 Miscellaneous Remarks on Prediction

(i) Do not be fooled by exaggerated claims

Prediction is a risky business. Perhaps this is one of the many reasons why it features so significantly in almost every ancient civilization and continues to occupy such a premier position in modern times. As far as gazing beyond the crystal balls and the oracle bones is concerned, fashions have come and gone even within the statistical community and now and then exaggerated claims have been made if only just to provide the occasional illusory excitement. In this respect, measures such as $\dot{F}_m(x), I_{1,m}(x), I_{2,m}(x)$ and others with their nonparametric estimates may be seen as timely moderators. It is now well known (e.g., Tong [1995]) that the predictability of a time series is reflected by the above measures which typically depend on x, the present state, and m, sometimes called the *predictive horizon*.

(ii) Never take comparative results on trust

Nobody can deny the value of a comparative study of different prediction methods. However, for such a study to have any scientific value, we maintain that it should be conducted on *genuine* post-sample data in the sense that the human operators really have no access to the post-sample data in any way at the time when the predictions are made. Despite the obvious truth of this statement, we observe that there are far more "fake" post-sample data predictions than genuine ones reported in the literature! The most recent large-scale example of genuine post-sample prediction seems to be the "Sante Fe Time Series Prediction and Analysis Competition, 1991" (results summarized in Weigend and Gershenfeld [1994]). Earlier large-scale examples were Makridakis and Hibon [1979] and Makridakis et al. [1984]. However, even in the above comparative studies, the "true answers" were actually known to the organisers (but not to the competitors). This state of affairs is not surprising in view of the necessary time constraint on the organisation. The situation may be compared with the 8-year waiting time for Ghaddar and Tong [1981], who in 1980 submitted their genuine out-of-sample predictions for the annual sunspot numbers for the years 1980 to 1987 inclusively. (For further discussion see Tong [1990], esp. Section 7.3.2.) They had to wait for almost 8 years before they knew if they had done a reasonable job! Of course, this was only a small scale exercise involving only one time series, albeit over a long time on a human scale.

Our experience convinces us that we must maintain a healthy scepticism when presented with comparative results purporting to demonstrate the superiority of a new method and yet only "historical" data are used.

We are of the opinion that when a method does show genuinely promising performance, it is most important to understand why, because no method will produce promising predictions all the time under all conditions. For example, Table 4 of Yao and Tong [1995, p. 84] shows that we should expect poorer prediction performance for the sunspot numbers over the "peak" years. Therefore, any claim of excellent performance over the "peak" years would be suspect.

Another example is the use of the neural network approach reported in Weigend and Gershenfeld [1994]. For the out-of-sample prediction of the laser data of the competition, the neural network approach produces the best entry as well as (almost) the worst entry. As neural networks tend to involve vast numbers of parameters to be tuned, different tunings can produce vastly different qualities of prediction. There is clearly a need for some systematic guidance in the tuning of the parameters.

(iii) Multiple-step models vs single-step models

Suppose we assume that model (2.2) is the true model. (This assumption could well be discovered to be invalid at a later stage leading to the problem of mis-specification.) Then the m-step ahead least-squares prediction of X_{t+m} based on observations up to now is simply the conditional mean of X_{t+m} given X_t. We can then use either the Chapman-Kolmogorov relation or the Monte Carlo method to evaluate it. (See, e.g., Tong [1995]). It is important to note that if F is *nonlinear* then

$$\begin{aligned} E[X_{t+2}|X_t] &= E[F(X_{t+1})|X_t] && (3.1)\\ &\neq F(E[X_{t+1}|X_t])\,. && (3.2) \end{aligned}$$

In other words, it is unwise to replace X_{t+1} by its 1-step ahead least-squares prediction obtained at time t in order to obtain the 2-step ahead least-squares prediction. The argument obviously applies to higher steps. The nonlinearity of F simply kills the commutativity of F and the conditional expectation operator. It is amazing how often people fall into this *commutativity-trap* and some of these are supposed to be nonlinear time series specialists! Linear prejudice dies hard.

Of course, we know that in practice F is rarely known and we replace it by its estimate, say $\hat{F}_1$. Typically, $\hat{F}_1$ is obtained as a result of minimizing an appropriate functional of the 1-step ahead predictors, say $\{X_{t+1} - \hat{F}_1(X_t),\ t = 1, \dots, N-1\}$. Clearly $\hat{F}_1$ is no longer the true model even if it is so asymptotically. Another complication which often exists in practice arises because we may be using a *mis-specified* model. For example, we may be using the model class (2.2) whilst the true model class is (2.6). In this case, it is pertinent to ask the question: if we want to predict m-steps ahead, is it better to first of all fit a multiple-step model of the form

$$X_{t+m} = F_m(X_t) + \epsilon_{t+m} \ , \tag{3.3}$$

by minimizing an appropriate functional of the m-step ahead prediction errors, say $\{X_{t+m} - \hat{F}_m(X_t),\ t = 1, \dots, N-m\}$, and then use $\hat{F}_m(X_t)$ as the m-step least-squares predictor of X_{t+m}? The comparison here is with the use of $\hat{F}_1$ of the single-step model coupled with the Chapman-Kolmogorov relation to produce an m-step least-squares prediction . (Of course, we will not commit the commutativity sin!) As far as we know, there is inadequate research on this question for the *nonlinear* case. However, R. L. Smith has presented some recent results at the Workshop reported in this volume. For the purely deterministic dynamical system, see, e.g., Farmer and Sidorowich [1987]. We conjecture that there is no uniform domination between the two approaches. In a sense the issue has to do with the robustness of $\hat{F}_1$.

For the linear case, there is some evidence which suggests that the multiple-step model may produce better prediction than the single-step model in *some* mis-specified situations. We summarize the findings of Tiao and Tsay [1994]. Specifically, let B denote the back-shift operator, namely $BX_t = X_{t-1}$. Suppose that the *true* model is

$$(1-B)^d X_t = \epsilon_t \ , \tag{3.4}$$

where $\{\epsilon_t\}$ is a sequence of independent identically distributed random variables each with a $\mathcal{N}(0,1)$ distribution, and $-0.5 < d < 0.5$. (It is known that models of this form have long-memory in the sense that the autocorrelation function is not absolutely summable.) Suppose that we do not know what the true model is and use instead a mixed autoregressive/moving average model to fit the data generated by (3.4); the *mis-specified* model takes the form

$$(1-\beta B)X_t = (1-\eta B)b_t \ , \tag{3.5}$$

with $|\beta| < 1$ and $|\eta| < 1$. We refer to this model by the usual acronym ARMA (1,1). For this ARMA (1,1) model, the standard m-step ahead least-squares predictor, $\hat{X}_t(m)$, and its corresponding error, $\hat{e}_t(m)$, are

$$\hat{X}_t(m) = \begin{cases} \beta X_t - \eta b_t & \text{for } m = 1 \ , \\ \beta^{m-1}\hat{X}_t(1) & \text{for } m > 1 \ , \end{cases} \tag{3.6}$$

and

$$\hat{e}_t(m) = \begin{cases} X_{t+1} - \hat{X}_t(1) & \text{for } m = 1\,, \\ X_{t+m} - \beta^{m-1}(\beta X_t - \eta b_t) & \text{for } m > 1\,. \end{cases} \tag{3.7}$$

Clearly, minimizing $Var[\hat{e}_t(m)]$ would produce different "least-squares estimates" of β and η for different m. (In practice, we would minimize the sample version of $Var[\hat{e}_t(m)]$.) Let $\beta(m)$ and $\eta(m)$ denote the minimizers of $Var[\hat{e}_t(m)]$ with respect to β and η. Then the multiple-step model would substitute $\beta(m)$ and $\eta(m)$ for β and η respectively in equation (3.5) to yield the m-step prediction. On the other hand, the single-step model would substitute $\beta(1)$ and $\eta(1)$ for β and η respectively in equation (3.5) to yield the m-step prediction. Tiao and Tsay [1994] have given Table 1 below to compare their prediction performance.

Table 1

Comparison of prediction performance by a multiple-step model and a single-step model. Min(m) denotes the theoretical variance of the prediction error when the true model is used. (This represents the minimum achievable value of the prediction variance for the different m.) Mℓ(m) denotes the ratio of the prediction variance for the multiple-step model related to Min(m). (The ratio is always greater than 1; the smaller the ratio the better the performance of the multiple-step model.) Similarly, Sℓ(m) denotes the variance ratio when the single-step model is used.

d	m	Min(m)	Mℓ(m)	$\beta(m)$	$\eta(m)$	Sℓ(m)
.45	1	1.000	1.044	.965	.554	1.000
	2	1.203	1.045	.982	.737	1.019
	3	1.309	1.043	.988	.813	1.041
	4	1.380	1.042	.991	.857	1.059
	5	1.433	1.040	.993	.885	1.073
	6	1.475	1.040	.994	.904	1.086
	7	1.509	1.039	.995	.917	1.098
	8	1.538	1.038	.996	.928	1.110
	9	1.564	1.037	.996	.936	1.121
	10	1.586	1.037	.997	.942	1.133
	15	1.670	1.034	.998	.961	1.193
	20	1.726	1.033	.998	.971	1.258
	50	1.896	1.028	.999	.989	1.572
	100	2.013	1.024	≈ 1	.994	1.716
	200	2.122	1.024	≈ 1	.996	1.674

d	m	Min(m)	Mℓ(m)	$\beta(m)$	$\eta(m)$	Sℓ(m)
.25	1	1.000	1.011	.835	.609	1.000
	2	1.063	1.009	.915	.784	1.005
	3	1.087	1.007	.944	.854	1.009
	4	1.101	1.006	.959	.890	1.011
	5	1.110	1.006	.967	.912	1.013
	6	1.116	1.005	.973	.927	1.015
	10	1.131	1.004	.984	.956	1.021
	15	1.141	1.003	.989	.971	1.024
	20	1.146	1.003	.992	.978	1.024
	50	1.159	1.002	.997	.991	1.017
	100	1.165	1.001	.998	.996	1.012
	200	1.170	1.001	.999	.997	1.008

The table shows that (i) for d near to 0.5, the single-step model leads to poor multiple-step ahead predictions and a substantial gain can result using the multiple-step model, and (ii) more importantly, the penalty of mis-specifying the model is not serious if the multiple-step model is used.

(iv) To transform or not to transform?

Critics of instantaneous transformation of the data prior to modelling have argued that it is difficult to undo the transformation if we want to use the model fitted to the transformed data to predict the future values of the untransformed data, unless the transformation is linear. It has been shown that whilst it is true that a naive back-transformation will lead to bias in the prediction, the bias can be removed substantially by introducing a bias correction. (See, e.g., Tong [1990], Section 6.2.4.) Specifically, let $y = f(z)$ denote the 1-1 smooth transformation from z to y. Let g denote the inverse of f. Let $\hat{Z}_t(m)$ and $\hat{Y}_t(m)$ denote the least-squares m-step ahead prediction at time t of Z_{t+m} and Y_{t+m} respectively. Then we may correct the bias by observing that

$$\hat{Z}_t(m) \approx g(\hat{Y}_t(m)) + \tfrac{1}{2}g''(\hat{Y}_t(m))Var[Y_{t+m}|Y_tY_{t-1},\dots,]. \tag{3.8}$$

(Cf. for $Z = e^Y$ where $Y \sim \mathcal{N}(\mu, \sigma^2), EZ = \exp\{\mu + \frac{1}{2}\sigma^2\}$.) For an example of a successful application, see, e.g., Tong [1990], p.422. Therefore, we would argue that prediction consideration alone does not constitute valid grounds for objection to instantaneous transformations.

(v) Nearly the same goodness of fit need not mean nearly the same prediction

It is recognised that sometimes different classes of models can give almost equally good fit to the same data set. However, it might not be as widely recognised that their predictions can sometimes be dramatically different. In fact, models which fit the same data set equally well could, in some circumstances, lead to dramatically different predictions. It is perhaps instructive to think of the fitting model as part of the initial condition of a *super dynamical system*, whose output is the prediction; this super dynamical system may be sensitively dependent on its initial condition including the fitting model.

Cox and Medley [1989] give a particularly striking illustration in the short-term prediction of the AIDS epidemic in the United Kingdom, which we summarize here.

As the issue is quite complex, we shall have to leave out some of the details. The basic model used is one of a series of point events occurring in continuous time in a Poisson process of rate $\lambda(t)$, to be called the *incidence function.* The point events represent patients newly diagnosed as having AIDS. Once diagnosed, the case is notified to the Communicable Disease Surveillance Centre (CDSC) of the Public Health Laboratories of the U.K. with a time delay (of between a few weeks and, in extreme circumstances, two or more years), called the notification delay. Cox and Medley [1989] used the AIDS data supplied by the CDSC for all cases up to 1 July 1988. These data give the calendar months of diagnosis and of the arrival of the report to CDSC. The purpose is to predict the actual (not just the notified) number of AIDS patients at any one time.

Cox and Medley [1989] have considered three different parametric forms of $\lambda(t)$, each coupled to a notification delay distribution which takes the form of a mixture of two first order Gamma distributions. The three parametric forms of $\lambda(t)$ are

$$\lambda_1(t) = \exp(\rho_1 + \rho_2 t - \rho_3 t^2), \tag{3.9}$$

$$\lambda_2(t) = \rho_3/[1 + \exp(\rho_1 - \rho_2 t)], \tag{3.10}$$

and

$$\lambda_3(t) = (\rho_1 + \rho_2 t)/[1 + \rho_3 \rho_1 \exp(-\rho_4 t)]. \tag{3.11}$$

The notification delay distribution is of the form

$$\int [\theta_1 \theta_2^2 x e^{-\theta_2 x} + (1 - \theta_1)\theta_3^2 x e^{-\theta_3 x}]dx, \tag{3.12}$$

where the integral is over the appropriate month.

Table 2

Maximum log likelihood estimates and maximized log likelihoods, ℓ, for different incidence functions and notification delay

ρ_1	ρ_2	ρ_3	ρ_4	θ_1	θ_2	θ_2	ℓ
$\lambda_1(t)$							
-2.57	1.55	0.05	–	0.57	10.65	1.52	5353.58
$\lambda_2(t)$							
8.22	0.97	1519	–	0.57	11.36	1.49	5348.09
$\lambda_3(t)$							
70.99	162.06	13.57	0.82	0.57	11.35	1.52	5350.80

Table 2 shows that over the period of the data there is no great difference between the $\lambda(t)$'s, and the maximized log likelihoods are quite close.

Cox and Medley [1989] have shown that despite the similar goodness of fit of the three models the differences between their predictions are major, even after a year or so. In fact, differences are qualitatively very significant. To proceed further it is wise that subject matter considerations be included in order to throw some light on the more plausible model and hence prediction of what is clearly an extremely pressing problem in human terms. Cox and Medley [1989] have invoked epidemiological theory to suggest that $\lambda_3(t)$ is the most plausible. Note that a standard black-box modelling approach based on the principle of parsimony, e.g., the use of model selection criteria (such as Akaike's information criterion, Schwartz's information criterion, Rissanen's minimizing description-length criterion, etc.) is here superceded by a substantive approach.

Acknowledgments

This paper was partially supported by the European Community grant ERB CHRX-CT 940693. I thank Professor Kung-sik Chan for discussion.

References

Cox, D. R. and Medley, G. F. [1989] *A process of events with notification delay and the forecasting of AIDS*, Philos. Trans. Roy. Soc. Lond., **B**, **325**, 135–45.

Farmer, J. D. and Sidorowich, J. J. [1987], *Predicting chaotic time series*, Phys. Rev. Lett., **59**, 845.

Ghaddar, D. K, and Tong, H. [1981], *Data transformation and self-exciting threshold autoregression*, Appl. Statist., **30**, 238.48.

Makridakis, S. and Hibon, M. [1979], *Accuracy of forecasting: an empirical investigation (with discussion)*, J. Roy. Statist. Soc., **A142**, 97–145.

Makridakis, S., Andersen, A., Carbone, R., Fildes, R., Hibon, M., Lewandowski, R., Newton, J., Parzen, E. and Winkler, R. [1984] *The Forecasting Accuracy of Major Time Series Methods.* John Wiley, New York.

Tiao, G. C. and Tsay, R. S. [1994], *Some advances in nonlinear and adaptive modeling in time series analysis*, J. Forecasting, **13**, 109–31.

Tong, H. [1990], *Nonlinear Time Series*, Oxford: Oxford Univ. Press.

Tong, H. [1995], *A personal overview of non-linear time series analysis from a chaos perspective (with discussion)*, Scand. J. Statist., **22**, 399–445.

Weigend, A. S. and Gershenfeld, N. A. [1994], *Time series prediction: forecasting the future and understanding the past*, Proceeding volume XV, Santa Fe Inst. Studies in the Sciences of Complexity, New York: Addison-Wesley.

Yao, Q. and Tong, H. [1995], *On prediction and chaos in stochastic systems*, in Chaos and Forecasting–Proceedings of the Royal Society Discussion Meeting, (H. Tong, ed.) Singapore: World Scientific.

Fields Institute Communications
Volume 11, 1997

A General Approach to Predictive and Fractal Scaling Dimensions in Discrete-Index Time Series

Colleen D. Cutler
Department of Statistics and Actuarial Science
University of Waterloo
Waterloo, Ontario, Canada N2L 3G1

Abstract. In this paper we establish properties of various fractal scaling dimensions (correlation dimension, information dimension, dimension of supports, etc.) for time series with a discrete time index. We also introduce the notion of *predictive dimension* and, through it, make precise the definitions of *stochastic* and *deterministic* as applied to stationary time series; specifically, we say that a stochastic time series is one possessing infinite predictive dimension. We develop results concerning the various notions of dimension, including their behaviours in time series which are functionals of finite-dimensional dynamical systems (f.d.d.s.). One key result is that even smooth functionals of f.d.d.s. may produce stochastic (rather than deterministic) time series; however, the fractal scaling dimensions of such a functional time series are always finite and bounded above by those of the underlying f.d.d.s.

1 Introduction

Modern analysis of experimental time series data has among its goals the separation of high-dimensional and stochastic dynamics from low-dimensional deterministic (in particular, chaotic) signals, estimation of system parameters (such as Lyapunov exponents and various fractal scaling dimensions), and, of course, prediction; see, for example, Broomhead and Jones [1989], Abarbanel et al. [1993], Ott, Sauer, and Yorke [1994], and Sugihara [1995]. A frequent assumption is that the observed time series X_n, $n = 1, 2, \ldots$ arises as a functional of an underlying finite-dimensional dynamical system, obtained by following some observable of the system as it evolves over time. Theorems of Takens [1981] and Sauer, Yorke, and Casdagli [1991] state, under certain mild assumptions, that for "almost all" choices of the observable, the structure of the asymptotics of such an underlying system can be recovered from the observed time series via the method of *time-delay embeddings*.

In the time-delay embedding method, successive observations (or, more generally, observations separated in time by some suitably chosen *delay parameter* $\tau \geq 1$) are grouped into d-dimensional vectors of the form $(X_{n\tau}, X_{(n+1)\tau}, \ldots, X_{(n+d-1)\tau})$,

1991 *Mathematics Subject Classification.* Primary 62M10, 58F13; Secondary 28A80.

This research supported by an operating grant from the Natural Sciences and Engineering Research Council of Canada. The author thanks an anonymous referee for useful comments.

the hope being that for sufficiently large d this *delay coordinate mapping* is in fact an *embedding* (see Definition 2.5 below) between the phase space of the original system and a subset of $\mathbb{R}^d$. (In this paper we will always take $\tau = 1$ for convenience.) The existence of an embedding means, among other things, that the distribution of the vector $(X_1, \ldots, X_d)$ (which is of course the d-dimensional joint distribution P_d of the time series) shares many of the properties of the invariant distribution of the hidden underlying system (many fractal dimension properties can be recovered with less than a full embedding; however this aspect will not be discussed in this paper). The rule, due to Sauer et al. [1991], is that generally d greater than twice the box-counting dimension of the underlying system attractor will suffice to obtain an embedding; the smallest value of d for which an embedding exists in a particular case is called the *minimal embedding dimension.* Various numerical and statistical techniques are employed to estimate the minimal embedding dimension from data and to efficiently reconstruct the phase space of the hidden dynamical system; see Casdagli [1989], Kennel, Brown, and Abarbanel [1992], Buzug and Pfister [1992], Gibson et al. [1992], Abarbanel et al. [1993], and Cheng and Tong [1995].

The goal of this paper is to study the dimension properties of the finite-dimensional (or joint) distributions of time series, with an emphasis on strictly stationary time series. We consider a collection of attributes we call *fractal scaling dimensions* (which include such typical quantities of interest as information dimension and correlation dimension) as well as a separate integer-valued index we call *predictive dimension,* which is based on the ability of a time series to predict itself using past observations. (Predictive dimension is closely related to minimal embedding dimension; see Remark 2.6.) Our approach differs from the scenario sketched in the first two paragraphs of this section in that we do not initially make any assumptions about the nature of the mechanism generating the time series and only later specialize to the situation where the time series is a functional of a finite-dimensional dynamical system. (A similar broad viewpoint is adopted in Cutler [1995a] in the case of stationary time series with a continuous time index.) One important reason for choosing this type of very general approach is to broaden our understanding of the distinction between *stochastic* and *deterministic* time series (in fact we offer precise definitions of these terms) and illuminate the extent to which notions of dimension and scaling will permit us to distinguish these two cases. For the reader already familiar with fractal dimension concepts in time series, the most interesting part of the paper is likely to be Example 3.3 (this example can be read independently of most of Section 2 provided the reader has seen Definitions 2.1 and 2.3 and has a basic understanding of fractal dimension terminology). Example 3.3 is a nontrivial example of a *stochastic* ergodic stationary time series in which all fractal scaling dimensions remain bounded. This example was motivated by the work of Osborne and Provenzale [1989] (abbreviated OP) who proposed that coloured noise with a power-law spectral density possessed finite correlation dimension and thus exhibited scaling behaviour analogous to that of a deterministic, rather than stochastic, signal. The OP example fails because true continuous-spectrum coloured noise in fact has infinite correlation dimension, while the discrete-spectrum simulation model used by OP disintegrates into uncountably many ergodic components each of which almost certainly has finite predictive dimension and can therefore be regarded as deterministic. (The predictive dimension of the components of the OP model will be the topic of a forthcoming paper.) See Cutler [1995a] for a rigorous analysis of correlation dimension in coloured noise, and Theiler [1991] for earlier motivating work.

In addition, Example 3.3 represents a stochastic time series which is also a smooth functional of a finite-dimensional dynamical system. We know that this is atypical behaviour for such functional time series since the results of Sauer et al. [1991] show that "almost all" (in the sense of prevalence) functionals result in an embedding between the time series and underlying finite-dimensional system. However, the study of atypical, or pathological, functionals may have more value than might initially be presumed. It is of course mathematically relevant in that some such functionals can be used to construct examples of stationary time series with unusual interesting behaviour, such as Example 3.3. But there may also be a practical relevance. For one thing, it is not clear that in practice functionals are chosen at random (to meet the prevalence criterion of Sauer et al. [1991]) or can always be sufficiently sensitive to perform as expected (see Remark 3.5 in this regard). More importantly, however, note that a given time series can be simultaneously regarded as a functional of a multitude of dynamical systems. (For example, a time series of temperatures taken at Waterloo can be regarded as a functional of the Waterloo climate, the Ontario climate, the Canadian climate, and even the global climate. While these systems are obviously connected, there is still the possibility of some self-contained local variables, particularly on a shorter time scale.) Thus, given a time series, we might turn the problem around and ask for which of a possible hierarchy of underlying dynamical systems an embedding exists. It might even be possible to express this in terms of notions such as genericity and prevalence (see Sauer et al. [1991] for definitions) on the entire collection of possible underlying dynamical systems.

The remainder of the paper is organized as follows: Section 2 of the paper provides precise definitions of all terms as well as statements of the main results, Section 3 is comprised of examples (with discussion) which are in fact counterexamples to some natural conjectures concerning dimension in time series, and Section 4 contains proofs of the results in Section 2.

2 Definitions, Notation, and Statement of Results

Throughout the following $\mathcal{X}$ will denote a complete separable metric space. In most applications we will have $\mathcal{X} = \mathbb{R}$ or $\mathcal{X} = \mathbb{R}^k$ (or some closed subset of one of those spaces) and the reader may simply regard $\mathcal{X}$ as a Euclidean space if desired. We will be concerned with time series X_n, $n = 1, 2, \dots$ taking values in $\mathcal{X}$. We will let P denote the distribution of the time series on the Borel sets of the infinite product space $\mathcal{X}^\infty = \times_{m=1}^{\infty}\mathcal{X}$ (equipped with the usual product topology) and let P_n denote the distribution of $(X_1, \dots, X_n)$ on the Borel sets of $\mathcal{X}^n = \times_{m=1}^{n}\mathcal{X}$. The collection P_n, $n = 1, 2, \dots$ comprise the finite-dimensional or *joint distributions* of the time series.

Definition 2.1 Let X_n, $n = 1, 2, \dots$ be a strictly stationary time series taking values in $\mathcal{X}$. We define the *predictive dimension* n_p to be

$$n_p = \min\{n \geq 1 \,|\, \text{there exists}\, T : \mathcal{X}^n \to \mathcal{X} \,\text{such that}\, X_{n+1} = T(X_1, \dots, X_n) \ \text{w.p.}\,1\}.$$

If no such function T exists for any $n \geq 1$, we set $n_p = \infty$. In the case that $n_p < \infty$, the function T corresponding to $n = n_p$ is called the *predictor function.*

If a predictor function T exists, it is clearly P_{n_p}*-a.s.* unique, and $X_{n_p+1} = T(X_1, \dots, X_{n_p})$ w.p. 1. Moreover, we have

Theorem 2.2 *Let X_n, $n = 1, 2, \dots$ be a strictly stationary time series with finite predictive dimension n_p and predictor function T. Then, for all integers*

$m \geq 0$

$$X_{m+n_p+1} = T(X_{m+1}, X_{m+2}, \dots, X_{m+n_p}) \quad \text{w.p. 1}. \tag{2.1}$$

The notion of predictive dimension leads to a precise definition of the terms *deterministic* and *stochastic* for a stationary time series.

Definition 2.3 We say that a strictly stationary time series X_n, $n = 1, 2, \dots$ is *deterministic* if $n_p < \infty$ and *stochastic* if $n_p = \infty$.

We now set some additional notation. A typical element of $\mathcal{X}^\infty$ will be denoted by the boldface $\boldsymbol{x} = (x_1, x_2, \dots)$, and the vector consisting of the first n coordinates of $\boldsymbol{x}$ will be denoted by $\boldsymbol{x}_{(n)} = (x_1, \dots, x_n)$. We let ρ denote the metric on $\mathcal{X}$ and assume throughout that the metric ρ_n on $\mathcal{X}^n$ is the "sup metric" given by

$$\rho_n(\boldsymbol{x}_{(n)}, \boldsymbol{z}_{(n)}) = \sup_{1 \leq j \leq n} \rho(x_j, z_j). \tag{2.2}$$

We will frequently need to simultaneously consider a second complete separable metric space $\mathcal{Y}$, and will use the notation ρ^* and ρ_n^* for the metrics on $\mathcal{Y}$ and $\mathcal{Y}^n$ respectively, where ρ_n^* is assumed to be the sup metric constructed from ρ^*.

Definition 2.4 A mapping $g : \mathcal{Y} \to \mathcal{X}$ between metric spaces is said to satisfy a *Lipschitz condition* if there exists a constant $k \geq 0$ such that

$$\rho(g(z), g(y)) \leq k\rho^*(z, y) \tag{2.3}$$

for all $z, y \in \mathcal{Y}$. The constant k is called a *Lipschitz constant.* If, in addition to (2.3), g is one-to-one and g^{-1} also satisfies a Lipschitz condition on its domain, then g is said to be *bi-Lipschitz.*

Definition 2.5 Let $\mathcal{Y}$ be a complete separable metric space. A Borel-measurable mapping $\varphi : \mathcal{Y} \to \mathcal{Y}$ is called a (discrete-time) *dynamical system* with *phase space* $\mathcal{Y}$, and in the case that $\mathcal{Y}$ is a compact subset of a finite-dimensional manifold (in particular, if $\mathcal{Y}$ is a compact subset of $\mathbb{R}^k$) we say that φ is a *finite-dimensional dynamical system (f.d.d.s.).* For each $y \in \mathcal{Y}$ the sequence of iterates $y, \varphi(y), \varphi^2(y), \dots$ forms an *orbit* of the dynamical system with *initial condition* y. We let μ_1 denote the distribution of the initial condition Y on $\mathcal{Y}$, and consider the generated time series $Y_n = \varphi^{n-1}(Y)$, $n = 1, 2, \dots$ where $\varphi^0(Y) = Y$. The joint distributions of this time series are denoted by μ_n, $n = 1, 2, \dots$ where μ_n is the distribution of $(Y_1, \dots, Y_n)$. The dynamical system is said to be *stationary* or *invariant* with respect to the distribution μ_1 (which is then called an *invariant distribution* of the system) if $\mu_1 \varphi^{-1} = \mu_1$. This condition is equivalent to stating that $Y_1, Y_2, \dots$ is strictly stationary with marginal distribution μ_1.

We note that the time series $Y_n = \varphi^{n-1}(Y)$, $n = 1, 2, \dots$ generated by a stationary dynamical system is of course deterministic with $n_p = 1$ and $T = \varphi$. In practice, however, we are often unable to view this complete system evolving on $\mathcal{Y}$ but rather observe a *functional* or *projection* $h : \mathcal{Y} \to \mathcal{X}$ of the system (typically $\mathcal{X} \subseteq \mathbb{R}$ or $\mathbb{R}^k$). This produces an *observed* time series X_n, $n = 1, 2, \dots$ (also stationary) taking values in $\mathcal{X}$, where $X_n = h(\varphi^{n-1}(Y))$.

The *delay coordinate mapping* $\Phi_d : \mathcal{Y} \to \mathcal{X}^d$ with embedding parameter d is given by

$$\Phi_d(y) = (h(y), h(\varphi(y)), \dots, h(\varphi^{d-1}(y))) \tag{2.4}$$

and is an attempt to identify the unknown initial condition Y of the original system with the observed delay coordinate vector $(X_1, \ldots, X_d)$ as $\Phi_d(Y) = (X_1, \ldots, X_d)$. We say that Φ_d is an *embedding* if it is a bi-Lipschitz map between $\mathcal{Y}$ and $\Phi_d(\mathcal{Y})$. Normally the definition of embedding includes a demand that differential structure also be preserved (see Sauer et al. [1991]) but here we are concerned only with preservation of dimension properties and do not even require differential structure on the underlying metric spaces. As noted in the Introduction, the *minimal embedding dimension* is the smallest value d_0 of d (if any) for which the delay coordinate map is an embedding. It follows from the one-to-one property of an embedding that any future value X_n of the observed time series is a deterministic function of the preceding d_0 values $X_{n-1}, \ldots, X_{n-d_0}$.

Remark 2.6 Predictive dimension, as we have defined it, is very close in spirit to the notion of minimal embedding dimension. However, there is a subtle but important difference between the two concepts. The notion of predictive dimension is based solely upon predictability properties of the given time series; there is no assumption made about the particular underlying mechanism generating the time series and no attempt to connect properties of the time series to properties of a particular dynamical system (other than the canonical one described below in Theorem 2.8). In the definition of minimal embedding dimension, however, the identification (up to an embedding) of the delay coordinate vectors of the time series with a specific presumed generating mechanism is of paramount importance, as the goal is to recover properties of that particular dynamical system. In Example 3.8 we exhibit a time series which is represented as a functional of two different non-equivalent dynamical systems φ and φ_1; the predictive dimension of the time series is 1 and corresponds to the minimal embedding dimension of φ_1. However, if we are actually interested in the more complex system φ and believe this to be the underlying mechanism, then a minimal embedding dimension does not exist as, due to poor choice of the functional h, the delay coordinate mapping does not produce an embedding for any value of d.

Remark 2.7 Also note that our definition of predictive dimension does not make any allowance for noise in the time series observations. The approach of Cheng and Tong [1995], using nonlinear autoregressive models, does incorporate noise, and the search for the minimal embedding dimension in their paper corresponds to a search for the predictive dimension of the deterministic skeleton of the process. Of course it is possible that the skeleton and the noise-contaminated process exhibit very different asymptotic properties.

The following theorem states that every deterministic time series can be viewed as the projection of a stationary dynamical system evolving on a finite product space; we call this the *canonical dynamical system* associated with the time series. The canonical system exhibits the same dimension properties as the functional time series (see Theorem 2.25 and Corollary 2.27) as the delay coordinate mapping produces an embedding (in fact the identity) between the phase space of the canonical system and $\mathcal{X}^{n_p}$.

Theorem 2.8 *Let X_n, $n = 1, 2, \ldots$ be a strictly stationary time series taking values in $\mathcal{X}$ and having joint distributions P_n, $n = 1, 2, \ldots$, finite predictive dimension n_p, and predictor function T. Then the mapping $\varphi_T : \mathcal{X}^{n_p} \to \mathcal{X}^{n_p}$ defined* *by*

$$\begin{aligned} \varphi_T(x) &= T(x) & \text{if } n_p = 1 \\ \varphi_T(x_1, \ldots, x_{n_p}) &= (x_2, \ldots, x_{n_p}, T(x_1, \ldots, x_{n_p})) & \text{if } n_p \geq 2 \end{aligned} \tag{2.5}$$

defines a dynamical system with invariant distribution P_{n_p}. Taking the initial condition to be $Y = (X_1, \ldots, X_{n_p})$ and defining $h : \mathcal{X}^{n_p} \to \mathcal{X}$ by $h(x_1, \ldots, x_{n_p}) = x_1$, the original time series is then recovered as the projection $X_n = h(\varphi_T^{n-1}(Y))$. Moreover, as $h(\varphi_T^{n-1}(\boldsymbol{x}_{(n_p)})) = x_n$ for each of $n = 1, \ldots, n_p$, the delay coordinate mapping

$$\Phi_{n_p}(\boldsymbol{x}_{(n_p)}) = (h(\boldsymbol{x}_{(n_p)}), h(\varphi_T(\boldsymbol{x}_{(n_p)})), \ldots, h(\varphi_T^{n_p-1}(\boldsymbol{x}_{(n_p)}))) \tag{2.6}$$

is in fact the identity map (hence an embedding) of $\mathcal{X}^{n_p}$ onto itself.

One goal of this paper is to examine the relationship between dimensions associated with the time series $Y_n = \varphi^{n-1}(Y)$, $n = 1, 2, \ldots$ generated by a stationary dynamical system, and the corresponding dimensions of an observed functional time series X_n, $n = 1, 2, \ldots$. Associated with any time series (not necessarily stationary) are a variety of scaling quantities which we will collectively call *fractal scaling dimensions.* Strictly speaking, these are scaling properties associated with the joint distributions P_n of the time series. We provide brief definitions of these quantities below (detailed descriptions and relationships between these definitions can be found in, for example, Cutler [1993].)

Definition 2.9 Given a probability measure ν, we let $\operatorname{Supp}\nu$ denote the (topological) *support* of ν i.e. the smallest closed set S such that $\nu(S) = 1$. Given any set E we let $\dim(E)$ and $\operatorname{Dim}(E)$ denote, respectively, the *Hausdorff* and *packing dimensions* of E. We also consider the (upper) *box-counting dimension*

$$\Delta(E) = \limsup_{\epsilon \to 0} \frac{\log N(\epsilon)}{\log 1/\epsilon}$$

where $N(\epsilon)$ is the minimum number of closed balls of diameter ϵ (or less) which cover E. If $N(\epsilon) = \infty$ for some $\epsilon > 0$ (which will occur if E is not totally bounded) then automatically $\Delta(E) = \infty$. We have the general inequality $\dim(E) \leq \operatorname{Dim}(E) \leq \Delta(E)$ for any subset E of any metric space; see Cutler [1995b].

In this paper we will be concerned with the Hausdorff, packing, and box-counting dimensions of the supports $\operatorname{Supp} P_n$ of the joint distributions of a time series.

Definition 2.10 Associated with any probability measure ν on the Borel sets of a metric space are auxiliary probability distributions $\hat{\nu}$ and $\tilde{\nu}$ on the Borel sets of the extended real line $[0, \infty]$ which, respectively, describe the manner in which ν-mass is distributed with respect to Hausdorff and packing dimension. Specifically

$$\hat{\nu}([0, c]) = \sup\{\nu(B) \mid \dim(B) \leq c,\ B \text{ Borel}\}.$$

The analogous definition holds for $\tilde{\nu}$ with "dim" replaced by "Dim." We call $\hat{\nu}$ (respectively, $\tilde{\nu}$) the *Hausdorff (respectively, packing) dimension distribution* of ν. In many typical cases we in fact have $\hat{\nu} = \tilde{\nu}$; however, in this paper we will not be concerned with the circumstances leading to equality. An important point of interest to us will be the behaviour of the dimension distributions $\widehat{P}_n$ and $\widetilde{P}_n$ of the joint distributions P_n.

The behaviour of the dimension distributions can be investigated through that of the *pointwise dimension map* (also called the *local Hölder exponent*) associated with ν

$$\alpha_\nu(x) = \lim_{\epsilon\to 0} \frac{\log \nu(B(x,\epsilon))}{\log \epsilon}$$

where $B(x,\epsilon)$ is the closed ball of radius ϵ centred at x. The above limit may not exist, in which case we consider the two maps α_ν^- and α_ν^+ defined by replacing "lim" with, respectively, "lim inf" and "lim sup". In fact α_ν^- is specifically connected with $\hat{\nu}$, and α_ν^+ is specifically connected with $\tilde{\nu}$, via the following

$$\begin{aligned} \hat{\nu}([0,c]) &= \nu(\{x \mid \alpha_\nu^-(x) \le c\}) \\ \tilde{\nu}([0,c]) &= \nu(\{x \mid \alpha_\nu^+(x) \le c\}). \end{aligned} \tag{2.7}$$

See Cutler [1990, 1993, 1995b] for details.

We will, for notational convenience, use the symbols α_n^-, α_n^+, and α_n in place of the more complicated $\alpha_{P_n}^-$, $\alpha_{P_n}^+$, and α_{P_n} (which have domain $\mathcal{X}^n$) for the joint distributions P_n. We will also find it convenient to be able to regard α_n^-, α_n^+, and α_n as simultaneously being defined on the infinite product space $\mathcal{X}^\infty$, and accomplish this by setting $\alpha_n^-(\boldsymbol{x}) = \alpha_n^-(\boldsymbol{x}_{(n)})$, with a similar interpretation of $\alpha_n^+(\boldsymbol{x})$ and $\alpha_n(\boldsymbol{x})$.

Definition 2.11 Global scaling properties of a distribution ν can be described by the *generalized Rényi dimensions* which, for each $q \neq 0$, are given by

$$d_\nu(q) = \lim_{\epsilon\to 0} \frac{\log \int \nu(B(x,\epsilon))^q \, \nu(dx)}{q \log \epsilon}.$$

If the limit does not exist we instead consider the "lim inf" and "lim sup" and define $d_\nu^-(q)$ and $d_\nu^+(q)$ accordingly. A special case of interest is $q = 1$, which describes the scaling behaviour of the average ν-mass in a ball of small radius ϵ. This quantity $d_\nu(1)$ is called the (spatial) *correlation dimension* of ν.

We also consider the *information dimension* of ν, given by

$$\sigma_\nu = \lim_{\epsilon\to 0} \frac{\int \log \nu(B(x,\epsilon)) \, \nu(dx)}{\log \epsilon}.$$

We consider the lower and upper limits σ_ν^- and σ_ν^+ if the limit defining σ_ν does not exist. Note that σ_ν plays the role of $d_\nu(0)$, although we may or may not have $\sigma_\nu = \lim_{q\to 0} d_\nu(q)$.

We will use the more convenient shorthand notation $d_n^-(q)$, $d_n^+(q)$, $d_n(q)$, σ_n^-, σ_n^+, and σ_n in connection with the joint distributions P_n.

Theorem 2.12 *Let X_n, $n = 1, 2, \ldots$ be a time series taking values in $\mathcal{X}$ and having joint distributions P_n. Then all fractal scaling dimensions of P_n, with the possible exceptions of* $\dim(\operatorname{Supp} P_n)$ *and* $\operatorname{Dim}(\operatorname{Supp} P_n)$, *increase with n. More precisely, for all n we have*

1. $d_n^-(q) \le d_{n+1}^-(q)$ *and* $d_n^+(q) \le d_{n+1}^+(q)$ *for each* $q \neq 0$
2. $\sigma_n^- \le \sigma_{n+1}^-$ *and* $\sigma_n^+ \le \sigma_{n+1}^+$
3. $\alpha_n^-(\boldsymbol{x}) \le \alpha_{n+1}^-(\boldsymbol{x})$ *and* $\alpha_n^+(\boldsymbol{x}) \le \alpha_{n+1}^+(\boldsymbol{x})$ *for each* $\boldsymbol{x} \in \mathcal{X}^\infty$
4. $\widehat{P}_n((c,\infty]) \le \widehat{P}_{n+1}((c,\infty])$ *and* $\widetilde{P}_n((c,\infty]) \le \widetilde{P}_{n+1}((c,\infty])$ *for each* $c \ge 0$
5. $\Delta(\operatorname{Supp} P_n) \le \Delta(\operatorname{Supp} P_{n+1})$
6. *if* $\operatorname{Supp} P_{n+1}$ *is compact, then we also have* $\dim(\operatorname{Supp} P_n) \le \dim(\operatorname{Supp} P_{n+1})$ *and* $\operatorname{Dim}(\operatorname{Supp} P_n) \le \operatorname{Dim}(\operatorname{Supp} P_{n+1})$.

Remark 2.13 If the assumption of compactness of the supports is removed, we may find $\dim(\operatorname{Supp} P_n) > \dim(\operatorname{Supp} P_{n+1})$ and $\operatorname{Dim}(\operatorname{Supp} P_n) > \operatorname{Dim}(\operatorname{Supp} P_{n+1})$ for some n, even for strictly stationary time series; see Example 3.1.

Remark 2.14 The increasing property of the fractal scaling dimensions of the joint distributions makes it possible, by taking limits as $n \to \infty$, to define a set of fractal scaling dimensions associated with the overall time series e.g. we could define the *upper information dimension of the time series* as $\sigma^+ = \lim_{n\to\infty} \sigma_n^+$. An alternative approach involves placing a suitable metric on the infinite product space $\mathcal{X}^\infty$ and defining the fractal scaling dimensions directly in terms of the overall distribution P; however, this approach seems unattractive because of the lack of a canonical choice for the metric and the possible inaccessibility of the resulting dimensions based on any finite sample $X_1, \ldots, X_n$ of the time series. (Such drawbacks are discussed in greater detail, albeit in the context of continuous-index time series, in Cutler [1995a].)

Corollary 2.15 *If X_n, $n = 1, 2, \ldots$ is a strictly stationary time series and* $\operatorname{Supp} P_1$ *is compact, then*

$$\text{dimension}(\operatorname{Supp} P_n) \le \text{dimension}(\operatorname{Supp} P_{n+1}) \tag{2.8}$$

for all $n \ge 1$, for dimension $= \dim, \operatorname{Dim}$, *or* Δ.

Proof It is straightforward to show that, for any time series,

$$\operatorname{Supp} P_n \subseteq \operatorname{Supp} Q_1 \times \operatorname{Supp} Q_2 \times \cdots \times \operatorname{Supp} Q_n \tag{2.9}$$

where Q_j is the distribution of X_j. (Note that $Q_1 = P_1$.) In the case of a strictly stationary time series we have $Q_j = P_1$ for every j, and so $\operatorname{Supp} P_n$ is compact for every n if and only if $\operatorname{Supp} P_1$ is compact. The rest then follows from Theorem 2.12. Example 3.1 shows that the assumption of compactness of $\operatorname{Supp} P_1$ cannot, in general, be removed. □

Remark 2.16 In subsequent theorems we will be assuming that certain mappings, e.g., predictor functions T, dynamical systems φ, functionals h, satisfy Lipschitz conditions over their entire domains. It might appear that we could possibly weaken these assumptions by assuming only, for example, that φ is Lipschitz μ_1-*a.s.* where μ_1 is the invariant initial distribution of the system. However, it is easy to show that if a function g is Lipschitz on a subset E of a complete metric space then g extends uniquely to a Lipschitz function (with the same Lipschitz constant) on the closure $\bar{E}$. Thus, if φ is μ_1-*a.s.* Lipschitz, then φ is automatically Lipschitz on $\operatorname{Supp} \mu_1$, and since $\varphi(\operatorname{Supp} \mu_1) \subseteq \operatorname{Supp} \mu_1$, we can simply restrict $\mathcal{Y} = \operatorname{Supp} \mu_1$. Similar arguments can be made in the case of predictors T and functionals h. Thus there is no loss of generality in making the simplifying assumption that the mappings are defined and Lipschitz over the entire given domains.

Definition 2.17 If $T : \mathcal{X}^{n_p} \to \mathcal{X}$ is the predictor function associated with a deterministic time series X_n, $n = 1, 2, \ldots$ we define, for each $n \ge n_p$, the vector-valued function $\boldsymbol{T}_n : \mathcal{X}^n \to \mathcal{X}^{n+1}$ by

$$\boldsymbol{T}_n(x_1, \ldots, x_n) = (x_1, \ldots, x_n, T(x_{n-n_p+1}, \ldots, x_n)).$$

Consequently, we have $(X_1, \ldots, X_{n+1}) = \boldsymbol{T}_n(X_1, \ldots, X_n)$ w. p. 1 for each $n \ge n_p$.

Theorem 2.18 *Let X_n, $n = 1, 2, \ldots$ be a strictly stationary time series taking values in $\mathcal{X}$ and having finite predictive dimension n_p. Suppose the predictor function T satisfies a Lipschitz condition on $\mathcal{X}^{n_p}$. Then, for $n \ge n_p$, all fractal scaling* *dimensions of the joint distributions P_n coincide with those of P_{n_p}. More precisely, for all $n > n_p$ we have*

1. $d_n^-(q) = d_{n_p}^-(q)$ *and* $d_n^+(q) = d_{n_p}^+(q)$ *for each* $q \neq 0$
2. $\sigma_n^- = \sigma_{n_p}^-$ *and* $\sigma_n^+ = \sigma_{n_p}^+$
3. $\alpha_n^-(\boldsymbol{x}) = \alpha_{n_p}^-(\boldsymbol{x})$ *and* $\alpha_n^+(\boldsymbol{x}) = \alpha_{n_p}^+(\boldsymbol{x})$ w. p. 1 *for* $\boldsymbol{x} \in \mathcal{X}^\infty$. *More specifically,* $\alpha_n^-(\boldsymbol{x}) = \alpha_{n_p}^-(\boldsymbol{x})$ *and* $\alpha_n^+(\boldsymbol{x}) = \alpha_{n_p}^+(\boldsymbol{x})$ *if* $\boldsymbol{x}$ *satisfies* $\boldsymbol{x}_{(n)} = \boldsymbol{T}_{n-1} \circ \cdots \circ \boldsymbol{T}_{n_p+1} \circ \boldsymbol{T}_{n_p}(\boldsymbol{x}_{(n_p)})$. *Otherwise* $\alpha_n^-(\boldsymbol{x}) = \alpha_n^+(\boldsymbol{x}) = \infty$.
4. $\widehat{P}_n = \widehat{P}_{n_p}$ *and* $\widetilde{P}_n = \widetilde{P}_{n_p}$
5. dimension(Supp P_n) = dimension(Supp P_{n_p}) *for* dimension = dim, Dim, *or* Δ.

Remark 2.19 The conclusions of the preceding theorem need not hold if the assumption that T is Lipschitz is eliminated; see Example 3.1 for a case where item 5. is violated. (We could obtain a weaker set of dimension inequalities if we were to assume, for example, that T satisfies a Hölder condition rather than a Lipschitz condition.) Note also that if T is Lipschitz we do not need to make any assumption about compactness of the supports in order to obtain item 5.

Definition 2.20 Let $\varphi : \mathcal{Y} \to \mathcal{Y}$. For each $n \geq 1$ we define the vector-valued function $\varphi_n : \mathcal{Y} \to \mathcal{Y}^n$ by

$$\varphi_n(y) = (y, \varphi(y), \ldots, \varphi^{n-1}(y)).$$

Note that φ_1 is the identity map on $\mathcal{Y}$, and that $\varphi_n(Y) = (Y_1, \ldots, Y_n)$.

Corollary 2.21 *Let $\varphi : \mathcal{Y} \to \mathcal{Y}$ be a dynamical system with invariant distribution μ_1 and suppose that φ satisfies a Lipschitz condition on $\mathcal{Y}$. Let Y be a random initial condition distributed according to μ_1. Then all fractal scaling dimensions of the joint distributions μ_n of the time series $Y_n = \varphi^{n-1}(Y)$, $n = 1, 2, \ldots$ coincide with those of μ_1. That is, for all $n > 1$ we have*

1. $d_{\mu_n}^-(q) = d_{\mu_1}^-(q)$ *and* $d_{\mu_n}^+(q) = d_{\mu_1}^+(q)$ *for each* $q \neq 0$
2. $\sigma_{\mu_n}^- = \sigma_{\mu_1}^-$ *and* $\sigma_{\mu_n}^+ = \sigma_{\mu_1}^+$
3. $\alpha_{\mu_n}^-(\boldsymbol{y}) = \alpha_{\mu_1}^-(\boldsymbol{y}_{(1)})$ *and* $\alpha_{\mu_n}^+(\boldsymbol{y}) = \alpha_{\mu_1}^+(\boldsymbol{y}_{(1)})$ w. p. 1 *for* $\boldsymbol{y} \in \mathcal{Y}^\infty$. *More specifically,* $\alpha_{\mu_n}^-(\varphi_n(y)) = \alpha_{\mu_1}^-(y)$ *and* $\alpha_{\mu_n}^+(\varphi_n(y)) = \alpha_{\mu_1}^+(y)$ *for all* $y \in \mathcal{Y}$.
4. $\hat{\mu}_n = \hat{\mu}_1$ *and* $\tilde{\mu}_n = \tilde{\mu}_1$
5. dimension(Supp μ_n) = dimension(Supp μ_1) *for* dimension = dim, Dim, *or* Δ.

Proof This follows directly from Theorem 2.18, with $\mathcal{X} = \mathcal{Y}$, $T = \varphi$, $n_p = 1$, $X_n = Y_n$, and $P_n = \mu_n$. □

Theorem 2.22 *Let $\varphi : \mathcal{Y} \to \mathcal{Y}$ be a dynamical system with invariant distribution μ_1 and suppose that φ satisfies a Lipschitz condition on $\mathcal{Y}$. Suppose $h : \mathcal{Y} \to \mathcal{X}$ also satisfies a Lipschitz condition and we observe the functional time series $X_n = h(\varphi^{n-1}(Y))$, $n = 1, 2, \ldots$ where Y is a random initial condition chosen according to μ_1. Then all fractal scaling dimensions of the joint distributions P_n, with the possible exceptions of* dim(Supp P_n) *and* Dim(Supp P_n), *are bounded above by the corresponding fractal scaling dimensions of μ_1. That is, for each $n \geq 1$ we have*

1. $d_n^-(q) \leq d_{\mu_1}^-(q)$ *and* $d_n^+(q) \leq d_{\mu_1}^+(q)$ *for each* $q \neq 0$
2. $\sigma_n^- \leq \sigma_{\mu_1}^-$ *and* $\sigma_n^+ \leq \sigma_{\mu_1}^+$

3. $\alpha_n^-(\boldsymbol{x}) \leq \inf\{\alpha_{\mu_1}^-(y) \mid \Phi_n(y) = \boldsymbol{x}_{(n)}\}$ *and* $\alpha_n^+(\boldsymbol{x}) \leq \inf\{\alpha_{\mu_1}^+(y) \mid \Phi_n(y) = \boldsymbol{x}_{(n)}\}$ w. p. 1 *for* $\boldsymbol{x} \in \mathcal{X}^\infty$. *More specifically,* $\alpha_n^-(\Phi_n(y)) \leq \alpha_{\mu_1}^-(y)$ *and* $\alpha_n^+(\Phi_n(y)) \leq \alpha_{\mu_1}^+(y)$ *for all* $y \in \mathcal{Y}$, *where* Φ_n *is the delay coordinate mapping of* $\mathcal{Y}$ *into* $\mathcal{X}^n$.
4. $\widehat{P}_n((c,\infty]) \leq \hat{\mu}_1((c,\infty])$ *and* $\widetilde{P}_n((c,\infty]) \leq \tilde{\mu}_1((c,\infty])$ *for each* $c \geq 0$
5. $\Delta(\operatorname{Supp} P_n) \leq \Delta(\operatorname{Supp} \mu_1)$
6. *if* $\operatorname{Supp} \mu_1$ *is compact or if* h *is a closed mapping, then we also have* $\dim(\operatorname{Supp} P_n) \leq \dim(\operatorname{Supp} \mu_1)$ *and* $\operatorname{Dim}(\operatorname{Supp} P_n) \leq \operatorname{Dim}(\operatorname{Supp} \mu_1)$.

Corollary 2.23 *Let* $\varphi : \mathcal{Y} \to \mathcal{Y}$ *be a finite-dimensional dynamical system, where* $\mathcal{Y}$ *is a compact subset of* $\mathbb{R}^k$. *Suppose that* φ *has invariant distribution* μ_1 *and satisfies a Lipschitz condition on* $\mathcal{Y}$. *Suppose* $h : \mathcal{Y} \to \mathcal{X}$ *also satisfies a Lipschitz condition and we observe the functional time series* $X_n = h(\varphi^{n-1}(Y))$, $n = 1, 2, \ldots$, *where* Y *is a random initial condition chosen according to* μ_1. *Then, for all* $n \geq 1$

1. $d_n^+(q) \leq k$ *for each* $q > 0$
2. $\sigma_n^+ \leq k$
3. $\alpha_n^+(\boldsymbol{x}) \leq k$ w. p. 1 *for* $\boldsymbol{x} \in \mathcal{X}^\infty$
4. $\widehat{P}_n((k,\infty]) = \widetilde{P}_n((k,\infty]) = 0$
5. $\Delta(\operatorname{Supp} P_n) \leq k$.

Proof For any distribution μ_1 on a compact subset of $\mathbb{R}^k$ it is well known that $\sigma_{\mu_1}^+ \leq k$, $\alpha_{\mu_1}^+(x) \leq k$ μ_1*-a.s.*, $d_{\mu_1}^+(q) \leq k$ for $q > 0$, and that the dimension distributions are supported on the interval $[0, k]$; see, for example, Cutler [1993]. It is also well known that the box-counting dimension of any compact subset of $\mathbb{R}^k$ (in this case $\operatorname{Supp} \mu_1$) is bounded above by k. The rest then follows from Theorem 2.22. □

Remark 2.24 The preceding theorem and corollary show that the fractal scaling dimensions of a smooth functional of an f.d.d.s. are always bounded. The same is not true of predictive dimension; Example 3.3 is a stochastic time series arising as a smooth functional of an f.d.d.s. The predictive dimension will be finite, however, if the delay coordinate mapping between the f.d.d.s. and the time series is an embedding for some d, and this is the focus of the remaining results in this section.

Theorem 2.25 *Let* $\varphi : \mathcal{Y} \to \mathcal{Y}$ *be a dynamical system with invariant distribution* μ_1, *and let* μ_n, $n = 1, 2, \ldots$ *be the joint distributions of the corresponding time series* $Y_n = \varphi^{n-1}(Y)$, $n = 1, 2, \ldots$ *where* Y *is a random initial condition with distribution* μ_1. *Suppose* $h : \mathcal{Y} \to \mathcal{X}$ *is a Borel-measurable mapping, and* $X_n = h(\varphi^{n-1}(Y))$, $n = 1, 2, \ldots$ *is the resulting functional time series with joint distributions* P_n, $n = 1, 2, \ldots$. *If the delay coordinate mapping* Φ_d *is an embedding for some* d *then the functional time series is deterministic with* $n_p \leq d$ *and, for each* $n \geq 1$, *all fractal scaling dimensions of* μ_n *coincide with those of* P_{d+n-1}. *That is*

1. $d_{\mu_n}^-(q) = d_{d+n-1}^-(q)$ *and* $d_{\mu_n}^+(q) = d_{d+n-1}^+(q)$ *for each* $q \neq 0$
2. $\sigma_{\mu_n}^- = \sigma_{d+n-1}^-$ *and* $\sigma_{\mu_n}^+ = \sigma_{d+n-1}^+$
3. $\alpha_{\mu_n}^-(\varphi_n(y)) = \alpha_{d+n-1}^-(\Phi_{d+n-1}(y))$ *and* $\alpha_{\mu_n}^+(\varphi_n(y)) = \alpha_{d+n-1}^+(\Phi_{d+n-1}(y))$ *for all* $y \in \mathcal{Y}$.
4. $\hat{\mu}_n = \widehat{P}_{d+n-1}$ *and* $\tilde{\mu}_n = \widetilde{P}_{d+n-1}$
5. $\operatorname{dimension}(\operatorname{Supp} \mu_n) = \operatorname{dimension}(\operatorname{Supp} P_{d+n-1})$ *for* dimension = dim, Dim, *or* Δ.

Corollary 2.26 *Suppose the hypotheses of Theorem 2.25 hold, and in addition* φ *satisfies a Lipschitz condition. Then the functional time series is deterministic with* $n_p \leq d$ *and, for all* $n \geq d$, *the fractal scaling dimensions of the joint distributions* P_n *coincide with those of* μ_1.

Proof This follows directly from Corollary 2.21 and Theorem 2.25. □

Corollary 2.27 *Let* $\varphi_T : \mathcal{X}^{n_p} \to \mathcal{X}^{n_p}$ *be the canonical dynamical system associated with a deterministic time series* X_n, $n = 1, 2, \ldots$ *having predictive dimension* n_p, *predictor function* T, *and joint distributions* P_n. *If* T *is Lipschitz, then all fractal scaling dimensions of the joint distributions* μ_n *of the time series* $Y_n = \varphi_T^{n-1}(Y)$, $n = 1, 2, \ldots$, *where* $Y = (X_1, \ldots, X_{n_p})$, *coincide with those of* P_{n_p}.

Proof The delay coordinate mapping in (2.6) is the identity and hence bi-Lipschitz, and so Theorem 2.25 implies that the fractal scaling dimensions of μ_n coincide with those of P_{n_p+n-1} for each $n \geq 1$. If in addition T is Lipschitz, Theorem 2.18 implies that P_{n_p+n-1} and P_{n_p} share the same dimensions for each n, yielding the desired result. □

3 Examples and Discussion

Example 3.1 In this example we construct a strictly stationary real-valued time series which is deterministic (with $n_p = 1$) having supports that satisfy $\dim(\operatorname{Supp} P_1) > \dim(\operatorname{Supp} P_n)$ for all $n \geq 2$. (See Theorems 2.12 and 2.18 and Corollary 2.15 in connection to this.)

We let $A_1 = \{2, 3, 4, \ldots\}$ and $A_2 = \mathbb{Q} \cap [0, 1]$, where $\mathbb{Q}$ denotes the rational numbers, and set $A = A_1 \cup A_2$. Now since both A_1 and A_2 are countably infinite, there exists a bijection $g : A_1 \to A_2$.

Define $T : A \to A$ by

$$T(x) = \begin{cases} g(x) & \text{if } x \in A_1 \\ g^{-1}(x) & \text{if } x \in A_2 \end{cases}.$$

Note that T is well defined (as A_1 and A_2 are disjoint sets) and is a bijection of A onto itself which in fact satisfies $T = T^{-1}$. T will be the predictor function of our time series which will take values in A and be defined recursively by $X_{n+1} = T(X_n)$. It remains only to specify the (necessarily discrete) marginal distribution P_1 of X_1 in such a way as to obtain stationarity; it suffices to ensure that $T(X_1)$ has the same distribution as X_1. This is accomplished by assigning positive probabilities to the points of A in any way such that the probabilities sum to 1 and satisfy $P_1(\{g(x)\}) = P_1(\{x\})$ for each $x \in A_1$. For example, we could define $P_1(\{g(x)\}) = P_1(\{x\}) = 2^{-x}$ for each of $x = 2, 3, \ldots$. As a result we get $P_1(\{T^{-1}(x)\}) = P_1(\{x\})$ for each $x \in A$, and the process is stationary.

Now since the closure $\bar{A} = A_1 \cup [0, 1]$, it follows that $\operatorname{Supp} P_1 = A_1 \cup [0, 1]$ and so $\dim(\operatorname{Supp} P_1) = \operatorname{Dim}(\operatorname{Supp} P_1) = 1$. The distribution P_2, however, is supported on the countable set $A^* = \{(x, T(x)) \mid x \in A\}$ which we will show is closed. Note that $\rho_2((x, T(x)), (z, T(z))) = \max(|x - z|, |T(x) - T(z)|)$. If at least one of x, z belongs to A_1, then $|x - z| \geq 1$, and if both x, z belong to A_2 then both $T(x), T(z)$ belong to A_1, giving $|T(x) - T(z)| \geq 1$. Thus any two distinct points $(x, T(x))$ and $(z, T(z))$ of A^* are at least distance 1 apart. Therefore A^* is closed, giving $\operatorname{Supp} P_2 = A^*$. Moreover, as A^* is countable, we have $\dim(\operatorname{Supp} P_2) = \operatorname{Dim}(\operatorname{Supp} P_2) = 0$. A similar argument shows $\dim(\operatorname{Supp} P_n) = \operatorname{Dim}(\operatorname{Supp} P_n) = 0$ for all $n > 2$.

We add that of course $\operatorname{Supp} P_1$ is not compact and the predictor function T is not Lipschitz (and in fact widely separates the points of A_2, which makes the example work).

Example 3.2 The simplest example of a stochastic time series with finite fractal scaling dimensions is a sequence of independent and identically distributed 0–1 variables. Let $0 < p < 1$ and set $P([X_n = 1]) = p$, $P([X_n = 0]) = 1 - p$. Then $\operatorname{Supp} P_n$ is a discrete set of 2^n points, and all fractal scaling dimensions of P_n are concentrated at the value 0, i.e., $\Delta(\operatorname{Supp} P_n) = 0$, $\widehat{P}_n = \widetilde{P}_n = \delta_0$ (where δ_x denotes the point mass at x), $d_n(q) = \sigma_n = 0$, and $\alpha_n(x) = 0$ P-*a.s.*

Example 3.3 Here we present a more sophisticated example in which the observed time series is stochastic but also a smooth functional of an underlying f.d.d.s. The joint distributions P_n have both discrete and continuous components, and the information dimension is strictly increasing in n and approaches a finite asymptote as $n \to \infty$.

Set $\mathcal{Y} = \mathcal{X} = [0,1]$ and let the f.d.d.s. be given by the tent map $\varphi : [0,1] \to [0,1]$ defined by

$$\varphi(y) = \begin{cases} 2y & \text{if } 0 \le y \le 1/2 \\ 2(1-y) & \text{if } 1/2 \le y \le 1 \end{cases} \tag{3.1}$$

and the functional $h : [0,1] \to [0,1]$ given by

$$h(y) = \begin{cases} 0 & \text{if } 0 \le y \le 1/2 \\ 2(y-1/2) & \text{if } 1/2 \le y \le 1 \end{cases} . \tag{3.2}$$

Note that φ and h are both Lipschitz continuous (see Remark 3.4 below for discussion of various smoother modifications of φ and h).

The original f.d.d.s. is invariant (in fact ergodic) with respect to Lebesgue measure $\mu_1 = \lambda$ on $[0,1]$, and all fractal scaling dimensions of the system are concentrated at the value 1. Specifically, $d_\lambda(q) = 1$ for all q, $\sigma_\lambda = 1$, $\alpha_\lambda(x) = 1$ for all $0 \le x \le 1$, $\hat{\lambda} = \tilde{\lambda} = \delta_1$, and $\operatorname{Supp} \lambda = [0,1]$ has Hausdorff, packing, and box-counting dimension 1.

To see that the functional time series $X_n = h(\varphi^{n-1}(Y))$, $n = 1, 2, \ldots$ has infinite predictive dimension, note that if the initial condition Y lies in the interval $[0, 2^{-n}]$ (an event having positive probability 2^{-n}) then $X_1 = \cdots = X_n = 0$ and the $(n+1)^{th}$ observation X_{n+1} cannot be predicted based on this information; in fact the conditional distribution of X_{n+1} given $X_1 = \cdots = X_n = 0$ coincides with the marginal P_1 of X_1, which is easily seen to be the discrete-continuous mixture

$$P_1 = \frac{1}{2}\delta_0 + \frac{1}{2}\lambda. \tag{3.3}$$

This is actually a very interesting example, because the future of the functional time series is completely determined once the first nonzero observation is encountered, so there is a certain type of predictability inherent in the process. However, as illustrated above, the length of time until occurrence of the first nonzero is random and can be arbitrarily long. Moreover, since the underlying f.d.d.s. is ergodic and will repeatedly visit small neighbourhoods of the origin, we will observe arbitrarily long strings of 0's occurring infinitely often in almost all realizations of the functional time series. (The functional time series, viewed as a shift dynamical system on the space of real sequences, is also ergodic as ergodicity is preserved under functionals.)

It is not difficult to see that the joint distribution P_n is a mixture of 2^n distributions in $\mathbb{R}^n$, one being the point mass at the origin $(0, \ldots, 0)$, and the remaining

$2^n - 1$ each being a continuous uniform distribution over a segment of a smooth 1-dimensional curve. The dimension distributions take the form

$$\widehat{P}_n = \widetilde{P}_n = \frac{1}{2^n}\delta_0 + (1 - \frac{1}{2^n})\delta_1 \tag{3.4}$$

and so vary continuously in n, in fact converging to $\hat{\lambda} = \delta_1$ as $n \to \infty$. The information dimensions, which may be computed as the means of the dimension distributions since each P_n has compact support, take the values

$$\sigma_n = 1 - \frac{1}{2^n} \tag{3.5}$$

and so approach the asymptote $\sigma_\lambda = 1$ as $n \to \infty$. The remaining Rényi dimensions, however, are constant in n and take the form

$$d_n(q) = \begin{cases} 0 & \text{for } q > 0 \\ 1 & \text{for } q < 0 \end{cases} . \tag{3.6}$$

The $q < 0$ region is determined by the uniform 1-dimensional components of P_n while the $q > 0$ region is always dominated by the discrete component of P_n. The target value $d_\lambda(q) = 1$ of the underlying f.d.d.s. is never actually approached by the functional time series for $q > 0$.

Remark 3.4 The choices of φ and h in the above example were motivated by ease of analysis of the resulting functional time series. Note that we could replace (3.1) with any smooth, even infinitely differentiable, system on [0,1] possessing an absolutely continuous invariant distribution with density function symmetric about the point 1/2 and bounded below by a positive constant; an example of such is given by the logistic map $\varphi(y) = 4y(1-y)$. We may also choose h to possess any number m of derivatives as desired by, for example, defining $h(y) = 0$ for $0 \le y \le 1/2$ and $h(y) = 2^{m+1}(y - 1/2)^{m+1}$ for $1/2 \le y \le 1$. We may even select h to be infinitely differentiable; an example is given by $h(y) = 0$ for $0 \le y \le 1/2$ and $h(y) = \exp(2 - (y - 1/2)^{-1})$ for $1/2 < y \le 1$.

All these choices produce a functional time series with the same qualitative behaviour as that of Example 3.3; the predictive dimension will be infinite, P_n will have an atom at $(0, \ldots, 0)$, and equations (3.4)–(3.6) will hold. The only real change is that the continuous 1-dimensional components of P_n will no longer be uniform and may possess infinite singularities. The point here is that smoothness of the underlying f.d.d.s. or of the projection h is not an impediment to the construction of functional time series which are stochastic.

Remark 3.5 It is also worth noting that h in (3.2) can be modified to be constant on only an arbitrarily small interval $[0, \epsilon]$, rather than on $[0, 1/2]$, without affecting the qualitative behaviour of the resulting functional time series in Example 3.3. This is of practical significance, as it raises the possibility of a functional being (for some reason, perhaps measurement limitations) insensitive to a small region of the phase space.

Remark 3.6 From the point of view of data analysis, the stochastic nature of a realization from Example 3.3 should reveal itself in the observation of strings of 0's of varying length followed by differing nonzero observations. However, a data analysis technique such as the false nearest neighbours method of Kennel et al. [1992] should also be effective in showing that the functional time series cannot be successfully embedded in a low-dimensional phase space. ▰

Remark 3.7 Example 3.3 also illustrates another interesting point about the joint distributions P_n of a projected time series. Note that the ergodic invariant measure λ of the f.d.d.s. has the property that its dimension distributions $\hat{\lambda}$ and $\tilde{\lambda}$ are each concentrated at a single point (in this case at the point 1). This property is called *exact dimensionality* and is always possessed by the *ergodic* invariant distributions of Lipschitz continuous systems (see Cutler [1990]). However, equation (3.4) shows that this property need not be inherited by the joint distributions of a functional time series.

Example 3.8 Here we illustrate the point that a deterministic time series can arise as a functional of many different, not necessarily equivalent, dynamical systems. Let K_1 and K_2 be compact subsets of $\mathbb{R}^2$, let $\varphi_1 : K_1 \to K_1$ be a smooth dynamical system with invariant distribution ν, and let $\varphi_2 : K_1 \times K_2 \to K_2$ be a smooth mapping. We now define a dynamical system $\varphi : K_1 \times K_2 \to K_1 \times K_2$ by

$$\varphi(x,y) = (\varphi_1(x), \varphi_2(x,y)) .$$

If we take the functional h to be projection on to the x coordinate, i.e., $h(x,y) = x$, then the canonical dynamical system associated with the functional time series is obviously φ_1, and we recover the fractal scaling properties of ν. These will generally differ from those of the invariant distribution of φ if the mapping φ_2 is sufficiently interesting. Of course this choice of h corresponds to a (rare) bad selection of functional for the f.d.d.s. φ.

4 Proofs of Theorems

Proof of Theorem 2.2 From stationarity it follows that for each $m \geq 0$

$$(X_{m+1}, \ldots, X_{m+n_p}, X_{m+n_p+1}) \stackrel{\mathcal{D}}{\sim} (X_1, \ldots, X_{n_p}, X_{n_p+1}) \tag{4.1}$$

where $\stackrel{\mathcal{D}}{\sim}$ denotes equality in distribution. Hence we have

$$(T(X_{m+1}, \ldots, X_{m+n_p}), X_{m+n_p+1}) \stackrel{\mathcal{D}}{\sim} (T(X_1, \ldots, X_{n_p}), X_{n_p+1}) . \tag{4.2}$$

It follows from the definition of predictive dimension that the right hand side of (4.2) lies w. p. 1 along the diagonal in $\mathcal{X}^2$; hence the left hand side of (4.2) must do the same, yielding $X_{m+n_p+1} = T(X_{m+1}, \ldots, X_{m+n_p})$ w. p. 1. □

Proof of Theorem 2.8 Let B be a Borel set in $\mathcal{X}^{n_p}$. From the stationarity of X_n, $n = 1, 2, \ldots$ we have

$$\begin{aligned} P_{n_p}(B) &= P([(X_1, \ldots, X_{n_p}) \in B]) \\ &= P([(X_2, \ldots, X_{n_p+1}) \in B]) \\ &= P([\varphi_T(X_1, \ldots, X_{n_p}) \in B]) \\ &= P_{n_p}(\varphi_T^{-1}(B)) \end{aligned} \tag{4.3}$$

yielding $P_{n_p}\varphi_T^{-1} = P_{n_p}$ as required. □

The proofs of the remaining theorems rest on the following lemmas.

Lemma 4.1 *Let $g : \mathcal{Y} \to \mathcal{X}$ be a mapping between metric spaces. If g satisfies a Lipschitz condition on $\mathcal{Y}$ then for every $E \subseteq \mathcal{Y}$ we have*

$$\text{dimension}(g(E)) \leq \text{dimension}(E) \tag{4.4}$$

for dimension = dim, Dim, *or* Δ. *If g is in fact bi-Lipschitz then for every* $E \subseteq \mathcal{Y}$ *we have the equality*

$$\text{dimension}(g(E)) = \text{dimension}(E) \tag{4.5}$$

for dimension = dim, Dim, *or* Δ.

Proof See Definition 2.4 for the definition of Lipschitz and bi-Lipschitz functions. Equations (4.4) and (4.5) are well known results; see, for example, Falconer [1990]. □

Lemma 4.2 *Let $g : \mathcal{Y} \to \mathcal{X}$ be a Lipschitz mapping. Then there exists a positive constant k such that, for all $y \in \mathcal{Y}$ and all $\epsilon > 0$*

$$B(y, \epsilon/k) \subseteq g^{-1}(B(g(y), \epsilon)) . \tag{4.6}$$

If g is in fact bi-Lipschitz, then there exist positive constants k_1 and k_2 such that, for all $y \in \mathcal{Y}$ and all $\epsilon > 0$

$$B(y, \epsilon/k_2) \subseteq g^{-1}(B(g(y), \epsilon)) \subseteq B(y, \epsilon/k_1) . \tag{4.7}$$

Proof If g is bi-Lipschitz there exist positive constants $k_1 \leq k_2$ such that

$$k_1 \rho^*(z, y) \leq \rho(g(z), g(y)) \leq k_2 \rho^*(z, y) \tag{4.8}$$

for all $z, y \in \mathcal{Y}$. Note that k_1^{-1} is a Lipschitz constant of g^{-1}. If g is only Lipschitz with Lipschitz constant k, then the right hand side of (4.8) holds with $k_2 = k$. The lemma follows. □

Lemma 4.3 *Let $g : \mathcal{Y} \to \mathcal{X}$ be a bi-Lipschitz map. Then g maps closed sets to closed sets (i.e., g is a closed mapping), and hence also Borel sets to Borel sets.*

Proof It follows, by simple standard analysis arguments, that the completeness of $\mathcal{Y}$ plus the left hand side of Equation (4.8) and continuity of g, imply that g maps closed sets to closed sets. Thus the domain $g(\mathcal{Y})$ of g^{-1} is closed and hence Borel. Now since g^{-1} is continuous and $g = (g^{-1})^{-1}$, it follows that g maps Borel sets to Borel sets. □

Lemma 4.4 *Let $g : \mathcal{Y} \to \mathcal{X}$ be Lipschitz, and suppose $\gamma = \nu g^{-1}$ where ν and γ are probability measures on the Borel sets of $\mathcal{Y}$ and $\mathcal{X}$ respectively. Then all fractal scaling dimensions of γ, with the possible exceptions of* $\dim(\operatorname{Supp}\gamma)$ *and* $\operatorname{Dim}(\operatorname{Supp}\gamma)$, *are bounded above by the corresponding fractal scaling dimensions of ν. That is*

1. $d_\gamma^-(q) \leq d_\nu^-(q)$ *and* $d_\gamma^+(q) \leq d_\nu^+(q)$ *for each* $q \neq 0$
2. $\sigma_\gamma^- \leq \sigma_\nu^-$ *and* $\sigma_\gamma^+ \leq \sigma_\nu^+$
3. $\alpha_\gamma^-(x) \leq \inf\{\alpha_\nu^-(y) \mid g(y) = x\}$ *and* $\alpha_\gamma^+(x) \leq \inf\{\alpha_\nu^+(y) \mid g(y) = x\}$ γ-*a.s. Alternatively,* $\alpha_\gamma^-(g(y)) \leq \alpha_\nu^-(y)$ *and* $\alpha_\gamma^+(g(y)) \leq \alpha_\nu^+(y)$ *for all* $y \in \mathcal{Y}$.
4. $\hat{\gamma}((c, \infty]) \leq \hat{\nu}((c, \infty])$ *and* $\tilde{\gamma}((c, \infty]) \leq \tilde{\nu}((c, \infty])$ *for each* $c \geq 0$
5. $\Delta(\operatorname{Supp}\gamma) \leq \Delta(\operatorname{Supp}\nu)$
6. *if* $\operatorname{Supp}\nu$ *is compact or if g is a closed mapping, then we also have* $\dim(\operatorname{Supp}\gamma) \leq \dim(\operatorname{Supp}\nu)$ *and* $\operatorname{Dim}(\operatorname{Supp}\gamma) \leq \operatorname{Dim}(\operatorname{Supp}\nu)$.

Proof Let k be a Lipschitz constant of g. From (4.6) and the relation $\gamma = \nu g^{-1}$ we obtain the inequality

$$\nu(B(y, \epsilon/k)) \leq \gamma(B(g(y), \epsilon)) \tag{4.9}$$

The proofs of items 1. and 2. are similar to one another, and follow from (4.9) and a change of variable. We will prove 2. Assuming for convenience that all limits exist (the same argument applies separately to each of "lim inf" and "lim sup"), we have

$$\begin{aligned}
\sigma_\gamma &= \lim_{\epsilon\to 0} \frac{\int_{\mathcal{X}} \log\gamma(B(x,\epsilon))\,\gamma(dx)}{\log\epsilon} \\
&= \lim_{\epsilon\to 0} \frac{\int_{\mathcal{X}} \log\gamma(B(x,\epsilon))\,\nu g^{-1}(dx)}{\log\epsilon} \\
&= \lim_{\epsilon\to 0} \frac{\int_{\mathcal{Y}} \log\gamma(B(g(y),\epsilon))\,\nu(dy)}{\log\epsilon} \quad \text{by change of variable} \\
&\leq \lim_{\epsilon\to 0} \frac{\int_{\mathcal{Y}} \log\nu(B(y,\epsilon/k))\,\nu(dy)}{\log\epsilon} \quad \text{using (4.9)} \\
&= \lim_{\epsilon\to 0} \left(\frac{\int_{\mathcal{Y}} \log\nu(B(y,\epsilon/k))\,\nu(dy)}{\log\epsilon/k}\right)\left(\frac{\log\epsilon/k}{\log\epsilon}\right) \\
&= \sigma_\nu
\end{aligned} \tag{4.10}$$

We now prove item 3. Again assuming that limits exist for convenience, we have

$$\begin{aligned}
\alpha_\gamma(g(y)) &= \lim_{\epsilon\to 0} \frac{\log\gamma(B(g(y),\epsilon))}{\log\epsilon} \\
&\leq \lim_{\epsilon\to 0} \frac{\log\nu(B(y,\epsilon/k))}{\log\epsilon} \quad \text{by (4.9)} \\
&= \lim_{\epsilon\to 0} \left(\frac{\log\nu(B(y,\epsilon/k))}{\log\epsilon/k}\right)\left(\frac{\log\epsilon/k}{\log\epsilon}\right) \\
&= \alpha_\nu(y).
\end{aligned} \tag{4.11}$$

Since $\gamma(g(\mathcal{Y})) = \nu(\mathcal{Y}) = 1$, item 3. follows. Now consider item 4. We have

$$\begin{aligned}
\hat{\gamma}([0,c]) &= \gamma(\{x \mid \alpha_\gamma(x) \leq c\}) \quad \text{using (2.7)} \\
&= \nu g^{-1}(\{x \mid \alpha_\gamma(x) \leq c\}) \\
&= \nu(\{y \mid \alpha_\gamma(g(y)) \leq c\}) \\
&\geq \nu(\{y \mid \alpha_\nu(y) \leq c\}) \quad \text{from (4.11)} \\
&= \hat{\nu}([0,c]) \quad \text{again using (2.7).}
\end{aligned} \tag{4.12}$$

Thus we must have $\hat{\gamma}((c,\infty]) \leq \hat{\nu}((c,\infty])$. An analogous argument holds for $\tilde{\gamma}$ and $\tilde{\nu}$. It remains to prove items 5. and 6. The key result needed here is the relation

$$\operatorname{Supp}\gamma = \overline{g(\operatorname{Supp}\nu)} \tag{4.13}$$

where the bar denotes closure. To prove (4.13), we first note that $\gamma(g(\operatorname{Supp}\nu)) = \nu(\operatorname{Supp}\nu) = 1$, yielding one direction $\operatorname{Supp}\gamma \subseteq \overline{g(\operatorname{Supp}\nu)}$. To see the opposite direction, let $\epsilon > 0$ and suppose $y \in g(\operatorname{Supp}\nu)$. Then, from the continuity of g, there exists $x \in \operatorname{Supp}\nu$ and $\delta > 0$ such that $g(B(x,\delta)) \subseteq B(y,\epsilon)$. Thus $\gamma(B(y,\epsilon)) \geq \gamma(g(B(x,\delta))) \geq \nu(B(x,\delta)) > 0$, giving $g(\operatorname{Supp}\nu) \subseteq \operatorname{Supp}\gamma$ and hence $\overline{g(\operatorname{Supp}\nu)} \subseteq$

$\operatorname{Supp}\gamma$. Now, using (4.13), the fact that the box-counting dimension of a set is the same as that of its closure, and (4.4), we get

$$\Delta(\operatorname{Supp}\gamma) = \Delta(\overline{g(\operatorname{Supp}\nu)}) = \Delta(g(\operatorname{Supp}\nu)) \le \Delta(\operatorname{Supp}\nu) \tag{4.14}$$

proving item 5. Now if g is a closed mapping or if $\operatorname{Supp}\nu$ is compact, then the image $g(\operatorname{Supp}\nu)$ is a closed set and we may remove the closure bar in (4.13) to get $\operatorname{Supp}\gamma = g(\operatorname{Supp}\nu)$. Item 6. then follows by applying (4.4). □

Lemma 4.5 *Let $g : \mathcal{Y} \to \mathcal{X}$ be bi-Lipschitz, and suppose $\gamma = \nu g^{-1}$ where ν and γ are probability measures on the Borel sets of $\mathcal{Y}$ and $\mathcal{X}$ respectively. Then all fractal scaling dimensions of γ coincide with those of ν. That is*

1. $d_\gamma^-(q) = d_\nu^-(q)$ *and* $d_\gamma^+(q) = d_\nu^+(q)$ *for each* $q \neq 0$
2. $\sigma_\gamma^- = \sigma_\nu^-$ *and* $\sigma_\gamma^+ = \sigma_\nu^+$
3. $\alpha_\gamma^-(x) = \alpha_\nu^-(g^{-1}(x))$ *and* $\alpha_\gamma^+(x) = \alpha_\nu^+(g^{-1}(x))$ γ*-a.s.*
   Alternatively, $\alpha_\gamma^-(g(y)) = \alpha_\nu^-(y)$ *and* $\alpha_\gamma^+(g(y)) = \alpha_\nu^+(y)$ *for all* $y \in \mathcal{Y}$.
4. $\hat{\gamma} = \hat{\nu}$ *and* $\tilde{\gamma} = \tilde{\nu}$
5. $\operatorname{dimension}(\operatorname{Supp}\gamma) = \operatorname{dimension}(\operatorname{Supp}\nu)$ *for* $\operatorname{dimension} = \dim, \operatorname{Dim}$, *or* Δ.

Proof As a consequence of the bi-Lipschitz property, Equation (4.7), and the relation $\gamma = \nu g^{-1}$, we have the inequalities

$$\nu(B(y, \epsilon/k_2)) \le \gamma(B(g(y), \epsilon)) \le \nu(B(y, \epsilon/k_1)). \tag{4.15}$$

Now the proofs of items 1.–4. are by change of variable as in the proof of Lemma 4.4, except that inequalities in both directions are obtained by using first the left hand side of (4.15) and then the right hand side of (4.15). To obtain item 5., it follows from (4.13) and Lemma 4.3 that $\operatorname{Supp}\gamma = g(\operatorname{Supp}\nu)$. Applying (4.5) then gives 5. □

Proof of Theorem 2.12 The usual relationship between joint distributions gives $P_n(B) = P_{n+1}(B \times \mathcal{X})$ for each Borel set B of $\mathcal{X}^n$, and so we may write $P_n = P_{n+1}\boldsymbol{\pi}_n^{-1}$ where $\boldsymbol{\pi}_n : \mathcal{X}^{n+1} \to \mathcal{X}^n$ is the coordinate mapping

$$\boldsymbol{\pi}_n(x_1, \ldots, x_{n+1}) = (x_1, \ldots, x_n). \tag{4.16}$$

We will write this in shorthand as $\boldsymbol{\pi}_n(\boldsymbol{x}_{(n+1)}) = \boldsymbol{x}_{(n)}$. The coordinate mapping is a contraction (hence Lipschitz) on $\mathcal{X}^{n+1}$ since $\rho_n(\boldsymbol{x}_{(n)}, \boldsymbol{z}_{(n)}) \le \rho_{n+1}(\boldsymbol{x}_{(n+1)}, \boldsymbol{z}_{(n+1)})$. Thus the theorem follows directly from Lemma 4.4 with $g = \boldsymbol{\pi}_n$. To clarify the notation in item 3., recall that we set $\alpha_n(\boldsymbol{x}) = \alpha_n(\boldsymbol{x}_{(n)})$ for $\boldsymbol{x} \in \mathcal{X}^\infty$, and so we have

$$\begin{aligned} \alpha_n(\boldsymbol{x}) = \alpha_n(\boldsymbol{x}_{(n)}) &= \alpha_n(\boldsymbol{\pi}_n(\boldsymbol{x}_{(n+1)})) \\ &\le \alpha_{n+1}(\boldsymbol{x}_{(n+1)}) \quad \text{using item 3. of Lemma 4.4} \\ &= \alpha_{n+1}(\boldsymbol{x}). \end{aligned} \tag{4.17}$$

The proof is complete. □

Remark 4.6 $\boldsymbol{\pi}_n$ is not a closed mapping, so if $\operatorname{Supp} P_{n+1}$ is not compact then the projection $\boldsymbol{\pi}_n(\operatorname{Supp} P_{n+1})$ need not be closed; an example is given by $A = \boldsymbol{\pi}_1(\operatorname{Supp} P_2)$ in Example 3.1.

Proof of Theorem 2.18 Let $n \ge n_p$ throughout the following and suppose that T has Lipschitz constant k. We will show that $P_{n+1} = P_n\boldsymbol{T}_n^{-1}$, where $\boldsymbol{T}_n$ is given in Definition 2.17, and that $\boldsymbol{T}_n$ is bi-Lipschitz. The theorem will then follow from Lemma 4.5 with $g = \boldsymbol{T}_n$.

We first verify that $\boldsymbol{T}_n$ is bi-Lipschitz with constants $k_1 = 1$ and $k_2 = \max(1, k)$. Note that $\boldsymbol{T}_n$ is obviously one-to-one and satisfies

$$\begin{aligned}
&\rho_{n+1}(\boldsymbol{T}_n(\boldsymbol{x}_{(n)}), \boldsymbol{T}_n(\boldsymbol{z}_{(n)})) \\
&\quad = \rho_{n+1}((x_1, \dots, x_n, T(x_{n-n_p+1}, \dots, x_n)), (z_1, \dots, z_n, T(z_{n-n_p+1}, \dots, z_n))) \\
&\quad = \max(\max_{1 \le j \le n} \rho(x_j, z_j), \rho(T(x_{n-n_p+1}, \dots, x_n), T(z_{n-n_p+1}, \dots, z_n))) \\
&\quad = \max(\rho_n(\boldsymbol{x}_{(n)}, \boldsymbol{z}_{(n)}), \rho(T(x_{n-n_p+1}, \dots, x_n), T(z_{n-n_p+1}, \dots, z_n))).
\end{aligned} \tag{4.18}$$

This immediately yields the left hand side inequality in (4.8) with $k_1 = 1$. Now

$$\begin{aligned}
\rho(T(x_{n-n_p+1}, \dots, x_n), T(z_{n-n_p+1}, \dots, z_n)) &\le k \max_{n-n_p+1 \le j \le n} \rho(x_j, z_j) \\
&\le k \quad \rho_n(\boldsymbol{x}_{(n)}, \boldsymbol{z}_{(n)})
\end{aligned} \tag{4.19}$$

and this and (4.18) give the right hand side inequality in (4.8) with $k_2 = \max(1, k)$. Now to see that $P_{n+1} = P_n \boldsymbol{T}_n^{-1}$, let B be a Borel set in $\mathcal{X}^{n+1}$ and note that

$$\begin{aligned}
P_{n+1}(B) &= P([(X_1, \dots, X_{n+1}) \in B]) \\
&= P([\boldsymbol{T}_n(X_1, \dots, X_n) \in B]) \\
&= P([(X_1, \dots, X_n) \in \boldsymbol{T}_n^{-1}(B)]) \\
&= P_n(\boldsymbol{T}_n^{-1}(B))
\end{aligned} \tag{4.20}$$

as claimed. Thus Lemma 4.5 applies. To clarify item 3., note that, similar to the proof of Theorem 2.12, we will have $\alpha_{n+1}(\boldsymbol{x}) = \alpha_n(\boldsymbol{x})$ for $\boldsymbol{x}$ satisfying $\boldsymbol{x}_{(n+1)} = \boldsymbol{T}_n(\boldsymbol{x}_{(n)})$. This is a set of probability one. If $\boldsymbol{x}_{(n+1)} \neq \boldsymbol{T}_n(\boldsymbol{x}_{(n)})$ then $\boldsymbol{x}_{(n+1)} \notin \operatorname{Supp} P_{n+1}$, giving $\alpha_{n+1}(\boldsymbol{x}) = \alpha_{n+1}(\boldsymbol{x}_{(n+1)}) = \infty$. □

Proof of Theorem 2.22 Let k be the Lipschitz constant associated with h and let μ_n, $n = 1, 2, \dots$ denote the joint distributions of the underlying time series $Y_n = \varphi^{n-1}(Y)$. Define $\boldsymbol{h}_n : \mathcal{Y}^n \to \mathcal{X}^n$ by $\boldsymbol{h}_n(y_1, \dots, y_n) = (h(y_1), \dots, h(y_n))$ and note that $\Phi_n = \boldsymbol{h}_n \circ \boldsymbol{\varphi}_n$ and that $\boldsymbol{h}_n$ is a closed mapping if h is a closed mapping. It is easy to check that $\boldsymbol{h}_n$ also has Lipschitz constant k, and since $(X_1, \dots, X_n) = \boldsymbol{h}_n(Y_1, \dots, Y_n)$ we have $P_n = \mu_n \boldsymbol{h}_n^{-1}$. Thus Lemma 4.4 applies with $g = \boldsymbol{h}_n$ and we conclude that the fractal scaling dimensions of P_n are bounded above by those of μ_n. Now from Corollary 2.21 it follows that the fractal scaling dimensions of μ_n coincide with those of μ_1, and the theorem is proved. Alternatively, one could use Lemma 4.4 in one step by noting that the composition $\Phi_n = \boldsymbol{h}_n \circ \boldsymbol{\varphi}_n$ is Lipschitz (and closed if h is closed) and that $P_n = \mu_1 \Phi_n^{-1}$. □

Proof of Theorem 2.25 To see that the functional time series has predictive dimension $n_p \le d$, note that $X_{d+1} = h(\varphi^d(Y)) = h \circ \varphi^d \circ \Phi_d^{-1}(X_1, \dots, X_d)$. Now let Φ_d have Lipschitz constant k, and let $\mathcal{E}_n = \{\boldsymbol{\varphi}_n(y) \mid y \in \mathcal{Y}\}$. We will show that the mapping $\boldsymbol{h}_{n,d} : \mathcal{E}_n \to \mathcal{X}^{d+n-1}$ defined by

$$\boldsymbol{h}_{n,d}(\boldsymbol{\varphi}_n(y)) = \Phi_{d+n-1}(y) \tag{4.21}$$

is a bi-Lipschitz map of $\mathcal{E}_n$ onto its range $\boldsymbol{h}_{n,d}(\mathcal{E}_n)$. We begin by constructing some auxiliary maps. First note that the metric on $(\mathcal{X}^d)^n$, derived from the metric ρ on $\mathcal{X}$ by using the sup metric throughout, coincides with ρ_{nd}, the metric on $\mathcal{X}^{nd}$. Let $\Theta_n : \mathcal{X}^{d+n-1} \to (\mathcal{X}^d)^n$ be defined by

$$\begin{aligned}
&\Theta_n(x_1, \dots, x_{d+n-1}) \\
&\quad = ((x_1, \dots, x_d), (x_2, \dots, x_{d+1}), \dots, (x_n, \dots, x_{d+n-1}))
\end{aligned} \tag{4.22}$$

and note that Θ_n is in fact an isometry between $\mathcal{X}^{d+n-1}$ and $\Theta_n(\mathcal{X}^{d+n-1})$, i.e., Θ_n is one-to-one and satisfies

$$\rho_{nd}(\Theta_n(\boldsymbol{z}_{(d+n-1)}), \Theta_n(\boldsymbol{x}_{(d+n-1)})) = \rho_{d+n-1}(\boldsymbol{z}_{(d+n-1)}, \boldsymbol{x}_{(d+n-1)}) \tag{4.23}$$

for all $\boldsymbol{z}_{(d+n-1)}, \boldsymbol{x}_{(d+n-1)} \in \mathcal{X}^{d+n-1}$. Now define $\bar{\boldsymbol{h}}_{n,d} : \mathcal{E}_n \to (\mathcal{X}^d)^n$ by

$$\bar{\boldsymbol{h}}_{n,d}(\varphi_n(y)) = (\Phi_d(y), \Phi_d(\varphi(y)), \ldots, \Phi_d(\varphi^{n-1}(y))). \tag{4.24}$$

Using the Lipschitz property of Φ_d, we obtain

$$\begin{aligned} \rho_{nd}(\bar{\boldsymbol{h}}_{n,d}(\varphi_n(y)), \bar{\boldsymbol{h}}_{n,d}(\varphi_n(z))) &= \sup_{1\le j\le n} \rho_d(\Phi_d(\varphi^{j-1}(y)), \Phi_d(\varphi^{j-1}(z))) \\ &\le k \sup_{1\le j\le n} \rho^*(\varphi^{j-1}(y), \varphi^{j-1}(z)) \qquad (4.25) \\ &= k\, \rho_n^*(\varphi_n(y), \varphi_n(z)) \end{aligned}$$

which shows that $\bar{\boldsymbol{h}}_{n,d}$ satisfies the same Lipschitz condition as Φ_d. A similar argument, using the fact that Φ_d^{-1} also satisfies a Lipschitz condition, shows that $\bar{\boldsymbol{h}}_{n,d}$ is bi-Lipschitz. This, plus (4.23), and the fact that $\boldsymbol{h}_{n,d}$ can be written as

$$\boldsymbol{h}_{n,d} = \Psi_n \circ \bar{\boldsymbol{h}}_{n,d} \tag{4.26}$$

where $\Psi_n = \Theta_n^{-1}$, proves that $\boldsymbol{h}_{n,d}$ is bi-Lipschitz. Now, since $(X_1, \ldots, X_{d+n-1}) = \boldsymbol{h}_{n,d}(Y_1, \ldots, Y_n)$, we have $P_{d+n-1} = \mu_n \boldsymbol{h}_{n,d}^{-1}$ and the proof follows from Lemma 4.5. □

References

Abarbanel, H. D. I., Brown, R., Sidorowich, J. J. and Tsimring, L. S. [1993] *The analysis of observed chaotic data in physical systems,* Rev. Modern Phys., **65,** 1331–1392.

Broomhead, D. S. and Jones, R. [1989] *Time series analysis,* Proc. Roy. Soc. London A, **423,** 103–122.

Buzug, T. and Pfister, G. [1992] *Comparisons of algorithms calculating optimal embedding parameters for delay time coordinates,* Physica D, **58,** 127–137.

Casdagli, M. C. [1989] *Nonlinear prediction of chaotic time series,* Physica D, **35,** 335–356.

Cheng, B. and Tong, H. [1995] *Orthogonal projection, embedding dimension, and sample size in chaotic time series from a statistical perspective,* Chaos and Forecasting (H. Tong, ed.), World Scientific, Singapore, 1–29. Condensed version in Philos. Trans. Roy. Soc. London A (1994), **348,** 325–339.

Cutler, C. D. [1990] *Connecting ergodicity and dimension in dynamical systems,* Ergodic Theory Dynamical Systems, **10,** 451–462.

Cutler, C. D. [1993] *A review of the theory and estimation of fractal dimension,* Dimension Estimation and Models (H. Tong, ed.), World Scientific, Singapore, 1–107.

Cutler, C. D. [1995a] *A theory of correlation dimension for stationary time series,* Chaos and Forecasting (H. Tong, ed.), World Scientific, Singapore, 31–56. Condensed version in Philos. Trans. Roy. Soc. London A [1994], **348,** 343–355.

Cutler, C. D. [1995b] *The density theorem and Hausdorff inequality for packing measure in general metric spaces,* Illinois J. Math., **39,** 676–694.

Falconer, K. J. [1990] *Fractal Geometry,* John Wiley & Sons, New York.

Gibson, J. F., Farmer, J. D., Casdagli, M. and Eubank, S. [1992] *An analytic approach to practical state space construction,* Physica D, **57,** 1–30.

Kennel, M. B., Brown, R. and Abarbanel, H. D. I. [1992] *Determining embedding dimension for phase space reconstruction using a geometrical construction,* Phys. Rev. A, **45,** 3403–3411.

Osborne, A. R. and Provenzale, A. [1989] *Finite correlation dimension for stochastic systems with power-law spectra,* Physica D, **35,** 357–381.

Ott, E., Sauer, T. and Yorke, J. A. (eds.) [1994] *Coping with Chaos,* John Wiley & Sons, New York.

Sauer, T., Yorke, J. A. and Casdagli, M. [1991] *Embedology,* J. Statist. Phys., **65,** 579–616.

Sugihara, G. [1995] *Prediction as a criterion for classifying natural time series,* Chaos and Forecasting (H. Tong, ed.), World Scientific, Singapore, 269–294. Condensed version in Philos. Trans. Roy. Soc. London A [1994], **348,** 477–495.

Takens, F. [1981] *Detecting strange attractors in turbulence,* Lecture Notes in Mathematics, **898,** Springer-Verlag, New York, 366–381.

Theiler, J. [1991] *Some comments on the correlation dimension of* $1/f^{\alpha}$ *noise,* Phys. Lett. A, **155,** 480–493.

Fields Institute Communications
Volume 11, 1997

Statistics for Continuity and Differentiability: An Application to Attractor Reconstruction from Time Series

Louis M. Pecora
Thomas L. Carroll
and
James F. Heagy
Code 6343, Naval Research Laboratory
Washington, DC, USA 20375

Abstract. Given a sampling of pairs of points from a function's domain and range, what can we say about the mathematical properties, if any, of that function? For example, were the point pairs related by a function that is continuous? Or differentiable? This is the same situation we are faced with when we apply standard statistical tools, like linear correlation, except here we are asking more fundamental questions about the properties of the functional relation, rather than about its exact form (linear, polynomial, etc.). We show here several statistics that provide likelihood or confidence measures of each functional property given samplings of point pairs. We apply these statistics to time-series reconstructions of attractors. An important observation is that many commonly asked questions about properties of such reconstructions are actually questions about more fundamental mathematical properties. Such questions include: determinism in a time series, general synchronization between attractors, the relation of a filtered or transformed attractor to the original one, and whether or not two time series come from the same dynamical system (dynamical interdependence).

1 Introduction

Statisticians are often faced with the problem of having pairs of points sampled from a process or experiment. The immediate question often is: what is the relationship, if any, between the point pairs? What kind of function, if any, relates the two data sets? A typical test is a calculation of the linear correlation coefficient. If this quantity is near ± 1, the data sets are considered linearly related. This test is generalizable to multivariate data, say $\boldsymbol{x} \in \mathbb{R}^n$ and $\boldsymbol{y} \in \mathbb{R}^m$, where $\boldsymbol{x} = \boldsymbol{A}\boldsymbol{y}$ is the functional relationship to be tested. (Other statistical quantities have also been developed to test for a possible linear relationship.) However, low linear correlation does not imply that other (nonlinear) relationships do not exist.

1991 *Mathematics Subject Classification.* Primary 58F13, 62M10; Secondary 62M20.

We propose a more general test than testing for one type of functional model. We show that it is possible to start with mathematical definitions of function properties and construct a statistic which can be calculated from a sampling of point pairs from the function's domain and range. The statistic would then indicate whether or not the points are related by a function with that property. For example, we construct a statistic to test for the existence of a *continuous* function. (Note that we assume no particular functional form.) We adhere strictly to the definition of continuity in constructing our statistic. A simple example of a situation where this test would be applied is shown in Figure 1. There we have a sampling of points from the domain (x) and range (y) of a possible function. We ask the question: what is the likelihood that these points were sampled from a continuous function $f : x \to y$?

We can also ask the same question about differentiability, namely, were these points sampled from a *differentiable* function? Again, we create a statistic by adhering to the mathematical definition of differentiability. We can generalize this approach to functions on higher dimensional manifolds. One place this becomes important (and the original motivation for this work) is in the analysis of time series reconstruction of attractors. Attractors exist on manifolds, and when we have two simultaneously measured time series we have the potential for their being related by a function. One attractor can be considered the domain of the function (we will refer to this as the X space) and the other the range of the function (the Y space). The importance of testing for continuity and differentiability comes from Takens' theorem (Takens [1981], Mané [1981], Sauer, Yorke, and Casdagli [1991]) and also from rephrasing common questions in mathematical terms. The former is a statement that one can reconstruct the trajectory of a dynamical system in phase space from a single time series of data from the system. In fact the theorem states that there is a diffeomorphism between the reconstructed trajectory and the system's actual trajectory in its physical phase space — a *diffeomorphism* is a continuous, differentiable mapping between geometric objects whose inverse is also continuous and differentiable. Thus, for example, if we have two simultaneously measured time series in an experiment we might want to see if there is a diffeomorphism between them. If so, they are measurements of the same dynamical system, by Takens' theorem. In this way we would have answered the "colloquial" question: are the time series from dynamically interacting parts of the experiment? Other colloquial questions include:

- Given a time series, does it come from a deterministic system? Such a system has the characteristic that points nearby in phase space stay nearby at later times (predictability). This is just a restatement of continuity in forward time. Other determinism questions (smoothness?) can also be raised. (For earlier related work on this topic see Kaplan [1994].)
- Does a transformation of a time series (e.g., filtering) preserve certain dynamical invariants? An example of this is fractal dimension, which is preserved under a C^1 transformation (Sauer and Yorke [1996]).
- Are two parts of a dynamical system in general synchronization? This would imply that a one-to-one (smooth, perhaps) transformation would exist between their trajectory reconstructions (Afraimovich et al. [1986], Rul'kov et al. [1995]). This is the same as asking if the trajectories are related by a continuous (and differentiable) and inversely continuous (and inversely differentiable) function.

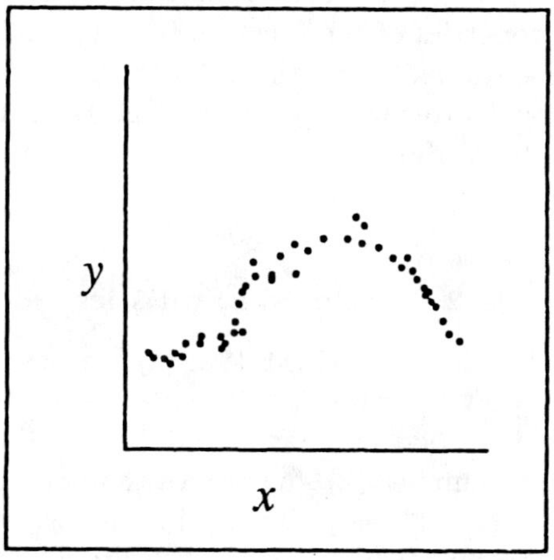

Figure 1 A sampling of points from a function.

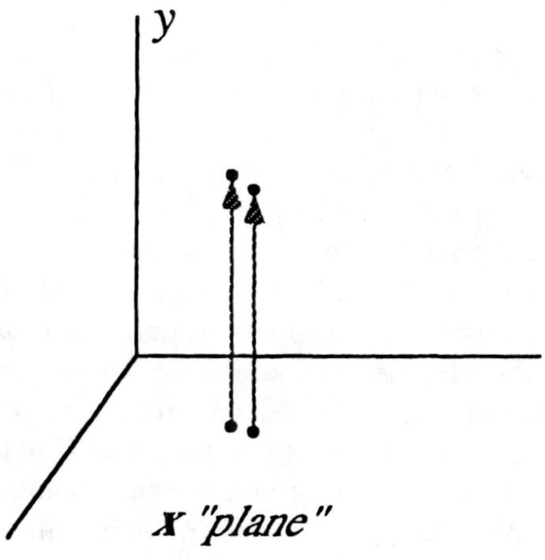

Figure 2 Schematic diagram showing predictable dependence of a variable y on other multivariate data x, the latter being represented by the flat plane. Nearby x points map to nearby y values.

In statistics we may ask the "colloquial" question: is a measurable variable y predictable from the value of some other (multivariate) variable, say $\boldsymbol{x}$? By this we usually mean that values of $\boldsymbol{x}$ which are close to one another are associated with values of y which are also (generally) close to one another. This is shown schematically in Figure 2, and is the *minimum* we would hope for in order to have any theory of data fitting. We note that although others have suggested statistical quantities which show relations between data sets and reconstructed attractors (Kaplan and Glass [1992], Prichard and Theiler [1995], Rul'kov et al. [1995]), none, with the exception of Kaplan [1994], attempts (as we do here) to determine the actual mathematical properties of the function (if any) relating the data sets.

In the next section we derive the statistics for testing for continuity and differentiability. We refer the reader to our longer paper (Pecora, Carroll, and Heagy [1995], hereafter Pecora et al. [1995]) for more details.

2 Topological Statistics

Our goal is to develop simple statistics that (1) are clearly connected to the one mathematical property they measure and (2) have a separate statistical hypothesis associated with them. Criterion number (1) guarantees that we are sure of what property we are testing. Number (2) guarantees that we know what we are using as a gauge of that property. These goals are, in a sense, reversed from those one usually finds in the literature, where tests which are an amalgam of mathematical properties and numerical algorithms are used to test for attributes which are derived. Although these numerical tests can be useful, it is not always clear what mathematical property is being tested.

We are testing for continuity (C^0) and differentiability (C^1). We give two separate statistical tests, one for each property.

2.1 Continuity Statistic. A test for continuity must start with the definition of continuity: the function f is continuous at a point $\boldsymbol{x}_0$ if $\forall\varepsilon > 0, \exists\delta > 0$ such that $\|\boldsymbol{x} - \boldsymbol{x}_0\| < \delta \implies \|f(\boldsymbol{x}) - f(\boldsymbol{x}_0)\| < \varepsilon$. In simpler terms, if we restrict ourselves to some local region around $f(\boldsymbol{x}_0)$, then there must exist a local region around $\boldsymbol{x}_0$ all of whose points are mapped into the $f(\boldsymbol{x}_0)$ region. From this definition we can generate an algorithm to apply to time series reconstructions.

Our reconstructions are matched in a functional relation (which we call f) in that for every point $\boldsymbol{x}$ we have a corresponding point $\boldsymbol{y}$; however, we do not know how points in local neighborhoods are paired. We choose an ε-sized set around a fiducial point $\boldsymbol{y}_0$, we also choose a δ-sized set around its preimage $\boldsymbol{x}_0$. We check whether all the points in the δ-set map into the ε-set. If not, we reduce δ and try again. We continue until we run out of points or all points from a small enough δ-set fall in the ε-set. We count the number of points in the ε-set (n_ε) and the δ-set (n_δ). We do not include the fiducial points $\boldsymbol{y}_0$ or $\boldsymbol{x}_0$, since they are present by construction. Generally $n_\varepsilon \geq n_\delta$.

We now choose the null hypothesis which helps us generate a probability that one would find n_ε and n_δ points in such an arrangement. Many null hypotheses are possible. We choose the simplest, namely that placements of the points on the $\boldsymbol{x}$ and $\boldsymbol{y}$ attractors are independent of each other. This null hypothesis is not trivial. It is typical of that which one would like to disprove early on in any attractor analysis, namely that the data sets have no relation to one another. We will see that it can show much more.

Given the null hypothesis, we approximate the probability of a point from the δ-set falling at random in the ε-set as $p = n_\varepsilon/N$, where N is the total number of points on the attractor. Then the probability that n_δ points will fall in the ε-set is p^{n_δ}. We obtain a likelihood that this will happen by taking the ratio of this probability to the probability p_{binmax} of the most likely event under the null hypothesis. The value of p_{binmax} is obtained by maximizing the binomial probability having parameters n_δ and p, i.e.,

$$p_{\text{binmax}} = \max_{0\le m\le n_\delta} \binom{n_\varepsilon}{m} p^m (1-p)^{n_\delta - m}. \tag{2.1}$$

We see that p^{n_δ} is simply the "tail-end" of the binomial distribution (i.e. $m = n_\delta$) while the maximum probability will generally occur for some intermediate number of points $m < n_\delta$. If $p^{n_\delta} << p_{\text{binmax}}$, then the null hypothesis is not likely and can be rejected.

We define the *continuity statistic* as $\Theta_{C^0} = 1 - p^{n_\delta}/p_{\text{binmax}}$. When $\Theta_{C^0} \approx 1$ we can confidently reject the null hypothesis. The points in the ε-set are behaving as though they are generated by a continuous function on the δ-set. When $\Theta_{C^0} \approx 0$ we cannot reject the null hypothesis and the points are behaving in a manner consistent with independence. Note that if we run out of points ($n_\delta = 0$), then we usually take the logical position that we cannot reject the null hypothesis and set $\Theta_{C^0} = 0$. Θ_{C^0} will depend on the resolution ε, and we will examine the statistic for a range of ε values. To get a global estimate of the continuity of f on the attractor we average Θ_{C^0} over the entire attractor, or over a random sampling of points from it. We present those averages here.

For inverse continuity (i.e., continuity of f^{-1}) we simply reverse the roles of the spaces (X and Y). Now we examine the mapping of δ-set points in Y into ε-sets in X. The statistical arguments are the same. We generate an inverse continuity statistic Θ_{I^0} which has the same meaning as Θ_{C^0}, except that it applies to $f^{-1}: Y \to X$.

2.2 Differentiability Statistic. The differentiability statistic is generated in the same vein as the continuity statistic. We start with the mathematical definition of a derivative and apply it locally to the two reconstructions. The generation of the linear map that approximates the derivative and the likelihood estimate associated with it are more complex than that for continuity. We show the details in our longer paper (Pecora et al. [1995]). We only outline the differentiability statistic here. The definition of a derivative at a point $\boldsymbol{x}_0$ is that a linear operator A exists such that $\forall\varepsilon > 0 \exists\delta > 0$ for which $\|\boldsymbol{x} - \boldsymbol{x}_0\| < \delta \implies \|f(\boldsymbol{x}_0) + A(\boldsymbol{x} - \boldsymbol{x}_0) - f(\boldsymbol{x})\| < \varepsilon\|\boldsymbol{x} - \boldsymbol{x}_0\|$. This means that there is a linear map that approximates the function at nearby points with an error in the approximation that is proportional to the distance between those points. Note that ε serves a purpose here different from continuity. Here it bounds the error in a linear mapping; it does not signify a distance.

The algorithm that we generate from this definition is to first choose an ε (error bound) and a δ. Then we find all the points in the local δ-set $\{\boldsymbol{x}_\imath\}$ and their $\boldsymbol{y}$ counterparts $\{\boldsymbol{y}_\imath\}$. We approximate the linear operator A as the least squares solution of the linear equations $A(\boldsymbol{x}_\imath - \boldsymbol{x}_0) = \boldsymbol{y}_\imath - \boldsymbol{y}_0$. The solution is accomplished by singular value decomposition (SVD) (Pecora et al. [1995]). The least-squares approach appears to be the best since it minimizes the errors in the approximation, which is just what ε is supposed to bound. We check if $\|\boldsymbol{y}_\imath - \boldsymbol{y}_0 - A(\boldsymbol{x}_\imath - \boldsymbol{x}_0)\| < \varepsilon\|\boldsymbol{x}_\imath - \boldsymbol{x}_0\|$ for each pair $(\boldsymbol{x}_\imath, \boldsymbol{y}_\imath)$. If not, we decrease δ and try again with fewer, but nearer, points. We continue this until we have success or we run out of points.

We choose the null hypothesis that the two sets of vectors $\{\boldsymbol{x}_\imath\}$ and $\{\boldsymbol{y}_\imath\}$ have zero correlation. We show (Pecora et al. [1995]) that this generates a likelihood that any two such sets will give the operator A "by accident" as $\exp[-\frac{1}{2}(n_\delta - r_x)(n_\delta - r_y)r^2\delta]$, where r^2 is the usual multivariate statistical correlation between $\{\boldsymbol{x}_\imath\}$ and $\{\boldsymbol{y}_\imath\}$, $\delta = \min(r_x, r_y)$, and r_x and r_y are the ranks of the $\boldsymbol{x}$ and $\boldsymbol{y}$ spaces that come out of the SVD (Pecora et al. [1995]). This is an asymptotic formula. The differentiability statistic Θ_{C^1} is given by one minus this likelihood. When $\Theta_{C^1} \approx 1$ we can reject the possibility that the points are accidentally related by a linear operator, a derivative. When $\Theta_{C^1} \approx 0$, we cannot reject the null hypothesis. As before, when we must shrink δ so small that no points other than $\boldsymbol{x}_0$ remain, we set $\Theta_{C^1} = 0$. Analogous to Θ_{C^0}, the statistic Θ_{C^1} depends on ε. We typically calculate Θ_{C^1} for a range of ε values and average over the attractor, or over a random sampling of points from it.

We now have two statistics Θ_{C^0} and Θ_{C^1} that we can use to test for continuity and differentiability in data sets. We will mostly concentrate on the continuity statistic, since the differentiability statistic follows it very closely in many cases, although we show one example for which this is not so. We give all ε and δ values in units of the standard deviation of the reconstructed attractor, so that, for example, $\varepsilon = 1.0$ is an ε-set with radius of one standard deviation. In all our studies so far, this seems to be a good normalization for set sizes.

Similar to Θ_{I^0}, we develop a statistic for the differentiability of the inverse function $f^{-1} : Y \to X$ by reversing the roles of X and Y. We now examine linear maps on δ-sets in Y which map to nearby points in X. The inverse differentiability statistic is called Θ_{I^1} and it has the same meaning as Θ_{C^1}.

2.3 Putting It All Together. We are now in a position to make statistical arguments for pairs of (multivariate) data points being related by a function which may have several different properties. It may simply be continuous. It may have a continuous inverse. Or both the function and its inverse may be continuous, in which case we can make an argument that it is a homeomorphism and hence all topological quantities will be invariant under it, i.e. the X data points and the Y data points share the same topological properties.

Similarly, should the function pass continuity tests, we can test for differentiability of the function or its inverse. Should both differentiability statistics be high (≈ 1), then we can argue that we have data points related by a diffeomorphism, i.e., local metric properties (like fractal dimension) are preserved.

We should always keep in mind that we are not making statements about the absolute continuity or differentiability of a function, but only the continuity or differentiability relative to the smallest ε's and δ's for which our statistics are good. That is, we always have a resolution limit associated with an argument for these properties. A good question is: if a function can be shown to be continuous or differentiable at some level of resolution, are all topological properties that can be calculated at or above that resolution invariant? We suspect the answer is "yes", but we cannot prove it.

3 Applications

3.1 Continuity vs. Linear Correlation in Lorenz Time Series. To help clarify the use of these statistics, we first present two cases, one in which we compare Θ_{C^0} to the usual statistical linear correlation and another in which we compare Θ_{C^0} to Θ_{C^1}.

Figure 3 shows a plot of x and z time series taken from a Lorenz system running in the chaotic regime ($\sigma = 10$, $b = 8/3$, and $\rho = 60$). If we calculate the linear correlation or cross correlation between these time series for various time delays (including zero) we will find that the correlation is very small. For example, for 5000 points it is less than 0.085 for any time lag. The multivariate linear correlation

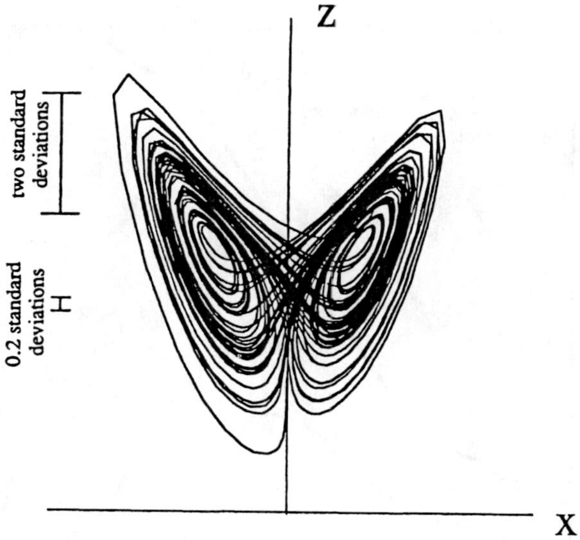

Figure 3 Plot of the x and z time series, along with two measure bars showing the size of the maximum ε-set used and the smallest ε-set at which the continuity statistic from x to z was still near 1.0.

calculated from time-delay reconstructions using the x and z series would show the same low correlation, as is easy to see from the form of the time-delay method. If we now calculate Θ_{C^0} using 4-dimensional reconstructions from the x and z series (5000 points), we see something entirely different. Figure 4 shows $\Theta_{C^0}(\varepsilon)$ vs. ε. The statistic is quite high (> 0.8) down to an ε of about 0.2 standard deviations. Figure 3 shows the size of ε-sets of 2 standard deviations in diameter and of 0.2 standard deviations in diameter. This shows that down to a fairly small resolution, with a small number of data points, we can establish that the mapping between the x attractor and the z reconstructed attractor is continuous, i.e., $f : x\ attractor \rightarrow z\ attractor$ is continuous. It turns out that the differentiability statistic Θ_{C^1} is also quite high, down to errors in the linear operator on the order of 0.2. On the other hand, the inverse map $f^{-1} : z\ attractor \rightarrow x\ attractor$ is not continuous (see Figure 4). This is not surprising since the x and y components of the Lorenz system have a two-to-one relation with the z component (the vector field is invariant to changing the signs of x and y). This means that $\pm x$ values will map to roughly the same point on the z attractor or, equivalently, each point on the z attractor maps back to two points on the x attractor. This is a discontinuity, and the statistic shows this. The statistic Θ_{I^0} is not zero since sets of finite δ size near the x origin will have points that can map into the same ε-set.

3.2 Differentiability of Sampling of Weierstrass Function. In our next test we examine a function f that is continuous but nowhere differentiable, the Weierstrass function. We show this function in Figure 5. The "time series" is just the pairs of points $(x, f(x))$ as x varies over the domain of f. Here there are no attractors to reconstruct from the time series; instead there is only a function $f : \mathbb{R} \rightarrow \mathbb{R}$. This shows that the statistics can be applied to any function given pairs of points in the domain and range. The continuity and differentiability statistics are shown in Figure 6. The continuity statistic, even for only 1000 points, starts high and degrades slowly with decreasing ε. As we add more points (up to 4000,

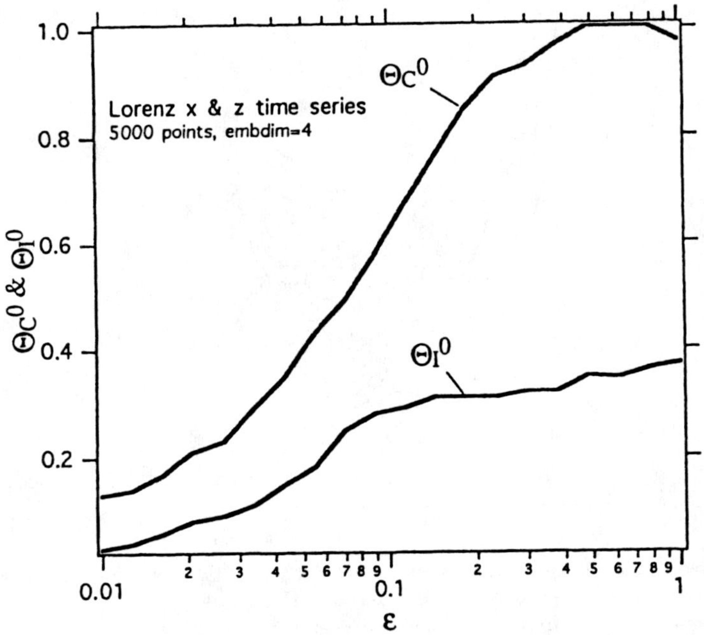

Figure 4 Continuity and inverse-continuity statistics for the x attractor to z attractor mapping.

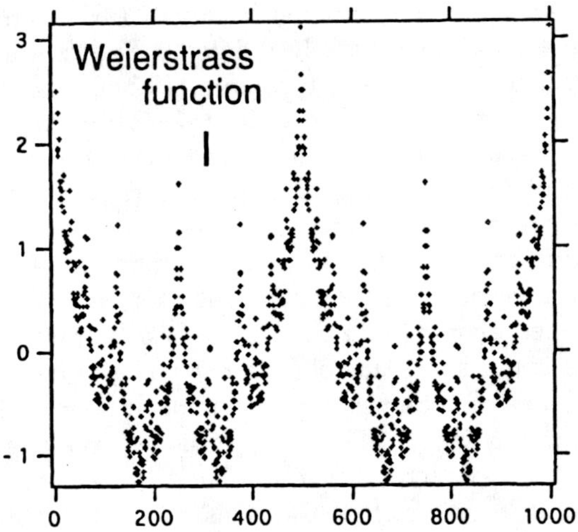

Figure 5 A plot of the Weierstrass function used to test the continuity and differentiability statistics.

in this case) the continuity improves. This is also a characteristic of a functional relation that is truly continuous. A function that has many discontinuities will not have Θ_{C^0} increase as the amount of data increases.

On the other hand, the differentiability statistic starts very low and quickly goes to zero as ε decreases. This indicates that the function is not differentiable (we cannot reject the null hypothesis for Θ_{C^1}). Thus, Θ_{C^0} and Θ_{C^1} do measure different properties.

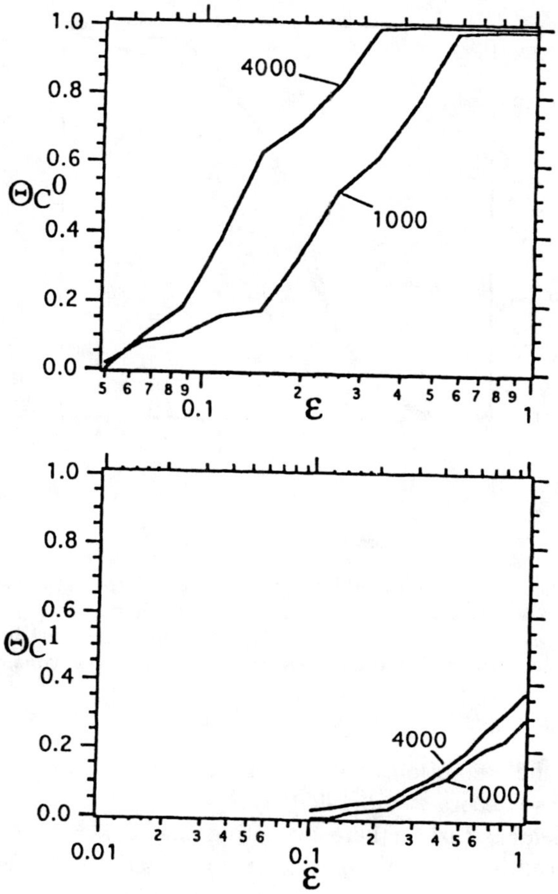

Figure 6 Continuity and differentiability statistics for the Weierstrass function.

3.3 Dynamical Interdependence. An interesting situation, especially regarding chaotic systems, is that of generalized synchronization (Afraimovich et al. [1986]). This occurs when two parts of a coupled dynamical system have the same dimension (same number of dynamical variables) and their respective trajectories or phase space points are related in a one-to-one smooth fashion. That is, there is a diffeomorphism between them. In the usual synchronization (Afraimovich et al. [1986], Carroll [1994], Carroll and Pecora [1991], Fujisaka and Yamada [1985], Heagy et al. [1994], Pecora and Carroll [1990, 1991], Rul'kov et al. [1994], Rul'kov et al. [1995]) the subsystems of a dynamical system are equal at each time step (e.g., $x = y$ in time series). In this case the function relating the parts of the dynamical system acts like the identity mapping, which is a diffeomorphism. However, Afraimovich showed that this can be relaxed to a situation as in the previous paragraph. Recent work by Rul'kov et al. [1995] has shown that one can devise a statistic (based on the false nearest neighbor method of Kennel et al. [1992], Kennel and Abarbanel [1993]) which can detect such general synchronization. We have shown that our continuity and differentiability statistics can do the same and at the same time specify the *type* of relationship (homeomorphism or diffeomorphism,

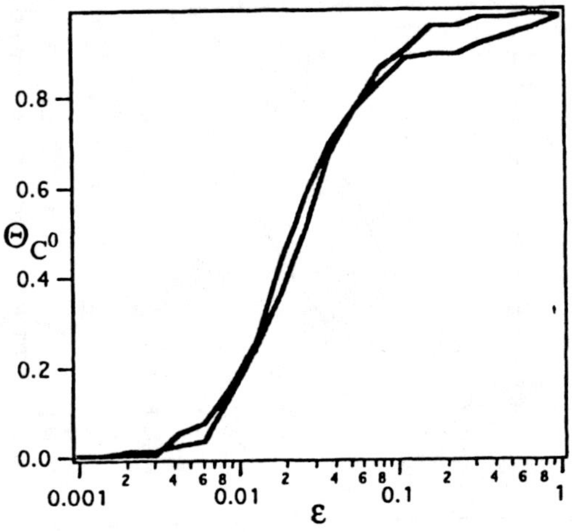

Figure 7 Continuity statistics for the mapping between attractors reconstructed from a Lorenz x and a simultaneously measured Lorenz y time series which, combined with Figure 8 shows that the two are related by a diffeomorphism and therefore the time series must come from the same dynamical system.

for example) and the resolution at which one can be confident that the function has these properties (Pecora et al. [1995]).

We take these ideas further here by noting that generalized synchronization, especially as determined from time series reconstructions, is really a particular type of *dynamical interdependence*. Since, generically, by Takens' theorem, we can measure two different dynamical variables, say x and y time series, and reconstruct the same attractor, up to a diffeomorphism, measuring any two dynamical variables in a system should yield attractors that have high Θ_{C^0}, Θ_{C^1}, Θ_{I^0}, and Θ_{I^1}. We note this is a generic statement and there are times it will fail, as in the above test between Lorenz x and z time series. We test the idea of dynamical interdependence on a Lorenz x and y time series (28,000 points), both of which can reproduce an equivalent Lorenz attractor. Figure 7 shows Θ_{C^0} and Θ_{I^0} for the attractors reconstructed from the Lorenz x and y time series. We see that the continuity and inverse continuity remain high down to ε radii of ≈ 0.07 standard deviations. As we saw above, these are small sets on the attractor and hence we can be confident down to fairly small resolutions that our attractors are related by a homeomorphism. Likewise the Θ_{C^1} and Θ_{I^1} statistics (Figure 8) remain high down to errors (ε) on the order of 0.1 standard deviation. This means our function relating the two attractors is smooth and the attractors are probably related by a diffeomorphism. If these were two time series taken simultaneously in an experiment, we would conclude that they came from the same dynamical system, i.e., they are dynamically interdependent.

3.4 Determinism. As our last example, we show how Θ_{C^0}, Θ_{C^1}, Θ_{I^0}, and Θ_{I^1} can be used to test for determinism in a time series. In this work we are very close in spirit and sometimes in actual practice (at least as far as forward, primitive determinism is concerned) to the paper by Kaplan [1994], which gives a

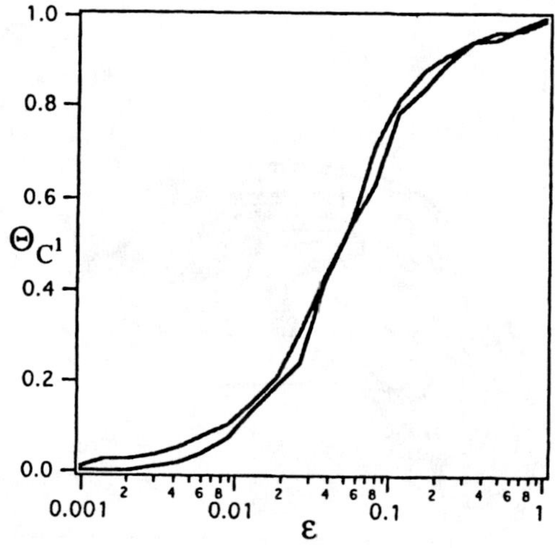

Figure 8 Differentiability statistics for the mapping between attractors reconstructed from a Lorenz x and a Lorenz y time series which, combined with Figure 7 shows that the two are related by a diffeomorphism and therefore the time series must come from the same dynamical system.

good introduction to the general idea of using continuity properties for determinism tests. We note here that determinism at its most primitive level simply requires phase-space points that are close in the present to be close in the future. This is forward determinism and is tested by calculating Θ_{C^0} for a time series reconstruction (X) and its forward, time-shifted version (Y). We can also test for forward, smooth determinism by calculating Θ_{C^1} on the same two time series. Invertible dynamics can be tested for by calculating Θ_{I^0} (primitive, reverse-time determinism) and Θ_{I^1} (smooth, reverse-time determinism). Thus, associated with each level of function property (where we restrict the function properties with each level) is a corresponding determinism.

Calculation of such levels of determinism could be useful if one were proposing to model a system for which experimental time series were available. This could dictate the level of model to use.

We used a Lorenz x time series to test for the various levels of determinism. We chose 100 points at random on the reconstructed attractor, then calculated the average of each statistic for points both forward and backward in time at time steps of $1, 2, 4, \ldots 2n$, $n = 6$. One time step was approximately one-quarter around the typical oscillation of the Lorenz system. Figure 9 (from Pecora et al. [1995]) shows the points on the attractor relative to one typical starting point. Figure 10 (from Pecora et al. [1995]) shows the statistics. We see that, given this time series, we would conclude that it was from a forward, smooth deterministic system. The reverse determinism is not as good, but is still present. The Lyapunov exponents determine the limits of forward and reverse determinism. The forward determinism is limited by the maximal positive exponent and the reverse determinism is limited by the minimal (in the sense of most negative) Lyapunov exponent. The most negative Lyapunov exponent of the Lorenz system is much larger in magnitude

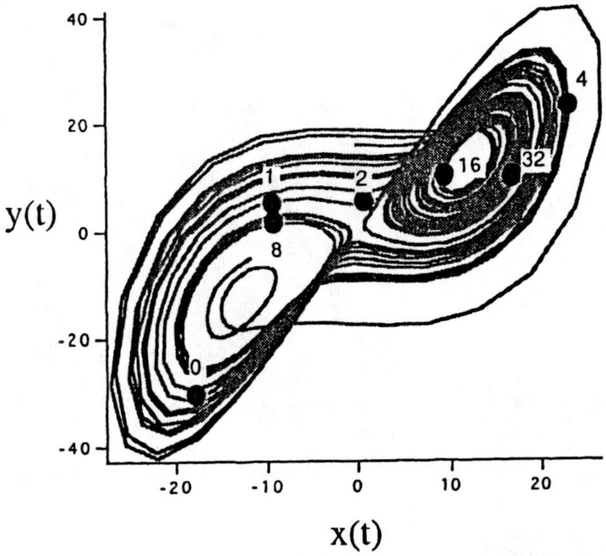

Figure 9 The points on the Lorenz attractor in forward time where $\Theta_{C^0}, \Theta_{C^1}, \Theta_{I^0}$, and Θ_{I^1} were calculated to check for various forms of determinism.

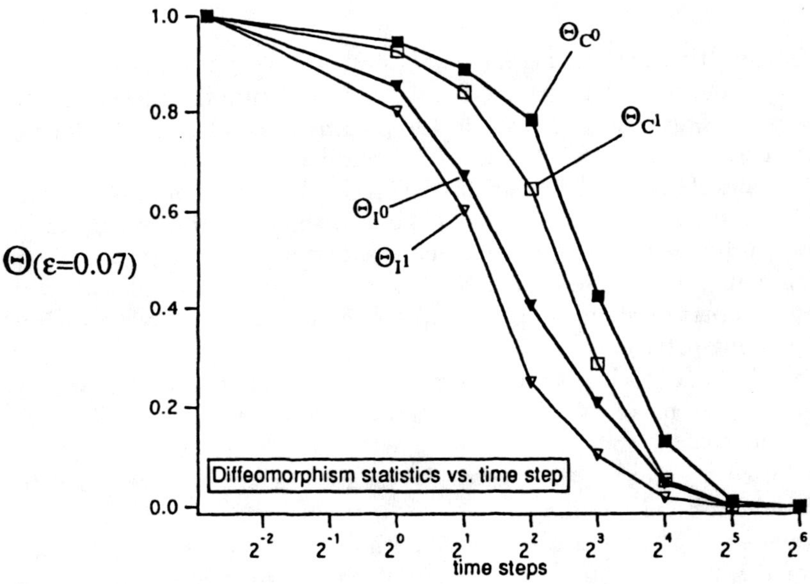

Figure 10 The statistics $\Theta_{C^0}, \Theta_{C^1}, \Theta_{I^0}$, and Θ_{I^1} which show the level of primitive determinism, smooth determinism in forward time and invertible primitive determinism and invertible smooth determinism in reverse time.

than the most positive. This probably accounts for the poorer values for the reverse determinism (Θ_{I^0} and Θ_{I^1}).

4 Conclusions and Remarks

We see that we can define statistics that address the question of whether data sets are related by functions with fundamental mathematical/analytical properties. This is a more general approach than asking whether or not the data are related by a particular type of function (e.g., linear map) or whether or not the data are distributed according to a particular type of probability distribution (e.g., Gaussian). In testing data sets, it would seem wise to also test for general properties to determine first whether or not there is even a tractable relation between data sets and, second, what level of function properties can be assigned to the relation (e.g., continuous? differentiable?).

Obviously, we can extend this work in two ways. One is to derive statistics for other mathematical properties, like twice differentiable or Lipschitz. The other way is to use other null hypotheses to generate our statistics. We have developed another continuity hypothesis, for example, which will appear elsewhere (Pecora and Carroll [1995]). A particularly challenging direction would be to create more robust versions of the statistics and/or null hypotheses, so that we can make conclusions in the presence of noise, much as is done now with linear correlations, ARMA modelling, and analyses of probability distributions. We have not investigated this possibility, but we do feel it would be very useful, and in some cases absolutely necessary, in order to use these statistics on actual data and experiments.

References

Afraimovich, V. S., Verichev, N. N. and Rabinovich, M. I. [1986] *Stochastic synchronization of oscillations in dissipative systems*, Inv. VUZ. Rasiofiz. RPQAEC. **29**, 795-803.

Carroll, T. L. [1994], *Synchronization of chaos*, Ciencia Hoje, **18**, 26.

Carroll, T. L. and Pecora, L. M. [1991] *Synchronizing chaotic circuits*, IEEE Trans. CAS. **38**, 453.

Fujisaka, H. and Yamada, T. [1983] *Stability theory of synchronized motion in coupled-oscillator systems*, Progress of Theoretical Physics **69(1)**, 32.

Fujisaka, H. and Yamada, T. [1985] *Stability theory of synchronized motion in coupled-oscillator systems IV*, Progress in Theoretical Physics **74**, 918.

Heagy, J. F., Carroll, T. L. and Pecora, L. M. [1994] *Synchronous chaos in coupled oscillator systems*, Physical Review. **E 50(3)**, 1874.

Kaplan, D. T. [1994] *Exceptional events as evidence for determinism*, Physica D, **73**, 38-48.

Kaplan, D. T. and Glass, L. [1992] *Direct test for determinism in a time series*, Physical Review Letters **68**, 427.

Kennel, M. B. and Abarbanel, H. D. I. [1993] *Local false nearest neighbors and the dynamical dimensions from observed chaotic data*, Physical Review. **E47**, 3057.

Kennel, M. B., Brown, R. and Abarbanel, H. D. I. [1992] *Determining embedding dimension for phase space reconstruction using the method of false nearest neighbors*, Physical Review. **A45**, 3403.

Mané, R. [1981] *Embeddings*, in Dynamical Systems and Turbulence (R. Rand and L. S. Young, eds.), Lecture Notes in Mathematics, vol. 898. Springer, Berlin, 230.

Pecora, L. M. and Carroll, T. L. [1990] *Synchronization in chaotic systems*, Physical Review Letters, **64**, 821.

Pecora, L. M. and Carroll, T. L. [1991] *Driving systems with chaotic signals*, Physical Review, **A44**, 2374.

Pecora, L. M. and Carroll, T. L. [1995] *Generalized synchronization and dynamical interdependence: statistical tests on time series*, in preparation.

Pecora, L. M., Carroll, T. L. and Heagy, J. F. [1995] *Statistics for mathematical properties of maps between time-series embeddings*, Physical Review **E52(4)**, 3420.

Prichard, D. and Theiler, J. [1995] *Generalized redundancies for time series analysis*, Physica D **84**, 476-493.

Rul'kov, N., Sushchik, M. M., Tsimring, L. S. and Abarbanel, H. D. I. [1995] *Generalized synchronization of chaos in directionally coupled chaotic systems*, Physical Review, **E51**, 980.

Rul'kov, N. F., Volkovskii, A. R., Rodriguez-Lozano, A., d. Rio, E. and Velarde, M. G. [1992] *Mutual synchronization of chaotic self-oscillators with dissipative coupling*, International Journal of Bifurcation and Chaos **2**, 669-676.

Rul'kov, N. F., Volkovskii, A. R., Rodriguez-Lozano, A., d. Rio, E. and Velarde, M. G. [1994] *Synchronous chaotic behavior of a response oscillator with chaotic driving*, Chaos, Solitons & Fractals. **4(2)**, 201-211.

Sauer, T. and Yorke, J. A. [1996] *Are the dimensions of a set and its image equal under typical smooth functions?*, Ergodic Theory and Dynamical Systems, to be published.

Sauer, T., Yorke, J. A. and Casdagli, M. [1991] *Embedology*, Journal of Statistical Physics **64**, 579.

Takens, F. [1981] *Detecting strange attractors in turbulence*, in Dynamical Systems and Turbulence (D. Rand and L.-S. Young (eds.), Warwick 1980, Lecture Notes in Mathematics, vol. 898, Springer, Berlin, 366.

Fields Institute Communications
Volume 11, 1997

Reconstruction of Integrate-and-Fire Dynamics

Tim Sauer
Department of Mathematical Sciences
George Mason University
Fairfax, VA, USA 22030

Abstract. The fact that the geometry of state space can be reconstructed from a time series of observations has led to innovative prediction and control algorithms for chaotic systems. In this article we show that state reconstruction of integrate-and-fire dynamics is possible from a series of interspike intervals. In particular, a chaotic spike train generated in this way can be predicted from its history. We show that it is possible to use system reconstruction based on spike trains alone to control unstable periodic trajectories of the system in two separate ways: by making small (subthreshold) changes in a system parameter, and by using an on-demand pacing protocol with large (superthreshold) pulses.

1 Introduction

The state of a system is a primitive concept that is fundamental to the understanding of dynamical behavior. The identification of state underlies all attempts at characterization, prediction, and control of deterministic systems.

The discovery of chaos in simple nonlinear systems prompted a radical change in dynamical system analysis from observations. The new ingredient due to chaos is the possibility that nonlinear systems with few degrees of freedom, while deterministic in principle, can create output signals that appear complex, and mimic stochastic signals from the point of view of conventional time series analysis. The reason for this is that trajectories that have nearly identical initial states will separate from one another at an exponentially fast rate. Exponential separation causes chaotic systems to exhibit much of the same long-term behavior as stochastic systems.

For a chaotic system, short-term prediction methods based on reconstructed states outperform the "optimal" linear predictor, if there are a reasonably low number of active degrees of freedom. To implement short-term prediction in practice requires global time series models. Geometric methods for reconstructing the state space of a nonlinear system from a time series of measurements began with the pioneering work of Packard et al. [1980] and Takens [1981]. (These efforts were

1991 *Mathematics Subject Classification.* Primary 58F13, 60G55; Secondary 62M20.

The author was supported in part by NSF Computational Mathematics Grant #DMS9305659.

related in spirit to the threshold-autoregressive models of Tong and Lim [1980] in the statistical literature. See also Aeyels [1981] for another version of Takens' theorem.) These articles and later work showed that vectors representing a number of delayed versions of some observable measurement can be put in a one-to-one correspondence with the states of the system being observed.

The fact that the current system state can be identified using a vector of time series measurements leads to a number of applications, since analysis of the geometry of the chaotic attractor underlying the time series can be performed in the state space reproduced from the delay-coordinate vectors. Applications include noise-filtering and prediction of chaotic time series, and control of unstable periodic orbits solely from a time series record (see Ott et al. [1994] for a broad sampling of applications).

In this article we explore the feasibility of applications such as prediction and control for a different type of dynamical data. Instead of using a vector of time series measurements to identify the state, we will use a vector of time intervals between firings of an integrate-and-fire dynamical system. We will show that in analogy with the time series case, there is a one-to-one correspondence between the states of the underlying dynamical system and the "interspike interval" vectors of sufficient length.

It is perhaps a surprising fact that, under the integrate-and-fire hypothesis, the intricate structure of the state space attractor is somehow contained in the spike train record. Together with the earlier results on time series, it hints at the multifaceted nature of faithful information transfer in complex systems. Neurophysiological systems in particular are widely assumed to process information coded in temporal patterns of action potential firings. See Schiff et al. [1994] for an application of interspike interval analysis to evoked and spontaneous neuronal population responses in rat hippocampal tissue.

Section 2 of this article explains how to reconstruct states of a dynamical system from a spike train. Section 3 discusses a practical consequence of the reconstruction: Spike trains of moderate length are predictable and controllable. We demonstrate that an unstable orbit of an attractor can be stabilized either with small or large perturbations, using knowledge of the system dynamics learned through spike trains. In Section 4, we present a sketch of the proof of an embedding theorem for interspike intervals.

2 State Representation from Spike Trains

Neurobiological systems are often marked by measurable pulses corresponding to a cell reaching a threshold potential, which triggers rapid depolarization, followed by repolarization, which restarts the cycle. The times of these discrete pulses can be recorded. This type of data differs from a time series of an observable measured at regular time intervals. Many hypotheses and models for the description of the time variability of these pulses have been proposed (see Tuckwell [1988]). A random process in which information is carried by a series of event timings is called a *point process.* See Lewis [1972] and Miller and Snyder [1991] for an introduction to the statistical literature on this subject. The goal of this article is to present an alternative method of analysis for spike timing data. The main question of this section is the following: If a point process is the manifestation of an underlying deterministic system, can the states of the system be uniquely identified from the information provided by the point process?

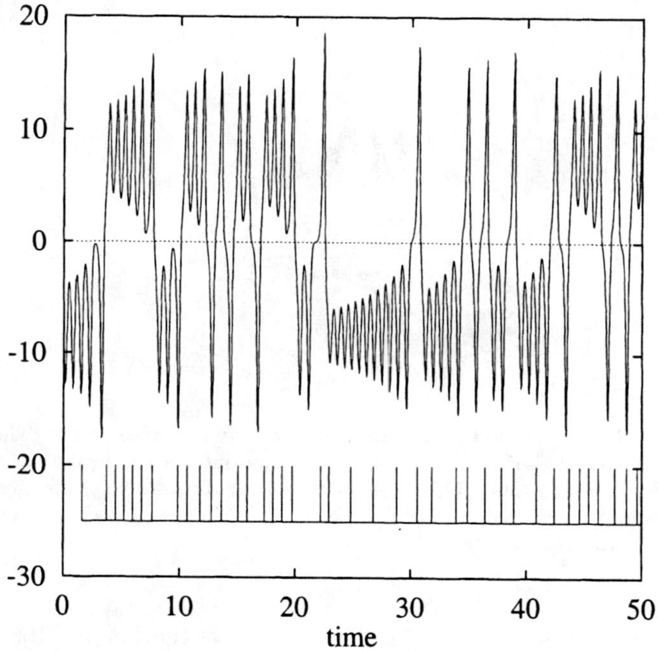

Figure 1 The upper trace is a solution of the Lorenz equations graphed as a function of time. The lower trace shows the times at which spikes are generated according to equation (2.1), with $S(t) = (x(t) + 2)^2$ and $\Theta = 60$.

We propose to use an integrate-and-fire model, a simple paradigm that shares qualitative characteristics with real systems, and that demonstrates a feasible way in which complicated information about system states can be communicated and transformed. This model allows us to study the linkage between continuous dynamics and the interspike intervals (ISI's) produced by them, rather than concentrating on the detailed mechanism of any single system.

The integrate-and-fire hypothesis connects an underlying continuous dynamical system to a point process in the following way. A time series from the dynamical system is integrated with respect to time; when it reaches a preset threshold, a spike is generated, after which the integration is restarted. In our simulations we hypothesize that the input to be integrated is a low-dimensional chaotic attractor. We use the Mackey-Glass equation and the Lorenz equations as simple representative examples.

Let $S(t)$ denote the signal produced by a positive function measured on the state space of a finite-dimensional dynamical system. Assume that the trajectories of the dynamical system are asymptotic to a compact attractor A. Let Θ be a positive number which represents the firing threshold. After fixing a starting time T_0 and initial condition x for the dynamical system, a series of "firing times" $T_1 < T_2 < T_3 < \ldots$ can be recursively defined by the equation

$$\int_{T_i}^{T_{i+1}} S(t)\, dt = \Theta. \tag{2.1}$$

From the firing times T_i, the interspike intervals can be defined as $t_i = T_i - T_{i-1}$.

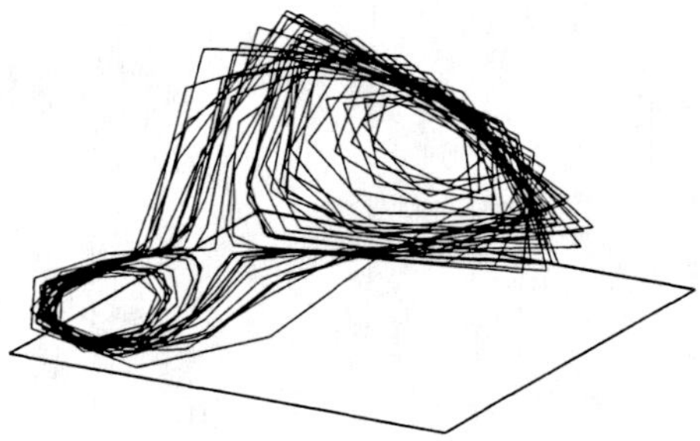

Figure 2 Reconstruction of Lorenz dynamics from a spike train. Three-dimensional vectors of successive interspike time intervals are plotted, connected by line segments. The series of time intervals used to make the reconstruction was created using integrate-and-fire model (2.1), with an input signal from the Lorenz attractor.

Figure 1 shows a time trace of the x-coordinate of the Lorenz [1963] equations

$$\begin{aligned} \dot{x} &= \sigma(y-x) \\ \dot{y} &= \rho x - y - xz \\ \dot{z} &= -\beta z + xy \end{aligned} \tag{2.2}$$

where the parameters are set at the standard values $\sigma = 10, \rho = 28, \beta = 8/3$. The positive "output function" to be integrated is $S(t) = (x(t)+2)^2$. In the lower part of the figure, the spiking times generated by (2.1) with threshold $\Theta = 60$ are shown as vertical line segments.

Under certain genericity conditions on the underlying dynamics, signal and threshold, the series $\{t_i\}$ of ISI's can be used to reconstruct the attractor A. Theorem 2.1 below states that there is a one-to-one correspondence between m-tuples of ISI's and attractor states, which associates each vector $(t_i, t_{i-1}, \dots, t_{i-m+1})$ of ISI's with the corresponding point $x(T_{i-1})$ on the attractor. In analogy with the generalization in Sauer et al. [1991] of the Takens [1981] theorem, the condition $m > 2d_0$ is sufficient, where d_0 is the box-counting dimension of the attractor A.

The one-to-one property is useful because the state of a deterministic dynamical system, and thus its future evolution, is completely specified by a point in its phase space. Suppose that when the system is in a given state x one observes the ISI-vector $G(x)$ in the reconstruction space, and that this is followed one second later by a particular event. If G is a continuous and one-to-one function, each appearance of the measurements represented by $G(x)$ will be followed one second later by the same event. This is because there is a uniquely-defined correspondence between the attractor states in phase space and their image vectors in reconstruction space. Thus there is predictive power in measurements that are matched to the system state x in a one-to-one manner.

Figure 2 shows a delay plot of interspike intervals produced by the x-coordinate of the Lorenz attractor. Since the box-counting dimension of the Lorenz attractor is slightly greater than 2, Theorem 2.1 implies that generic reconstructions in $m = 5$ dimensions are topologically equivalent to the original Lorenz attractor, although its orientation in five dimensional Euclidean space could be rather twisted. Figure 2 can be viewed as a projection from five dimensional space to three dimensional space of this topologically equivalent attractor.

To conclude this section we give a precise statement of the one-to-one continuous correspondence referred to above. A sketch of the proof of Theorem 2.1 is provided in Section 4. Assume that the underlying dynamical system is represented by the autonomous system of differential equations $\dot{x} = f(x)$ where x is a k-dimensional variable. The state space for this dynamical system is $\mathbb{R}^k$. Let A be a compact attractor for the dynamics in $\mathbb{R}^k$, and denote by $d_0(A)$ the box-counting dimension of A.

As the trajectory $x(t)$ moves along the attractor A, a signal $S(t) = h(x(t))$ is defined from an *output function* $h : \mathbb{R}^k \to (0, \infty)$ on the current state. If we identify a state x_0 at time T_0, then the interspike interval vector $(T_1 - T_0, \dots, T_m - T_{m-1})$ generated according to (2.1) can be associated to that state, resulting in the function $G_h : \mathbb{R}^k \to \mathbb{R}^m$ defined by

$$G_h(x_0) = (T_1 - T_0, \dots, T_m - T_{m-1}). \tag{2.3}$$

It is easy to check that the function G_h is continuous and differentiable as long as the positive-valued function h is continuous and differentiable.

A point x_0 in the state space R^k of the autonomous system $\dot{x} = f(x)$ is an *equilibrium point* if $f(x_0) = 0$, implying that the unique solution through x_0 is $x(t) = x_0$. Theorem 2.1 says that if there are only finitely many equilibria, and if $m > 2d_0(A)$, then for typical positive-valued output functions h, the ISI reconstruction function G_h is one-to-one on the attractor A.

Theorem 2.1 *Let $\dot{x} = f(x)$ be an autonomous system of differential equations on $\mathbb{R}^k$ with compact invariant set A of box-counting dimension $d_0(A)$. Assume that A contains at most a finite number of equilibrium points. Then there is a dense set of positive-valued output functions h for which the ISI-vector function G_h is one-to-one on A. In other words, for any smooth function $h_0 : \mathbb{R}^k \to \mathbb{R}$ which is positive on A, and for any $\epsilon > 0$, there exists an h within ϵ of h_0 in the C^1-metric such that G_h is one-to-one on A.*

Theorem 2.1 implies that the set of positive-valued h for which the reconstruction function G_h is one-to-one on A is dense in the C^1-topology. The proof, found in Section 4, reaches a somewhat stronger, measure-theoretic conclusion. A finite-dimensional family q_c of perturbing functions for h_0 is constructed such that $G_{h_0+q_c}$ is one-to-one for almost every c, in the sense of Lebesgue measure on Euclidean space. (The family contains the zero-function, so that the $h_0 + q_c$ include arbitrarily small perturbations of h_0.) Therefore the function G_h is one-to-one for a prevalent subset of h, in the sense of Hunt et al. [1992].

3 Prediction and Control From Interspike Intervals

A practical consequence of Theorem 2.1 is that the ISI vectors $(t_i, \dots, t_{i-m+1})$ can be used to reconstruct the attractor sufficiently to make measurements of dynamical invariants possible, and to do short-term prediction on the series $\{t_i\}$. Thus the possibility exists of predicting future interspike intervals from past history, and moreover to use this knowledge to control the system.

To explain the meaning of short-term prediction, it is instructive to recall the time series case. For chaotic time series, the length of time horizon for which prediction is effective depends on the separation time of nearby trajectories, and so inversely on the largest Lyapunov exponent. For ISI series, the horizon depends both on the Lyapunov exponents of the underlying process and on the threshold Θ. As Θ in (2.1) is increased, the effect on the interspike interval reconstruction is analogous to increasing the sampling interval in the case of time series reconstruction. We expect the quality of the reconstruction to degrade when the threshold is high enough that the intervals become comparable to the reciprocal of the largest Lyapunov exponent, the decorrelation time of the dynamics. Whether predictability can be found in a particular set of experimental data may depend on this issue.

Sauer [1994] explored the possibility of predicting a spike train from its history. By comparing predictability of the spike train with predictability of surrogate versions of itself constructed to share the main statistical properties (Theiler et al. [1992]), it was shown that it is feasible to extract information about nonlinearities of the underlying system from the spike train. Those findings indicated that data sets of lengths accessible in physiological settings may be sufficient in principle to distinguish deterministic chaos from random processes.

Stabilization of unstable periodic orbits in chaotic systems using small controls was described in Ott, Grebogi, and Yorke [1990] and is commonly referred to as OGY control. Several experimental examples of the success of this type of stabilization are documented in Ott et al. [1994]. The fundamental idea is to determine some of the low period unstable orbits embedded in the attractor, and then use small feedback to control the system. The OGY method is model-free, in that it relies on time series information to reconstruct the linear dynamics in the vicinity of the unstable orbit.

In this section we would like to prove the viability of this concept when the dynamical information is carried by interspike intervals instead of a time series. We will demonstrate two different types of control protocols. The first will be the analogue of OGY control for spike trains. Using small changes in a system parameter, an unstable orbit will be stabilized. This type of control is called *subthreshold control,* since the control signal is not large enough on its own to make the system fire. For this demonstration we will use spike intervals produced by an observable from the Lorenz system, and stabilize the figure-8 trajectory.

The second protocol will use superthreshold pulses to stabilize an unstable equilibrium. This technique, often called *on-demand pacing,* will be shown to work for deterministic as well as random integrate-and-fire systems. Therefore, although the success of subthreshold control can be viewed as evidence of deterministic dynamical behavior in an unknown system, the same cannot be said for superthreshold control.

We begin with subthreshold control. The Lorenz attractor of (2.2) exhibits trajectories cycling around two unstable equilibrium points. The number of times a typical trajectory cycles around one point before moving to the other is known to be unpredictable on moderate to long time scales. We will choose to stabilize the unstable periodic trajectory which cycles exactly once around one equilibrium, then exactly once around the other, before returning to the first and repeating the process. This "figure-eight" trajectory of the Lorenz attractor is unstable, in the sense that a trajectory started near this behavior will tend to move away from it.

In this feasibility study we used the Lorenz equations (2.2) to generate an input signal $S(t) = x(t) + 25$. The parameters $\sigma = 28$ and $\rho = 10$ are set at the standard values, and the parameter β is nominally set at the standard value $8/3$, but will be used as control parameter.

Using the reconstruction of Lorenz dynamics solely from the interspike intervals produced by (2.1), local linear models for a pair of points along the figure-eight orbit

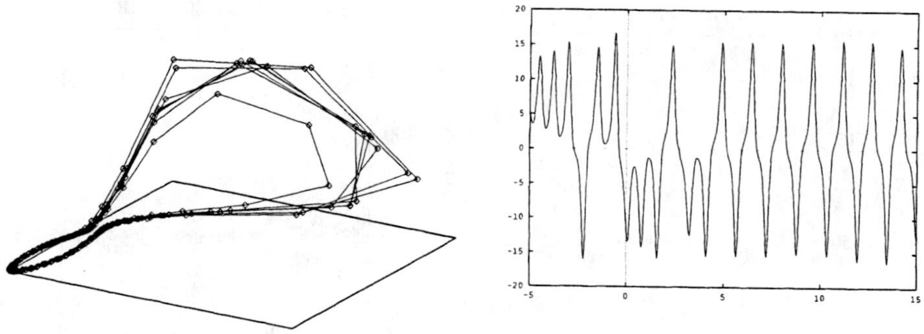

(a) (b)

Figure 3 Subthreshold control of the unstable "figure-eight" orbit of the Lorenz attractor, using the interspike interval history only. (a) A plot in reconstructed state space. The three dimensions of the space are three consecutive time intervals. After an initial transient, the system becomes trapped in the figure-eight orbit. (b) A plot of the time series from the Lorenz attractor, which is being integrated to form the spikes. The vertical line shows the time at which the control protocol is turned on.

were constructed numerically. This step involves careful use of interpolation, since as Figure 3(a) shows, any given point on the reconstructed orbit may be reached only rarely by an interspike interval vector.

Figure 3(a) shows the result of the control procedure. Using small perturbations to the nominal value of β, after an initial transient (the single inside loop), the stabilization is achieved, and the system will continue the figure-eight behavior indefinitely. A time trace of the x-coordinate of the underlying Lorenz attractor being used to drive the integrate-and-fire model is shown in Figure 3(b). It was not used in the controlling process, and is shown here simply to verify success of the control procedure.

Superthreshold control of spike train dynamics is conceptually simpler. In this scenario, the goal is to change an erratic firing pattern to a regular pattern. The controller has the capability of causing the system to fire at any time, say through a large pulse applied to the system. We choose a desired fixed interspike interval and then ask the controller to apply superthreshold spikes to regulate the system to spiking rhythmically at that interval. In the on-demand pacing protocol, the controller applies the external spike whenever the interspike interval has exceeded the desired value. Obviously, this will cause the spike intervals to be capped at the desired level. What we will show is that in addition, under realistic assumptions on the dynamics, the spike times will be bounded from below as well, resulting in an evenly-timed spike sequence.

To make a model which is perhaps closer to a neurophysiological process, we add to the integrate-and-fire model (2.1) some relative refractoriness. For this experiment we use a time series from the Mackey-Glass equation as the signal to be integrated. The Mackey-Glass [1977] delay differential equation is

$$\dot{x}(t) = -0.1x(t) + \frac{0.2x(t-\Delta)}{1+x(t-\Delta)^{10}}. \tag{3.1}$$

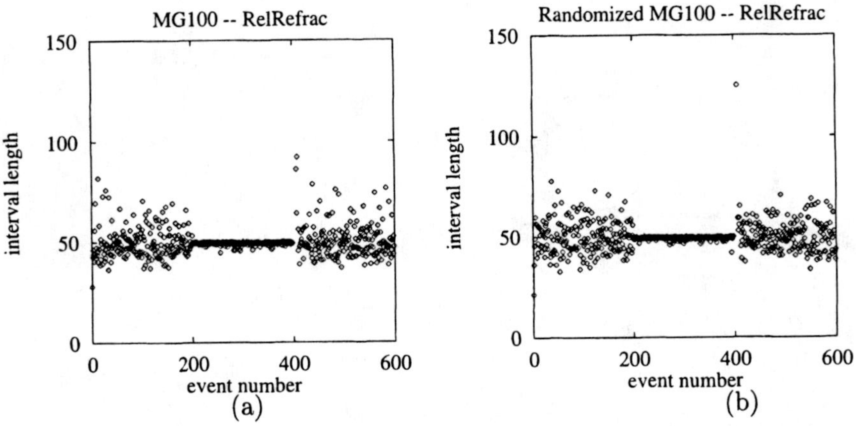

Figure 4 Application of superthreshold pulses to control an integrate-and-fire model with relative refractoriness. (a) The input signal to the integrator is a Mackey-Glass time series. Intervals are determined according to (2.1), where Θ is represented by a leaky potential. On-demand pacing control is applied between event numbers 200 and 400 only. (b) Input signal to the integrator is random noise with the same power spectrum as in (a). Similar control can be achieved for this random spike train.

When $\Delta = 30$, the attractor for this system has dimension around 3.5, according to numerical estimation. For $\Delta = 100$, the correlation dimension is approximately 7.1. (See Ding et al. [1993].)

The relative refractoriness is included by changing the constant threshold Θ in the integrate-and-fire hypothesis (2.1) to a simple "leaky potential" function. The threshold Θ will decrease at a constant rate ($\dot{\Theta} = -0.5$ in our example) until firing, at which time Θ is replaced by $\Theta + 50$, representing the repolarization stage.

The results of control are shown in Figure 4(a). A signal from equation (3.1) with $\Delta = 100$ is integrated until the threshold is met as in (2.1), with the assumption that the threshold Θ evolves as discussed above. For spike (event) numbers 1 to 200, the interspike interval lengths of the free-running system is graphed on the vertical axis. Since the integrated signal is 7-dimensional, no pattern can be ascertained by visual inspection. After event 200, the on-demand control protocol is turned on. The system immediately relaxes into a rhythmical behavior, with interspike intervals held near a fixed value. Note that not only are there no intervals above the nominal value (which is trivially enforced by on-demand pacing), but intervals below the nominal value have been largely eliminated as well. After spike 400, when the on-demand pacing is suspended, the system re-establishes its prior erratic behavior.

The same effect can be seen when the deterministic Mackey-Glass input is replaced by a random noise input. We manufacture the noise by creating a random-phase surrogate of the Mackey-Glass time series. That is, the phases of a Fourier-transformed version of the time series are randomized, and then the inverse Fourier transform is applied, resulting in a random signal with the same spectral characteristics as the deterministic Mackey-Glass signal used above. Figure 4(b) shows the same on-demand control technique as applied above. The control has an effect similar to that in the deterministic case.

4 An Embedding Theorem for Interspike Intervals

In this section we state an embedding theorem in the spirit of Sauer, Yorke, and Casdagli [1991], but for ISI vectors from a spike train instead of delay coordinate vectors from a time series. The details of the proof of this theorem are substantially more complicated than in the time series case—we will only sketch the proof in this article. We begin with a few lemmas.

Lemma 4.1 *Let k, m and p be integers, C an bounded open domain in $\mathbb{R}^p$, and let $\{G_c : \mathbb{R}^k \to \mathbb{R}^m | c \in C\}$ be a smooth family of smooth functions. Assume that there is an integer n and constants $\rho, \sigma > 0$ such that the nth largest singular value of the $m \times p$ Jacobian matrix*

$$\frac{\partial[G_c(x) - G_c(y)]}{\partial c} \tag{4.1}$$

is at least $\min\{\rho, \sigma|x-y|\}$ for all x, y lying in some bounded subset A of $\mathbb{R}^k$ and for all $c \in C$. Let $d_0(A)$ be the box-counting dimension of A. If $n > 2d_0(A)$, then G_c is one-to-one on A for (Lebesgue) almost every $c \in C$.

Proof Consider pairs $x, y \in \mathbb{R}^k$ separated by at least δ. Since the nth largest singular value of the Jacobian matrix (4.1) is at least $\min\{\sigma\delta, \rho\}$, there is a uniform constant B (depending only on δ) such that the point $G_c(x) - G_c(y)$ lies within ϵ of the origin for c contained in a subset of p-dimensional Lebesgue measure at most $B\epsilon^n$.

This means that two ϵ-balls in R^k separated by at least δ can have intersecting images under G_c for c in a set of measure at most $B\epsilon^n$. Since there are ϵ^{-d_0} balls covering A, there are at most ϵ^{-2d_0} such pairs. The set of c for which there is some pair of ϵ-balls whose images intersect under G_c has measure at most ϵ^{n-2d_0}. By hypothesis this expression tends to 0 as ϵ tends to 0, so we have shown that G_c is one-to-one on the pairs separated by at least δ, except for c belonging to a Lebesgue measure zero subset S_δ. By repeating the argument for $\delta = 1/i$, $i = 1, 2, 3, \ldots$, we find that each G_c for which c is not in the (measure zero) union of the $S_{1/i}$ is one-to-one on A. □

Lemma 4.2 *Let M be a $m \times k$ matrix. Assume that for some $\epsilon > 0$, the image of the unit ball $B(1)$ in $\mathbb{R}^k$ under M contains ϵe_i for each unit coordinate axis vector e_i. Then the mth largest singular value of the matrix M is at least $\epsilon/\sqrt{m}$.*

Proof The ellipsoid $MB(1)$ is centered at the origin and contains ϵe_i, so must contain an m-dimensional ball centered at the origin of radius $\epsilon/\sqrt{m}$. □

Lemma 4.3 *Assume that g is a positive real-valued continuous function of the two variables c, t. Let $\Theta > 0$ be a fixed constant, and define the function $T(c)$ by*

$$\int_{t_0}^{T(c)} g_c(t)\, dt = \Theta. \tag{4.2}$$

Assume also that $t_0 < t_1 < T(c)$, and that $g_c(t)$ is independent of c for $t_1 < t < T(c)$. If there is a constant $K > 0$ such that

$$\frac{d}{dc}\int_{t_0}^{t_1} g_c(t)\, dt \geq K, \tag{4.3}$$

then

$$\frac{dT}{dc} \leq -\frac{K}{g_c(T(c))}. \tag{4.4}$$

Proof Equation (4.2) gives

$$\Theta = \int_{t_0}^{t_1} g_c(t)\, dt + \int_{t_1}^{T(c)} g_c(t)\, dt. \tag{4.5}$$

Differentiating with respect to c yields

$$\begin{aligned} 0 &= \frac{d}{dc}\int_{t_0}^{t_1} g_c(t)\, dt + g_c(T(c))\frac{dT}{dc} \\ &\geq K + g_c(T(c))\frac{dT}{dc}, \end{aligned} \tag{4.6}$$

which implies the result. □

Using these lemmas, we will sketch a proof of Theorem 4.4. To simplify matters, we assume that the invariant set A contains no equilibrium points. The result holds as long as the number of equilibrium points in A is finite, although the proof is more complicated.

Let

$$\dot{x} = f(x) \tag{4.7}$$

be an autonomous system of k ordinary differential equations, with $x \in \mathbb{R}^k$. The system is said to satisfy a *Lipschitz condition* on the set A if there exists a constant $L > 0$ such that $|f(x) - f(y)| \leq L|x - y|$ for each x, y in A.

Assuming that a Lipschitz condition is satisfied, we can define the *flow* $F_t(x)$ as the unique solution at time t of the system (4.7) with initial value x. In addition, such a system satisfies continuous dependence on initial conditions; in fact, there exists a constant $\lambda > 1$ such that

$$\frac{|x-y|}{\lambda^t} < |F_t(x) - F_t(y)| < \lambda^t|x - y| \tag{4.8}$$

for each pair of points x, y in A. A set A is called *invariant* under the dynamics of (4.7), if $F_t(A)$ is contained in A for all $t \geq 0$. Recall the definition of the ISI-reconstruction function G_h from (2.3).

Theorem 4.4 *Let (4.7) be a system of differential equations with a Lipschitz condition on a compact invariant subset A of $\mathbb{R}^k$. Assume that m is an integer greater than $2d_0(A)$, and that A contains no equilibrium points of (4.7). Then there exists an open domain C in some $\mathbb{R}^p$ and a smooth family $\{q_c : \mathbb{R}^k \to \mathbb{R} | c \in C\}$ of smooth functions such that for each smooth positive real-valued function $h : \mathbb{R}^k \to \mathbb{R}$, the ISI-reconstruction function G_{h+q_c} is one-to-one on A for (Lebesgue) almost every $c \in C$.*

Proof (Sketch) The philosophy of this proof is that since the trajectories of the system are fairly straight on small scales, we can perturb h in carefully selected small regions in order to tease apart images of points in $\mathbb{R}^k$ that have mapped together under G_h. To do this trajectory-straightening, it helps to notice that since the compact set A contains no equilibrium points, there exists a minimum speed limit $s > 0$ for trajectories; that is, $|f(x)| > s$ for all $x \in A$. On the other hand, the Lipschitz constant L puts an upper bound on the spatial rate of change of f: Within a cube of width δ in $\mathbb{R}^k$, the velocity f can change by at most δL. We will choose δ small enough that $\delta < .01s/L$; then the velocity of the flow can change by at most 1% within a δ-cube. Trajectories move almost in straight lines through such cubes.

Since $|\dot{x}| > s$, a trajectory can spend at most $2\sqrt{k}\delta/s$ time units in any δ-cube before exiting. The reason is that $\sqrt{k}\delta$ is the longest straight segment contained in a δ-cube; since the trajectory may deviate by only 1% from a straight line, the distance traveled will be less than $2\sqrt{k}\delta$, and the speed is at least s.

Let $b = \min\{|h(x)| : x \in A\} > 0$ and $B = \max\{1 + |h(x)| : x \in A\}$. Clearly $\tau = \Theta/bm$ is an upper bound for the time needed to create m interspike intervals. In addition to the previous condition on δ, we require δ small enough to satisfy

$$3\frac{2\sqrt{k}\delta}{s}B < \Theta,$$

where Θ is the firing threshold. This requirement ensures that the trajectory must pass through at least 3 different δ-cubes between each firing.

On a δ-cube, there are k smooth "bump functions" $q(x)$, each equal to zero for points outside of the cube, and such that for two trajectories $F_t(x), F_t(y)$ passing through the cube, at least one of the q satisfies

$$\int [q(F_t(x)) - q(F_t(y))]\, dt \geq \frac{D}{2\sqrt{k}}, \tag{4.9}$$

where D is the minimum distance between $F_t(x)$ and $F_t(y)$ in the δ-cube. Note that (4.8) shows that $D > |x - y|/\lambda^\tau$ for $0 < t < \tau$. The collection of all linear combinations $\sum c_i q_i, |c_i| < 1$ of these bump functions on a grid of δ-cubes covering A is the smooth family of smooth functions we need.

Finally we suggest how to apply Lemma 4.1 to finish the proof. Let x, y be a pair of points in $\mathbb{R}^k$. Assume first that the trajectories of x and y do not come within $2\sqrt{k}\delta$ of one another during time τ. (After this amount of time, m interspike intervals have already occurred and the proximity of $F_t(x)$ and $F_t(y)$ are no longer important.) Since the δ-cubes the trajectories travel through are disjoint, we can add a bump function to the trajectory of x between the ith and $(i+1)$st firing, and subtract one between the $(i+1)$st and $(i+2)$nd so that the derivative

$$\frac{d[G_{h+c_1q}(x) - G_{h+c_1q}(x)]}{dc_1} \tag{4.10}$$

is a vector in the direction of the ith coordinate axis in $\mathbb{R}^m$, of length at least $\delta/(2B\sqrt{k})$. Here we have used Lemma 4.3 with $g_c(t) = (h+cq)(x(t)) - (h+cq)(y(t))$ and $K = \delta/(2\sqrt{k})$. Since this can be achieved for each $i = 1, \ldots, m$, Lemma 4.2 implies that the mth largest singular value of

$$\frac{\partial[G_c(x) - G_c(y)]}{\partial c} \tag{4.11}$$

must be at least $1/\sqrt{m}$ times this, or $\delta/(2B\sqrt{km})$. In the case that the trajectories of x and y become close during the interval of τ time units, we can use the same strategy, using $K = |x - y|/(2\lambda^\tau\sqrt{k})$. Then the derivative (4.10) is a vector of length at least $|x - y|/(2B\lambda^\tau\sqrt{k})$, so that the mth largest singular value of (4.11) is at least $|x - y|/(2B\lambda^\tau\sqrt{km})$. Now Lemma 4.1 can be applied with $\rho = \delta/(2B\sqrt{km})$ and $\sigma = 1/(2\lambda^\tau B\sqrt{km})$ to conclude the proof. □

5 Summary

We have suggested methods of studying and exploiting interspike interval series recorded from experiment, in the absence of known model equations. If they are generated according to hypothesis (2.1), the ISI vectors can be used to represent states of the underlying attractor in a similar manner as delay coordinate vectors from amplitude time series.

Although Equation (2.1) was chosen to be as free of assumptions as possible, it is a hypothesis, and the primary features of any particular physiological system may not be well represented by this hypothesis. This study is therefore a kind of feasibility study. Further work with other integrate-and-fire models for specific contexts, such as the leaky integrator of Section 3, is needed.

One strong motivation for this work is to explore modes of information processing in neural systems. The possibility exists that much of the communication and processing that transpires in the central nervous system, for example, is coded in the temporal firing patterns of neural circuits. Our approach is designed to provide fundamental knowledge toward understanding the so-called temporal coding problem.

This study shows the theoretical possibility of using the OGY control procedure, with obvious modifications, for systems measured only by spike trains. In systems where deterministic but aperiodic spike trains are measured, and where a global parameter is accessible for small time-dependent perturbations, this technique may be used to entrain the system into periodic spiking behavior.

Moreover, it was shown that the on-demand control procedure can be used to enforce equilibrium spiking behavior in a neurophysiologically motivated spiking model system. This capability was shown to exist for a deterministic system as well as a similar system that was principally random. We conclude that ability to control is not necessarily a reliable diagnostic for deterministic dynamical mechanisms.

References

Aeyels, D. [1981] *Generic observability of differentiable systems*, SIAM J. Control and Opt. **19**, 595-603.

Ding, M., Grebogi, C., Ott, E., Sauer, T. and Yorke, J. A. [1993] *Estimating correlation dimension from a chaotic time series: when does plateau onset occur?* Physica D **69**, 404.

Hunt, B., Sauer, T. and Yorke, J. A. [1992] *Prevalence: A translation-invariant "almost every" on infinite dimensional spaces.* Bull. A.M.S. **27**, 217-238.

Lorenz, E. [1963] *Deterministic nonperiodic flow*, J. Atmos. Sci. **20**, 130 - 141.

Mackey, M. C. and Glass, L. [1977] *Oscillations and chaos in physiological control systems*, Science **197**, 287.

Lewis, P. A. W. [1972] *Stochastic Point Processes: Statistical Analysis, Theory, and Applications*(ed). Wiley-Interscience, New York.

Miller, M. and Snyder, D. [1991] *Random Point Processes in Time and Space.* Springer-Verlag, New York.

Ott, E., Grebogi, C. and Yorke, J. A. [1990] *Controlling chaos*, Phys. Rev. Lett. **64**, 1196.

Ott, E., Sauer, T. and Yorke, J. A. [1994] *Coping with Chaos: Analysis of Chaotic Data and the Exploitation of Chaotic Systems.* Wiley Interscience, New York.

Packard, N., Crutchfield, J., Farmer, J. D. and Shaw, R. [1980] *Geometry from a time series*, Phys. Rev. Lett. **45**, 712.

Sauer, T. [1994] *Reconstruction of dynamical systems from interspike intervals*, Phys. Rev. Lett. **72**, 3811.

Sauer, T., Yorke, J. A. and Casdagli, M. [1991] *Embedology*, J. Stat. Phys. **65**, 579.

Schiff, S. J., Jerger, K., Chang, T., Sauer, T. and Aitken, P. G. [1994] *Stochastic versus deterministic variability in simple neuronal circuits. II. Hippocampal slice*, Biophys. J. **67**, 684.

Takens, F. [1981] *Detecting strange attractors in turbulence*, Lecture Notes in Math. **898**, Springer-Verlag, Berlin.

Theiler, J., Eubank, S. and Longtin, A., Galdrakian, B. and Farmer, J. D. [1992] *Testing for nonlinearity in time series: the method of surrogate data,* Physica D **58**, 77.

Tong, H. and Lim, K. S. [1980] *Threshold autoregression, limit cycles and cyclical data,* J. Roy. Stat. Soc. B **42**, 245-292.

Tuckwell, H. [1988] *Introduction to Theoretical Neurobiology*, vols 1,2. Cambridge University Press.

Fields Institute Communications
Volume 11, 1997

On the Validity of the Method of Surrogate Data

Kung-Sik Chan
Department of Statistics & Actuarial Science
University of Iowa
Iowa City, Iowa, USA 52242

Abstract. We consider the validity of the method of surrogate data from the viewpoint of whether or not the nominal false rejection rate of a test statistic calibrated by the method of surrogate data (approximately) equals the true false rejection rate. Some asymptotic properties of the surrogate data method are derived. We review the classical approach of reducing a composite hypothesis to a simple hypothesis via conditioning upon a sufficient statistic. Using this approach, we show that the method of surrogate data is exactly valid for any test statistic if the null hypothesis assumes the class of stationary Gaussian circular processes. We construct a test statistic for which the method of surrogate data is incorrect when the true model is a stationary Gaussian AR(1) process. However, it is shown that for a statistic that is asymptotically independent of the "end" condition of the data, the distribution is asymptotically unaltered when the underlying stationary Gaussian AR(1) process is changed into a sequence of approximating stationary Gaussian circular processes.

1 Introduction

The problem of testing whether or not a data set is generated by some chaotic system is generally rather difficult. One direct approach to answering the above question is to test if the maximum Lyapunov exponent is positive or not. However, the sampling properties of the estimator of the maximum Lyapunov exponent are not well understood. Among the indirect approaches, some of the more popular test statistics are based on (1) nonlinear prediction, (2) correlation dimension and (3) information. See Theiler et al. [1992], Yao and Tong [1996] and the references in these two papers. However, these test statistics also suffer from the problem of having intractable sampling distributions.

Instead of deciding whether the dynamics are chaotic, a more manageable problem is to test whether the data are generated by some linear stochastic process. (A process $\{X_t\}$ is *linear* if the best linear predictor of X_t given the past observations is the best predictor in mean square sense; Hannan [1979].) We shall assume that the process is stationary. In fact, we consider a more restricted null hypothesis,

---

1991 *Mathematics Subject Classification.* Primary 62G09, 58F13; Secondary 62M10, 60G15.

Work of K. S. Chan was supported in part by NSF Grant DMS-9504798. I thank Professors Colleen Cutler, Howell Tong and an anonymous referee for their careful reading of a previous draft of the paper, and their valuable comments.

namely, H_0: the process (possibly after instantaneous transformation) is a stationary Gaussian process with fast decaying auto-correlations, that is, $\sum_{j=1}^{\infty} j\gamma_j^2 < \infty$, where γ_j is the jth-lag auto-correlation of the process. Here, the null hypothesis is composite in the sense that the true model is still not specified; we only know that the true model is determined by its first two moments with fast decaying auto-correlations. It is at first not clear how to find the null sampling distribution for any test statistic. Theiler et al. [1992] suggested the innovative idea of *surrogate data* as a way to calibrate the (statistical) significance of any test statistic. See also Kennel and Isabelle [1992]. Briefly, the method of surrogate data consists of the following steps:

1. Randomly draw a time series of the same size and whose sample mean and periodogram are identical to that of the observed data; the time series so obtained is called a *surrogate series.*
2. Compute the test statistic, say $T = T_N$, for the surrogate data obtained in step (1), where N is the sample size.
3. Repeat steps (1) and (2) a large number of times to build up a reference null distribution for T.
4. Use the reference null distribution to assess the significance of the observed value of T.

Step (1) is carried out by the method of Fast Fourier Transform (FFT). (Details are given later.) Step (4) could be implemented in several ways. For example, if small T-values indicate evidence against the null hypothesis, one can compute the approximate p-value as the percentile of the observed T-value with respect to the reference distribution. If $t_1, t_2, \ldots, t_m$ are the T-values for the surrogate data obtained in step (3), then the approximate p-value is equal to the fraction of t_i's not greater than t_{obs}, i.e., $\#\{t_i \leq t_{\text{obs}}\}/m$. One-sided upper p-values and two-sided p-values can be similarly defined. Theiler et al. [1992] also suggested calculating the t-ratio, the number of standard deviations the observed T-value deviates from the mean of the reference null distribution, and using the standard normal approximation to assess the significance of the ratio. Theiler et al. [1992] suggested that, by preserving the sample mean and the periodogram, the surrogate data preserve the linear structure of the observed data, and hence provide an appropriate method for calibrating the significance of any test statistic testing for nonlinearity.

In the extreme case that the test statistic is determined by the sample mean and the periodogram, its distribution over the surrogate data degenerates to a constant, and the method of surrogate data becomes useless. Thus, implicitly assumed in the method of surrogate data is that the test statistic is what may be called a *nonlinear test statistic*, i.e., it is *not* asymptotically determined by the sample mean and the periodogram. This restriction is assumed here also, and it seems natural because only nonlinear test statistics will have any discriminatory power for testing for nonlinearity.

We consider the validity of the method of surrogate data from the viewpoint of whether or not the nominal false rejection rate (also called the *size*) of a test statistic calibrated by the method of surrogate data (approximately) equals the true false rejection rate. For example, we may reject the null hypothesis of linearity if and only if the lower 1-sided p-value of a test statistic, based on the method of surrogate data, is smaller than, say, 5%. Does the actual false rejection rate of the above rejection criterion (approximately) equal 5%?

We now outline the content of this paper. In Section 2, the method of surrogate data is described in some detail. It is shown that, in terms of finite dimensional distributions, the surrogate data are asymptotically Gaussian with population auto-covariances equal to the corresponding observed circular auto-covariances. This result suggests that the surrogate data introduce spurious correlation among the first and the last few surrogate data points.

In Section 3, we discuss a classical approach to constructing test statistics whose sampling distributions are invariant under the null hypothesis. (Much of this material is classical, but is included here for completeness and ease of reference.) The basic idea is as follows. Suppose that under the null hypothesis, a (possibly vectorial) sufficient statistic, say denoted by S_N, exists. Then the conditional distribution of any test statistic, say T_N, given $S_N = s$, is invariant under the null hypothesis. Moreover, the conditional distribution of T_N given $S_N = s$ may be used to assess the statistical significance of the observed value of T_N. For example, rejection criterion using p-values based on the conditional distributions is exactly valid: the nominal false rejection rate is identical to the actual false rejection rate. Fancier valid rejection rules may be constructed by gluing together rejection regions (of the same size) based on the conditional distributions. The conditional distribution of T_N given $S_N = s$ is ordinarily analytically intractable, but may be simulated via a Monte Carlo method:

1. draw realizations Y_N of the time series under the null hypothesis that preserve the observed sufficient statistic, i.e., $S_N(Y_N) = s$.
2. use the empirical distribution of $T_N(Y_N)$ to approximate the conditional distribution of T_N given $S_N = s$.

The link between this classical approach and testing with the method of surrogate data lies in establishing that the sample mean and the sample periodogram are sufficient for the parameters of Gaussian processes with fast decaying auto-covariances. Unfortunately, except for the class of stationary Gaussian *circular* processes (defined below), the sample mean and the periodogram are *not* exactly sufficient for Gaussian processes.

We say that a vector of observations $\mathbf{X}_N = (X_1, X_2, \ldots, X_N)$ comes from a *circular process* $X_1, X_2, \ldots$ *of index* N if $X_{j+kN} = X_j$ for each $1 \le j \le N$ and positive integer k. Such a process is called Gaussian if the (multivariate) distribution of $\mathbf{X}_N$ is Gaussian. Stationarity (in the usual sense) of a Gaussian circular process of index N is equivalent to requiring that the means of the X_j's be identical and that the covariance matrix of $\mathbf{X}_N$ be a *circular matrix* (see Section 4 for the definition of a circular matrix, and Hannan [1962] for more discussion). In Section 4, we show that the method of surrogate data is valid for any test statistic if the data $\mathbf{X}_N$ is generated by a stationary Gaussian circular process of matching index N (this unlikely assumption makes it of dubious value directly for data analysis; however, circular processes can be used effectively as approximations under certain circumstances — see the discussion at the end of the next paragraph, as well as Section 5).

In Section 5, we assume that the underlying process is a stationary Gaussian AR(1) process. As mentioned earlier, the surrogate data does a poor job in that its auto-covariance function deviates from the true auto-covariance function in the tail. This suggests that the method of surrogate data may not work for test statistics which depend heavily on the first and last few observations. We show that this is so (see Example 5.1) by exhibiting a test statistic which essentially depends on the first and last observations, and for which the method of surrogate data fails; the actual false rejection rate diverges sharply from the nominal false rejection rate when the AR(1) coefficient is close to 1 in magnitude. Thus, it is clear that some restriction must be placed on the test statistic in order for the method of surrogate data to work. We then assume that the test statistic satisfies the conditions that under the null hypothesis, (1) it is asymptotically independent of the end effects of the observations, i.e., asymptotically independent of any finite number of observations in the beginning and end, and (2) after suitable normalization if necessary, it is asymptotically distributed as some (possibly non-Gaussian) non-degenerate distribution. Condition (1) is needed to avoid the kind of pathology exhibited by Example 5.1. Condition (2) is ordinarily satisfied by many test statistics, and will be explained further in Section 5. It suffices that such normalization can be done theoretically to produce a non-degenerate distribution for the test statistic. In practice, the normalization need not be carried out because its implementation only induces a re-scaling which will have no effect on the p-values or the standard deviations, based on the surrogate data, of the observed test statistic. It is shown that under these regularity conditions, the asymptotic distribution of the test statistic is unaltered when the underlying stationary Gaussian AR(1) process is changed into a sequence of approximating stationary Gaussian circular processes. The discussions in Section 4 then suggest that the method of surrogate data is asymptotically valid for such test statistics. However, it is desirable to strengthen the result as follows: Let Q_N be the $\alpha \times 100$ percentile of the conditional distribution of T_N given the sample mean and the periodogram. We conjecture that, under suitable conditions, the true false rejection rate $\text{Prob}(T_N \leq Q_N | H_0)$ converges to α as $N \to \infty$.

2 The Method of Surrogate Data

Recall that the null hypothesis H_0 specifies that the data $\mathbf{X} = \mathbf{X}_N = (X_1, X_2, \ldots, X_N)^T$ is generated by a stationary Gaussian process with fast decaying autocorrelations. Denote the mean of X_t by μ. We first introduce some notions from frequency domain analysis. (See Priestley [1981] for an introduction to spectral analysis.) The k-th lag auto-covariance of (X_t) is $\gamma(k) = \gamma_k = E((X_t - \mu)(X_{t+k} - \mu))$, where $E()$ denotes taking the expectation of the expression enclosed by parentheses. We assume that the spectral density function, h, exists so that $\{\gamma_k\}_{k \in \mathbf{Z}}$ and h form a Fourier pair:

$$\begin{aligned} \gamma_k &= \int_{-\pi}^{\pi} \exp(ik\omega) h(\omega) d\omega, \text{and} \\ h(\omega) &= \frac{1}{2\pi} \sum_{k=-\infty}^{\infty} \gamma_k \exp(-ik\omega), \quad -\pi \leq \omega \leq \pi \end{aligned}$$

where $i = \sqrt{-1}$. Given the data $\mathbf{X}_N$, its discrete Fourier transform (DFT) is defined as

$$\zeta(\omega) = \zeta_{\mathbf{X}_N}(\omega) = \frac{1}{\sqrt{2\pi N}} \sum_{t=1}^{N} X_t \exp(-i\omega t), \quad -\pi \leq \omega \leq \pi. \tag{2.1}$$

The (modified) periodogram is an estimate of the spectral density function, and is defined by the following formula:

$$I(\mathbf{X},\omega) = I(\mathbf{X}_N,\omega) = I(\omega) = |\zeta_{\mathbf{X}_N}(\omega)|^2 = \frac{1}{2\pi}\sum_{k=-(N-1)}^{N-1} r'_k \exp(-ik\omega) \qquad (2.2)$$

where $r'_k = \frac{1}{N}\sum_{t=1}^{N-k} X_t X_{t+k}$ is the k-th lag (uncentered) sample auto-covariance. The centered k-th lag sample auto-covariance is denoted by $r_k = \sum_{t=1}^{N-k}(X_t - \bar{X})(X_{t+k} - \bar{X})/N$.

The DFT is often only computed at a set of equally distributed angular frequencies: $\omega_j = 2\pi j/N, j = 1,2,\ldots,N$. Note that $\zeta_N = (\zeta(\omega_1),\zeta(\omega_2),\ldots,\zeta(\omega_N))^T = \frac{1}{\sqrt{2\pi}}P_N\mathbf{X}_N$ where the j-th row of P_N is equal to $\frac{1}{\sqrt{N}}(\exp(-i\omega_j), \exp(-i2\omega_j),\ldots,$ $\exp(-iN\omega_j))$. P_N is an orthonormal matrix with its complex conjugate P_N^* being its inverse; hence $\mathbf{X}_N$ can be recovered from the DFT of ζ_N:

$$X_t = 2\pi \times \frac{1}{\sqrt{2\pi N}}\sum_{j=1}^{N}\zeta(\omega_j)\exp(ij\omega_t), \quad t = 1,2,\ldots,N. \qquad (2.3)$$

Because the X's are real numbers, its DFT enjoys the property that $\zeta(\omega_j) = \zeta^*(\omega_{N-j}), \forall j$, where for any complex number $c = a+bi$, its conjugate is $c^* = a - bi$. Hence $\zeta(\omega_0) = \zeta(\omega_N)$ is a real number, and so is $\zeta(\omega_{\frac{N}{2}})$ whenever N is an even number; otherwise the ζ's are generally complex. Written in the polar form, $\zeta(\omega_j) = \sqrt{I(\omega_j)}\exp(i\theta_j)$. The phases θ satisfy the relationship $\theta_j = -\theta_{N-j}, \forall j$. If N is even, $\theta_{N/2}$ must be either 0 or π because $\zeta(\omega_{N/2})$ is real. Equation (2.3) can be written as follows:

$$X_t = \sqrt{\frac{2\pi}{N}}\sum_{j=1}^{N}\sqrt{I(\omega_j)}\exp(i(\omega_t j + \theta_j)), \quad t = 1,2,\ldots,N. \qquad (2.4)$$

This equation shows that assuming N is an odd number, $\mathbf{X}_N$ can be determined from the sample mean, the periodogram values $I(\omega_j), j = 1,2,\ldots,(N-1)/2$ and the phases $\theta_1,\theta_2,\ldots,\theta_{(N-1)/2}$. In fact, we have

$$X_t = \bar{X} + \sqrt{\frac{2\pi}{N}}\sum_{j=1}^{(N-1)/2} 2\sqrt{I(\omega_j)}\cos(\omega_t j + \theta_j). \qquad (2.5)$$

Hence, $\eta = (\bar{X}, \sqrt{I(\mathbf{X},\omega_1)}, \sqrt{I(\mathbf{X},\omega_2)},\ldots,\sqrt{I(\mathbf{X},\omega_{(N-1)/2})},\theta_1,\ldots,\theta_{(N-1)/2})^T$ and $\mathbf{X}_N$ are one-to-one functions of each other.

Claim 2.1 The Jacobian $J = \det(\frac{\partial \mathbf{X}_N}{\partial\eta})$ is proportional to $\sqrt{\prod_1^{(N-1)/2} I(\omega_j)}$.

The proof of this claim is deferred to Appendix 1. When N is even, the RHS of (2.5) has to be augmented by the term $\sqrt{\frac{2\pi}{N}I(\mathbf{X},\omega_{N/2})}\cos(\pi t + \theta_{N/2})$.

The method of surrogate data generates fictitious data $\mathbf{Y} = \mathbf{Y}_N = (Y_1, Y_2,\ldots,$ $Y_N)^T$ that preserves the observed sample mean and periodogram: $\forall 1 \le t \le N$, and $N = 2m+1$,

$$Y_t = \bar{X} + \sqrt{\frac{2\pi}{N}}\sum_{j=1}^{m} 2\sqrt{I(\mathbf{X},\omega_j)}\cos(\omega_t j + \theta_j) \qquad (2.6)$$

where if $N = 2m+1$, $\theta_1,\theta_2,\ldots,\theta_m$ is iid $U[0,2\pi]$; and if $N = 2(m+1)$, the RHS of (2.6) has to be augmented by the term $\sqrt{\frac{2\pi}{N}I(\mathbf{X},\omega_{N/2})}\cos(\pi t + \theta_{N/2})$ and

$\theta_1, \ldots, \theta_m$ are iid $U[0, 2\pi]$ and independent of $\theta_{N/2}$ (which is equal to 0 or π with probability 0.5 each). Let $S = S_N = S(X_N)$ denote the sample mean $\bar{X}$ and the periodogram values $I(\mathbf{X}, \omega_j), j = 1, 2, \ldots, N$. Let $\mathcal{X}_s$ be the set of all $\mathbf{Y}$'s such that $S(\mathbf{Y}_N) = s$. (Here, we assume that s is the observed value of S.) The distribution of $\mathbf{Y}$ depends on $\mathbf{X}$ only through $s = S(\mathbf{X})$. Below, we sometimes write $\mathbf{Y}(S)$ for $\mathbf{Y}$ to indicate the dependence of $\mathbf{Y}$ on S.

Let $T = T_N$ be a test statistic. Theiler et al. [1992] suggested to calibrate an observed value $t = T(\mathbf{X})$ by the distribution of $T(\mathbf{Y}(s))$ where $s = S(\mathbf{X})$, i.e. $\mathbf{Y}$ is generated according to (2.6) with $\bar{X}$ and the periodogram values set to the corresponding observed values. Ordinarily, the analytic form of the distribution of $T(\mathbf{Y})$ is not available. However, the distribution of $T(\mathbf{Y})$ can be readily simulated:

(1) simulate iid samples $\mathbf{Y}^{(1)}, \ldots \mathbf{Y}^{(m)}$ according to (2.6);
(2) compute $T(\mathbf{Y}^{(1)}), \ldots, T(\mathbf{Y}^{(m)})$. Use the empirical cumulative distribution $F_{T,m}(t) = \#\{j : T(\mathbf{Y}^{(j)}) \leq t\}/m$ to approximate the cumulative distribution of $T(\mathbf{Y})$.

Step (1) can be done efficiently via the method of fast Fourier transform. For $S = s$, the distribution of $T(\mathbf{Y})$ is invariant for $\mathbf{X} \in \mathcal{X}_s$. Thus the method of surrogate data is valid generally if and only if the conditional distribution of $T(\mathbf{X})$ given $S = s$ is equal to the distribution of $T(\mathbf{Y}(s))$, for *all* s and *all* test statistics T (and for all N). The latter condition is equivalent to the condition that the conditional distribution of $\mathbf{X}$ given $S = s$ is equal to the distribution of $\mathbf{Y}(s)$ for all s, because of the assumed arbitrariness of T.

Remark 2.2 It may be argued that rather than requiring that the surrogate data method be valid for all test statistics, we need only be concerned with its validity for the test statistic of interest. However, this only raises the questions of (1) characterizing $\mathcal{C}$, the class of test statistics for which the method of surrogate data is valid, and (2) how to ascertain whether or not the test statistic of interest belongs to $\mathcal{C}$. Both in Theiler et al. [1992] and Kennel and Isabelle [1992], it seems that the method of surrogate data is proposed for calibrating *any* test statistic. Hence, we shall at first regard the surrogate data method as valid if and only if it is valid for all test statistics. Later on, we shall see that this requirement is too stringent, and must be relaxed.

The following result shows that $\forall s$, the conditional distribution of $\mathbf{Y}(s)$ is uniform over $\mathcal{X}_s$. In other words, a data vector can be randomly drawn from $\mathcal{X}_s$ by generating $\mathbf{Y}$ according to (2.6) with uniformly distributed phases.

Theorem 2.3 *For all s, the (surrogate data) distribution of $\mathbf{Y}(s)$ is uniform over $\mathcal{X}_s$.*

Proof We first consider the case that $N = 2m+1$ is odd. $\mathcal{X}_s$ is a differentiable manifold with a global co-ordinate system: $\mathbf{Y} = \mathbf{Y}(\theta_1, \ldots, \theta_m)$ defined by (2.6). The range of the θ's can be identified with the product of m circles. (Here, we identified 0 with 2π so that $[0, 2\pi]$ becomes a circle.) Below, we need to use some formula for integrating functions over a differentiable manifold. See Section 26 in

Dubrovin et al. [1984] for proofs of these formulas.

$$\begin{aligned} dY_t &= \sum_{j=1}^{m} \frac{\partial Y_t}{\partial \theta_j} d\theta_j \\ &= -\sqrt{\frac{2\pi}{N}} \sum_{j=1}^{m} 2\sqrt{I(\mathbf{X}, \omega_j)} \sin(\omega_t j + \theta_j) d\theta_j. \end{aligned}$$

Thus, $d\mathbf{Y} = -\sqrt{4\pi} \sum_{j=1}^{m} \sqrt{I(\mathbf{X}, \omega_j)} \mathbf{u}_j d\theta_j$ where $\mathbf{u}_j = (u_{1j}, u_{2j}, \ldots, u_{Nj})^T$ and $u_{tj} = \sin(\omega_t j + \theta_j)\sqrt{\frac{2}{N}}$. It can be readily checked that the $\mathbf{u}$'s are orthogonal to each other and are of unit length. Hence,

$$\sum (dY_t)^2 = d^T\mathbf{Y} d\mathbf{Y} = 4\pi \sum_{1}^{m} I(\mathbf{X}, \omega_j)(d\theta_j)^2. \tag{2.7}$$

Equation (2.7) implies that the joint pdf of $\mathbf{Y}$ over $\mathcal{X}_s$ is equal to (see formula (2) on p.249 of Dubrovin et al. [1984])

$$\begin{aligned} f_{\mathbf{Y}}(Y_1, \ldots, Y_N) &= \left((4\pi)^m \prod_{j=1}^{m} I(\mathbf{X}, \omega_j) \right)^{-1/2} f_{\theta}(\theta_1, \ldots, \theta_m) \\ &= \text{constant}, \end{aligned}$$

because the joint pdf of the θ's is constant.

For the case that N is even, note that

$$\sum (-1)^t Y_t = \cos(\theta_{N/2}) \sqrt{2\pi I(\mathbf{X}, \omega_{N/2})/N}$$

which is positive or negative depending on whether or not $\theta_{N/2}$ is 0 or π. Therefore, $\mathcal{X}_s$ is a disconnected manifold consisting of two connected manifolds over each of which $\theta_{N/2}$ is constant. The proof for the case of an odd N can be adapted to show that $\mathbf{Y}(s)$ is uniformly distributed over each of the two connected manifolds. Because $\theta_{N/2}$ is 0 or π with equal probability, $\mathbf{Y}(s)$ is uniformly distributed over $\mathcal{X}_s$. □

The above theorem provides a simple necessary and sufficient condition for the validity of the method of surrogate data.

Corollary 2.4 *The method of surrogate data is exactly valid for any test statistic if and only if the conditional distribution of* $\mathbf{X}_N$ *given* $S = s$ *(the sample mean and the sample periodogram) is uniform over* $\mathcal{X}_s$.

We now summarize some properties of the surrogate data that will be needed later on. See also Press et al. [1986] for an introduction to DFT and some of its properties. By construction, the surrogate data $\mathbf{Y}$ preserve the observed sample mean and periodogram, i.e., $\bar{Y} = \bar{X}$ and $I(\mathbf{Y}, \omega_j) = I(\mathbf{X}, \omega_j), \forall j = 1, 2, \ldots, N$. As a consequence, $\mathbf{Y}$ preserves the data circular auto-covariances, i.e., $\forall k, \sum_{t=1}^{N}(Y_t - \bar{Y})(Y_{t+k} - \bar{Y})/N = \sum_{t=1}^{N}(X_t - \bar{X})(X_{t+k} - \bar{X})/N = r_{k,c}$, where $Y_{t+N} = Y_t$ and $X_{t+N} = X_t$ for all $t \geq 0$. Below, we assume that $S = s$ is fixed and $\mathbf{Y}$ is generated according to (2.6). Assume also that $N = 2m + 1$ is odd. Then, (henceforth in this section, all expectations and covariances are calculated according to the surrogate data distribution),

$$\begin{aligned} E(Y_t) &= \bar{X} + \sqrt{\frac{2\pi}{N}} \sum_{j=1}^{m} 2\sqrt{I(\mathbf{X}, \omega_j)} E(cos(\omega_t + \theta_j)) \\ &= \bar{X} \end{aligned}$$

because $\cos(\omega_t j + \theta_j)$ is of zero mean and variance equal to $1/2$. We now derive the auto-covariances for the Y's. Let $p \geq q$ be two positive integers.

$$\begin{aligned} \operatorname{cov}(Y_p, Y_q) &= E((Y_p - \bar{X})(Y_q - \bar{X})) \\ &= \frac{2\pi}{N} \sum_{j=1}^{m} 4I(\mathbf{X}, \omega_j) E(\cos(\omega_p j + \theta_j) \cos(\omega_q j + \theta_j)) \\ &\quad + \sum_{j \neq l} 4\sqrt{I(\mathbf{X}, \omega_j) I(\mathbf{X}, \omega_l)} E(\cos(\omega_p j + \theta_j)) E(\cos(\omega_q l + \theta_l)) \\ &= \frac{4\pi}{N} \sum_{j=1}^{m} I(\mathbf{X}, \omega_j) E(\cos(\omega_p j + \omega_q j + 2\theta_j) + \cos((\omega_p - \omega_q) j))) \\ &= \frac{2\pi}{N} \{ \sum_{j=0}^{N-1} I(\mathbf{X}, \omega_j) \exp(i2\pi(p-q)j/N) - I(\mathbf{X}, \omega_0) \} \\ &= \sum_{j=0}^{N-1} \frac{1}{N} \{ \sum_{k=-N+1}^{N-1} r'_k \exp(-ik\omega_j + i2\pi(p-q)j/N)) \} - (\bar{X})^2 \\ &= \sum_{k=-N+1}^{N-1} \frac{r'_k}{N} \sum_{j=0}^{N-1} \exp\{-i2\pi(jk - (p-q)j)/N\} - (\bar{X})^2 \\ &= r'_{p-q} + r'_{N-(p-q)} - (\bar{X})^2 \\ &= r_{p-q,c}. \end{aligned}$$

This shows that the population auto-covariance function of $\mathbf{Y}$ is exactly equal to the circular auto-covariance function of the data. In particular, $\operatorname{cov}(Y_1, Y_N) = r_{1,c} = r_1 + r_{N-1}$. Hence, the sample auto-covariance function of the surrogate data might not resemble that of $\mathbf{X}$ in the tail. These results are now summarized in the following theorem.

Theorem 2.5 *For fixed N and s, i.e., for fixed sample size, fixed sample mean and periodogram values at $\omega_j, j = 1, \ldots, N$, we have (1) $\forall t, E(Y_t) = 0$ and (2) $\operatorname{cov}(Y_p, Y_q) = r_{p-q,c} = r_{p-q} + r_{N-(p-q)}$.*

Later, we shall need to determine the asymptotic joint distribution of Y_1 and Y_N for large sample sizes. Let p be a positive integer. Note that $Y_p - \bar{X}$ is a sum of independent random variables. It can be expected that $(Y_1, Y_N)^T$ is asymptotically normal, as is rigorously established in the following result.

Theorem 2.6 *Assume that (X_t) is a linear Gaussian process:*

$$X_t = \mu + \sum_{j=0}^{\infty} g_j a_{t-j}, \forall t,$$

where (a_t) is iid $N(0,\sigma_a^2)$;$g_0 = 1$ and the g's are constants such that $\sum_{j=1}^{\infty} |g_j| j^{\alpha} < \infty$ for some $\alpha > 1/2$. Also assume that the spectral density function of (X_t), h, is bounded away from zero and infinity. Then for any integer p, Y_p is asymptotically $N(\mu, \gamma_0)$. Also, $(Y_1, Y_N)^T$ is asymptotically normal with mean vector $(\mu, \mu)^T$ and covariance matrix equal to

$$\Sigma = \begin{pmatrix} \gamma_0 & \gamma_1 \\ \gamma_1 & \gamma_0 \end{pmatrix},$$

as $N \to \infty$, where μ is the population mean of X and the γ's are the population auto-covariances of the X's.

Proof For any p, $\cos(\omega_p j + \theta_j)$ is of zero mean and its variance is equal to 1/2. Now, $\bar{Y}_p - \bar{X}$ is a summand of independent variables. The variance of the j-th summand is $4\pi I(\mathbf{X}, \omega_j)/N$. The variance of $\bar{Y}_p - \bar{X}$ is then equal to r_0. A sufficient (Lyapunov) condition for the asymptotic normality of $Y_p - \bar{X}$ is (assuming $N = 2m + 1$)

$$\frac{\sum_{j=1}^{m} I^{3/2}(\mathbf{X}_N, \omega_j)}{(\sum_{j=1}^{m} I(\mathbf{X}_N, \omega_j))^{3/2}} \to 0. \tag{2.8}$$

See Corollary 2 on p. 331 of Shiryayev [1984], where we set $\delta = 1$. (2.8) holds if

$$\max_{1 \le j \le m} \frac{I(\mathbf{X}_N, \omega_j)}{\sum_{j=1}^{m} I(\mathbf{X}_N, \omega_j)} \to 0. \tag{2.9}$$

To prove (2.9), let

$$T_q = \frac{\sum_{j=1}^{q} I(\mathbf{X}_N, \omega_j)/h(\omega_j)}{\sum_{j=1}^{m} I(\mathbf{X}_N, \omega_j)/h(\omega_j)}.$$

Formula (6.2.171) in Priestley [1981] states that

$$\max_{1 \le q \le m} |T_q - q/m| = O_p(1/\sqrt{N}).$$

Because h is bounded below from zero and above from infinity, the LHS of (2.9) is of the order of $\max_{1 \le q \le m} |T_q - T_{q-1}| = O_p(1/\sqrt{N})$, where $T_0 = 0$. It follows from Corollary 2 and Theorem 6 on respectively p. 207 and 210 in Hannan [1970] that $\bar{X} \to \mu$ and $r_0 \to \gamma_0$ almost surely. Hence, the CLT implies that Y_p is asymptotically $N(\mu, \gamma_0)$.

We now sketch how to verify the asymptotic normality of $(Y_1, Y_N)^T$. Let α and β be two constants. Define $Z_N = \alpha(Y_1 - \bar{X}) + \beta(Y_N - \bar{X})$. Then Z_N is a finite sum of independent random variables, and its j-th summand is $\sqrt{\frac{8\pi}{N}}\sqrt{I(\mathbf{X}, \omega_j)}(\alpha\cos(\omega_1 j + \theta_j) + \beta\cos(\omega_N j + \theta_j))$. The variance of Z_N is equal to $(\alpha^2 + \beta^2) r_0 + 2\alpha\beta(r_1 + r_{N-1})$ The Lyapunov condition for the CLT for (Z_N) is just (2.8) and hence it is satisfied. Now, Theorem 6 on p.210 in Hannan [1970] implies that $r_1 \to \gamma_0$ and since $\gamma_{N-1} = O_p(1/N)$, the variance of Z_N approaches $(\alpha^2 + \beta^2)\gamma_0 + 2\alpha\beta\gamma_1$. Upon invoking the Cramer-Wold device, we obtain the asymptotic normality result for (Y_1, Y_N) stated in the theorem. □

Remark 2.7 It is clear from the proofs of Theorem 2.6 and Theorem 2.5 that, in terms of finite dimensional distributions, the surrogate data behave like a Gaussian process whose auto-covariance function matches asymptotically the observed circular auto-covariances of the data.

3 Composite Null Hypothesis, Sufficiency and Similar Tests

In this section, we describe a general approach to testing with a composite null hypothesis, which sheds useful insight on when the method of surrogate data may work. Much of the material in this section is classical but is included here for the sake of completeness; (Cox and Hinkley [1974]). Let $f(x|\theta)$ be the pdf of the data $X = x$, where θ denotes the unknown true parameter, possibly of infinite dimension. (Here, X and θ are unrelated to those in the previous sections.) Suppose the null hypothesis H_0 specifies that $\theta \in \Omega$. A test is implemented by defining a critical region C, a subset of the sample space, and H_0 is rejected if and only if the observed data lies in C. The *size* of the test is defined as $\alpha(\theta) = P(X \in C|\theta), \theta \in \Omega$, which is generally dependent on the true θ. A *similar* test requires that the size of the test be constant for all $\theta \in \Omega$, and the constant size is then called the size of the similar test.

Under some regularity conditions, any similar test has the so-called *Neyman structure.* To explain the concept of the Neyman structure, we first briefly review the concept of sufficiency. A statistic (possibly vector-valued) $S = S(X)$ is called a *sufficient statistic* (under H_0) if the conditional pdf of X given S is independent of $\theta \in \Omega$, i.e., $p(x|S;\theta) \equiv p(x|S)$ is free of θ. Intuitively, a sufficient statistic summarizes all the information contained in the data on the unknown parameter $\theta \in \Omega$. This can be seen from the following well known argument. Consider drawing the data X in the following two steps:

(1) first draw $S = s$ according to the pdf $p(s|\theta)$;
(2) then draw $X = x$ from the conditional pdf $p(x|s)$.

The second step does not involve θ, and hence provides no information on θ. Therefore, the sufficient statistic S contains all the relevant information on $\theta \in \Omega$ that is available in the data. In other words, summarizing the data by a sufficient statistic involves no loss of information, as long as the null hypothesis is valid. The whole data set is, of course, sufficient, but we generally look for the most economical summary of information, which is embodied in the notion of minimal sufficiency. A sufficient statistic S is called a *minimal sufficient statistic* if it is a (measurable) function of any other sufficient statistic, i.e., if S' is a sufficient statistic, $S = g(S')$ for some (measurable) function g. Hence, a minimal sufficient statistic is unique in the sense that any two minimal sufficient statistics are one-to-one functions of each other.

Sufficient statistics can be recognized via the factorization criterion:

$$p(x|\theta) = w(x)g(S(x), \theta), \theta \in \Omega. \tag{3.1}$$

$S = S(X)$ is sufficient (for all $\theta \in \Omega$) if and only if Equation (3.1) holds for some w and g. The factorization criterion is often formulated in terms of the log likelihood function which is $\ell_x(\theta) = \log p(x|\theta)$. Then (3.1) becomes

$$\ell_x(\theta) = w_1(x) + g_1(S(x), \theta), \forall\theta \in \Omega, \tag{3.2}$$

for some functions w_1 and g_1.

It is easily seen from (3.1) that $p(s|\theta)$, the pdf of S, is proportional to $g(s, \theta)$. Let $\mathcal{X}_s = \{x : S(x) = s\}$, the set of all samples with the sufficient statistic S taking the value s. Given $S = s$, X lies in $\mathcal{X}_s$. It can be readily seen that given $S = s$, the conditional pdf of X is proportional to $w(x)$ which is independent of θ. If $w(x) \equiv$ a constant, then $\forall s$, conditional on $S = s$, X is uniformly distributed over $\mathcal{X}_s$.

Example 3.1 Let $X = (X_1, X_2, X_3, \ldots, X_N)^T$ be a vector of N iid real valued random variables with unknown common pdf f. Here, the parameter space Ω consists of all pdf's on the real line. Then the joint pdf of $X_1 = x_1, X_2 = x_2, \ldots, X_N = x_N$ is $f(x_1)f(x_2)\cdots f(x_N) = \prod_1^N f(x_{(j)})$, where $x_{(1)} \leq x_{(2)} \leq \cdots \leq x_{(N)}$ is a rearrangement of the data, often called the order statistics. From the factorization criterion, we see that the order statistics $X_{(1)} \leq X_{(2)} \leq \cdots \leq X_{(N)}$ are jointly sufficient. In fact, the order statistics can be shown to be minimal sufficient. Given the order statistics, X is a random permutation of the order statistics.

Minimal sufficient statistics can be readily determined for the class of pdfs called the *exponential family,* which includes the Normal (or Gaussian) distribution, the Poisson distributions, etc. See Lehmann [1983] for proofs of some properties of the exponential family stated below. A pdf, $p(x|\theta)$, is said to belong to the exponential family if

$$p(x|\theta) = \exp(\sum_{j=1}^{k} \eta_j(\theta)S_j(x) - B(\theta))h(x), \quad \theta \in \Omega. \tag{3.3}$$

We shall assume that (1) the random variables $\{S_1(X), \ldots, S_k(X)\}$ are not linearly dependent, and (2) $\{(\eta_1(\theta), \ldots, \eta_k(\theta))^T, \theta \in \Omega\} \subseteq \mathbf{R}^k$ contains a non-empty k-dimensional open rectangle. Then it can be shown that $S = (S_1(X), \ldots, S_k(X))^T$ is a minimal sufficient statistic. Moreover, S is *complete*: there exists no nontrivial unbiased estimator of zero in terms of S, i.e., $E(h(S)|\theta) \equiv 0, \forall\theta \in \Omega$ implies that h is identically equal to zero. The existence of a complete sufficient statistic implies that certain optimal statistical procedures are unique. Its relation with similar tests will be discussed shortly.

Example 3.2 Suppose that $X_1, X_2, \ldots, X_N$ are iid $N(\mu, \sigma^2)$ with unknown mean $-\infty < \mu < \infty$ and unknown variance $\sigma^2 > 0$. Let $X = (X_1, X_2, \ldots, X_N)^T$. Its pdf is equal to

$$\begin{aligned} p(X|\mu, \sigma^2) &= \exp(-\sum_{j=1}^{N} \frac{(X_j - \mu)^2}{2\sigma^2} - \frac{N}{2}\log(2\pi\sigma^2)) \\ &= \exp(-\sum \frac{X_j^2}{2\sigma^2} + \sum X_j \frac{\mu}{\sigma^2} - \frac{\mu^2}{2\sigma^2 N} - \frac{N}{2}\log(2\pi\sigma^2)). \end{aligned}$$

Here $k = 2, S_1 = \sum X_j^2, S_2 = \sum X_j, \eta_1 = \frac{-1}{2\sigma^2}, \eta_2 = \frac{\mu}{\sigma^2}$ and $B(\theta) = \frac{\mu^2}{2\sigma^2 N} + \frac{N}{2}\log(2\pi\sigma^2))$. Then $0 > \eta_1$ and $-\infty < \eta_2 < \infty$, and hence the allowable range of η_1, η_2 forms an open rectangle. Therefore, $S = (\sum X_j, \sum X_j^2)^T$ is a complete minimal sufficient statistic. We may also adopt the sample mean $\bar{X}$ and the sample variance $\hat{\sigma}^2 = \sum(X_t - \bar{X})^2/(N-1)$ as the jointly complete minimal sufficient statistics as they are clearly one-to-one function of S. Given $\bar{X}$ and $\hat{\sigma}^2$, the data X is distributed uniformly on the sphere centered at $\bar{x}$, and with radius equal to $(N-1)\hat{\sigma}^2$.

We now consider how to construct similar tests when a sufficient statistic is available. Recall that given the sufficient statistic $S = s$, the conditional pdf of X is proportional to $w(x)$ over $\mathcal{X}_s = \{x : S(x) = s\}$. We shall assume that for each s, one could define a critical sub-region $C_{s,\alpha} \subset \mathcal{X}_s$ so that $(\int_{C_{s,\alpha}} w(x)dx)/(\int_{\mathcal{X}_s} w(x)dx) = \alpha$, where α is the size of the test. Moreover, we shall assume that for all s, $C_{s,\alpha} \subset C_{s,\alpha'}$ whenever $\alpha < \alpha'$. (This requirement ensures that if the test is significant at a certain significance level say α, it is significant at all significance levels higher than

α.) Now, define the overall critical region $C_\alpha = \cup_s C_{s,\alpha}$. (Given the observed data x, we compute s and reject H_0 if and only if x falls inside $C_{s,\alpha}$.) The critical region C_α is of size α because, for all $\theta \in \Omega$,

$$\begin{aligned} \text{Prob}(X \in C_\alpha|\theta) &= E(\text{Prob}(X \in C_{s,\alpha}|S = s;\theta)|\theta) \\ &= E(\alpha|\theta) = \alpha. \end{aligned} \tag{3.4}$$

A test constructed as above is called a test with *Neyman structure.* It is interesting to note that when a complete sufficient statistic S exists, any similar test is a test with Neyman structure.

Provided that $\mathcal{X}_s$ is a sufficiently large set, there will be ample freedom in choosing the critical region $C_{s,\alpha}$. For a parametric case, there are various likelihood-based approaches leading to the optimal choices of $C_{s,\alpha}$. However, in a non-parametric setting, these likelihood based methods are not available. Suppose a test statistic T is deemed appropriate in testing the validity of H_0, which need not induce a similar test. Also, suppose that small values of T are more likely to occur under some alternative hypothesis than under the null hypothesis. Then, one can set $C_{s,\alpha}$ by including points in $\mathcal{X}_s$ with small T-values until the size of the set equals α, i.e., $C_{s,\alpha} = \{x \in \mathcal{X}_s : T(x) \le t_0\}$ with its probability equal to α under the conditional distribution of X on $\mathcal{X}_s$. Equivalently, one computes the (conditional) p-value which is defined as $\text{Prob}(T \le T_{\text{obs}}|S = s_{\text{obs}})$, and rejects H_0 if and only if the p-value is less than α. Under the null hypothesis, the p-value so computed is U$[0,1]$ distributed which follows essentially from (3.4). The exact computation of the p-value is generally difficult. In some cases, asymptotic approximation of the p-value is available. It could also be approximated via Monte Carlo by (1) drawing a large number of independent $X_1, X_2, \dots, X_m$ with pdf proportional to $w(x)$ over $\mathcal{X}_s$, and (2) approximate the p-value by $\hat{p} = \frac{1}{m}\#\{T(X_i) < t_{\text{obs}}\}$. The Monte Carlo procedure is only limited by the ease of drawing samples from the appropriate conditional distribution over $\mathcal{X}_s$.

We shall see in the next section that under the null hypothesis that the underlying process is a stationary Gaussian circular process of matching index N, the sample mean and the periodogram are jointly sufficient, and $w(x)$ in (3.1) is constant so that the conditional distribution of the data given the sample mean and the periodogram is always uniformly distributed. The method of surrogate data then simply provides an efficient scheme of drawing a random sample from the uniform distribution over $\mathcal{X}_s$. And the method of surrogate data is exactly valid for any test statistic.

Example 3.1 (continued) A suitable statistic for testing the iid assumption of $X_1, X_2, \dots, X_N$ is the first lag sample auto-correlation:

$$T = \frac{\sum_{j=1}^{N-1}(X_j - \bar{X})(X_{j+1} - \bar{X})}{\sum_{j=1}^{N}(X_j - \bar{X})^2}.$$

Recall that the order statistics are minimal sufficient, and given their values, the data is a random permutation (shuffling) of the N order statistic values. Suppose, for simplicity, the order values are $x_1 \le x_2 \le \cdots \le x_N$. Then, given the order statistics,

$$
\begin{aligned}
T &= \frac{\sum_{j=1}^{N-1}(x_{\pi(j)} - \bar{x})(x_{\pi(j+1)} - \bar{x})}{(N-1)\hat{\sigma}^2} \\
&= \frac{\sum_{j=1}^{N-1}(x_j - \bar{x})(x_{\pi(\pi^{-1}(j)+1)} - \bar{x})}{(N-1)\hat{\sigma}^2},
\end{aligned}
$$

where $\pi : \{1, 2, \ldots, N\} \to \{1, 2, \ldots, N\}$ is a random permutation, $\bar{x} = \sum x_j/N$ and $\hat{\sigma}^2$ is the sample variance. But $\pi(\pi^{-1}(\cdot) + 1)$ is just a random permutation of $\{1, 2, \ldots, N\}$. Hence, conditional on the order statistics, the distribution of T is identical to the random variable defined by $\sum_{j=1}^{N_1}(x_j - \bar{x})(x_{\pi(j)} - \bar{x})/[(N-1)\hat{\sigma}^2]$, which can be studied via simulation. However, for large N, the permutation Central Limit Theorem (see Sen [1985]) entails that under suitable conditions on the x's, the conditional distribution of T is asymptotically Gaussian with zero mean and variance equal to $1/N$.

Example 3.2 (continued) To test if the data are independent and identically normally distributed with the same mean and variance against the alternative that the common distribution is skewed, we may employ the skewness test statistic

$$
T = \frac{\sum(X_j - \bar{X})^3}{\hat{\sigma}^3}.
$$

Notice that T is invariant w.r.t. location-scale changes. Therefore,

$$
\begin{aligned}
&\text{Prob}(T \le t|\mu, \sigma^2) \\
=\;& \text{Prob}(T(\sigma X + \mu) \le t|\mu = 0, \sigma = 1) \\
=\;& \text{Prob}(T(X) \le t|\mu = 0, \sigma = 1),
\end{aligned}
$$

and hence T is ancillary: its distribution is independent of the true parameter under the null hypothesis. Because $(\bar{X}, \hat{\sigma}^2)$ is a complete sufficient statistic and T is ancillary, Basu's theorem (see Cox and Hinkley [1974]) implies that they are conditionally independent for all $\theta \in \Omega$. Consequently, the conditional distribution of T given $\bar{X}$ and S^2 is equal to its unconditional distribution which is asymptotically normally distributed.

To conclude this section, we mention that despite the elegant framework of similar tests, this approach may suffer from several problems. The principal problem is that an exactly sufficient statistic with a not too small $\mathcal{X}_s$ may not be available. We might see that, in some cases, the existence of an asymptotically sufficient statistic may partially circumvent the preceding problem. Another potential problem with similar tests is that the conditional distribution of T given the sufficient statistic may remain the same under some alternative hypotheses for which the similar test is then not useful in detecting.

4 Stationary Gaussian Circular Processes

In this section, we aim to show that under some strong conditions on the null hypothesis, the method of surrogate data is exactly valid for any test statistic. The justification of the method of surrogate data will be based on the sufficiency principle discussed in Section 3. The null hypothesis specifies that the process (X_t) is a stationary Gaussian process with unknown mean μ and spectral density function $h(\omega), -\pi \le \omega \le \pi$. Below $\theta = (\mu, h)$ stands for the unknown parameter.

Let the data be $\mathbf{X}_N = \mathbf{X} = (X_1, X_2, \ldots, X_N)^T$ and $B_{h,N} = B_h = B$ denote the covariance matrix of $\mathbf{X}$. The logarithm of the pdf of $\mathbf{X}$ is

$$\begin{aligned}\ell_N &= \log p(X_1, X_2, \ldots X_N|\theta)\\ &= -\frac{1}{2}(N\log(2\pi) + \log(|B_h|) + (\mathbf{X} - \mu\mathbf{1})^T B_h^{-1}(\mathbf{X} - \mu\mathbf{1})) \qquad (4.1)\\ &\approx -\frac{N}{2}(\log(2\pi) + \frac{1}{2\pi}\int_{-\pi}^{\pi}\log(2\pi h(\omega))d\omega + \frac{1}{2\pi}\{\int_{-\pi}^{\pi}\frac{I(\mathbf{X},\omega)}{h(\omega)}d\omega -\\ &\quad - \frac{2\mu}{h(0)}\bar{X} + \frac{\mu^2}{h(0)}\}), \qquad (4.2)\end{aligned}$$

where $|B_h|$ is the determinant of B_h, and $\mathbf{1}$ is an $N \times 1$ vector consisting of all 1's. Whittle [1952] first suggested the above approximation with $\mu = 0$. For $\mu = 0$, it is shown in Dzhaparidze [1986] that assuming (1) the spectral density function of (X_t) is bounded away from zero and infinity, and (2) $\sum j\gamma_j^2 < \infty$, the error of the approximation is $o_p(N^{-1/2})$. The above approximation suggests that the sample mean and the periodogram are asymptotically jointly sufficient. The key to the above approximation is that $B_{h,N}$ is approximately diagonalizable as $B_{h,N} \approx P_N^* \Lambda_N P_N$ where P_N is as defined in Section 2 and Λ_N is a diagonal matrix with its diagonal elements being $2\pi(h(\omega_1), h(\omega_2), \ldots, h(\omega_N))^T$.

We now consider a special case under which the sample mean and the periodogram are *exactly* jointly sufficient. We assume that the data $\mathbf{X}_N$ are generated by a stationary Gaussian circular process of index N, as defined in the introduction. As a result of the definition, the auto-covariances of a stationary circular process satisfy the relationship that $\gamma_j = \gamma_{N-j}, j = 0, 1, 2, \ldots, N$. The covariance matrix B then becomes a *circular* matrix, i.e., each of its rows is obtained by shifting the preceding row one element to the right, with the last element of a preceding row becoming the first element of the next row:

$$B = \begin{bmatrix} \gamma_0 & \gamma_1 & \gamma_2 & \cdot & \cdot & \cdot & \gamma_{N-1}\\ \gamma_{N-1} & \gamma_0 & \gamma_1 & \cdot & \cdot & \cdot & \gamma_{N-2}\\ \vdots & \vdots & \vdots & \vdots & \vdots & \vdots & \vdots\\ \gamma_1 & \gamma_2 & \gamma_3 & \cdot & \cdot & \cdot & \gamma_0 \end{bmatrix}$$

B is exactly diagonalizable with eigenvalues $\widetilde{h}_N(\omega_j) = \sum_{k=0}^{N-1}\gamma_k \exp(ik\omega_j)$ and the jth column of P_N^* as the corresponding eigenvector. In matrix form, we have $B = P^*\Lambda P$, where $\Lambda = \Lambda_N$ is a diagonal matrix whose diagonal elements are $\widetilde{h}_N(\omega_j)$. Note that $\widetilde{h}_N(\omega_N) = \widetilde{h}_N(0)$. Then, the log likelihood of $\mathbf{X}$ becomes (see Appendix 1 for a derivation)

$$\ell_N = -\frac{N}{2}(\log(2\pi) + \sum_{j=1}^{N}\log\widetilde{h}_N(\omega_j)/N + \{\sum_{j=1}^{N}\frac{2\pi I_N(\mathbf{X};\omega_j)}{N\widetilde{h}_N(\omega_j)} - 2\bar{X}\frac{\mu}{\widetilde{h}_N(0)} + \frac{\mu^2}{\widetilde{h}_N(0)}\}). \qquad (4.3)$$

Thus, the log likelihood function depends on the data through the sample mean and the periodogram values $I_N(\mathbf{X}, \omega_j), j = 1, 2, \ldots, [N/2]$. Recall that we write $S = S_N$ for the sample mean and the above periodogram values. The factorization criterion (3.2) implies that S is sufficient. Moreover, because w_1 in (3.2) is constant, the conditional distribution of $\mathbf{X}$ given S is uniformly distributed over $\mathcal{X}_S$. Hence, the method of surrogate data is valid if the null hypothesis assumes that the process is a stationary Gaussian circular process (of index matching the sample size).

5 Stationary Gaussian AR(1) processes

We have seen from the previous section that the method of surrogate data is valid if we impose the circular stationarity assumption on the underlying process. However, this is a rather strong assumption which is not likely to be satisfied in practice. In Section 4, it is shown that the sample mean and the sample periodogram (at $\omega_j, j = 1, 2, \ldots, N$) are asymptotically jointly sufficient for a stationary Gaussian process with fast decaying auto-correlations. This raises the hope that perhaps the method of surrogate data is asymptotically valid for the null hypothesis that the process is a stationary Gaussian process with fast decaying autocorrelations.

In this section, we study the validity of the method of surrogate data assuming that the true model is a stationary Gaussian AR(1) process:

$$(X_t - \mu) - \phi(X_{t-1} - \mu) = a_t,$$

where $|\phi| < 1$, μ is the mean, and (a_t) is iid $N(0, \sigma_a^2)$. The stationary distribution of X_t is $N(\mu, \sigma_a^2/(1-\phi^2))$.

Example 5.1 We now construct a test statistic for which the method of surrogate data fails. Let $T_N = \frac{X_1 - X_N}{\sqrt{2}\hat{\sigma}_N}$ where $\hat{\sigma}_N^2$ is the sample variance. Under the stationary AR(1) assumption, T_N is asymptotically $N(0,1)$ because X_1 and X_N are asymptotically independent and identically distributed as $N(0, \sigma^2)$. T_N may be useful for detecting whether or not X_1 and X_N are highly correlated. If X_1 and X_N are positively (negatively) correlated, then T_N tends to be more (less) concentrated at zero than if they are independent. T_N is probably not useful for detecting nonlinearity. However, the point is to show that the surrogate data method fails for this statistic, and hence is not valid for *every* test statistic under the AR(1) assumption. The method of surrogate data calibrates T_N according to the surrogate data distribution of $\frac{Y_1 - Y_N}{\sqrt{2}\hat{\sigma}_N}$ where $(Y_1, Y_2, \ldots, Y_N)^T$ is the surrogate data preserving the observed sample mean and periodogram. It follows from Theorem 2.6 that $(Y_1, Y_N)^T$ is asymptotically bivariate Normal with zero mean vector and covariance matrix equal to

$$\Sigma = \sigma^2 \begin{pmatrix} 1 & \phi \\ \phi & 1 \end{pmatrix}.$$

Hence, $\frac{Y_1 - Y_N}{\sqrt{2}\hat{\sigma}_N}$ is asymptotically $N(0, (1-\phi))$. Notice that this distribution is independent of S_N, the sample mean and the periodogram. Let z_α be the $\alpha \times 100$ percentile of the standard Normal distribution. According to the surrogate data distribution, the $\alpha \times 100$ percentile of $T_N(\mathbf{Y}_N) = \frac{Y_1 - Y_N}{\sqrt{2}\hat{\sigma}_N}$ is approximately equal to $\sqrt{1-\phi}z_\alpha$. Suppose we reject the linear hypothesis if the observed T_N value is too small based on the surrogate data. Specifically, we reject the linear hypothesis if $T_N(\mathbf{X}_N) \le \sqrt{1-\phi}z_\alpha$. Then the nominal rate of false rejection of the linear hypothesis is $\alpha \times 100\%$. But, the actual rejection rate is $\text{Prob}(T_N(\mathbf{X}_N) \le \sqrt{1-\phi}z_\alpha) = \Phi(\sqrt{1-\phi}z_\alpha)$, where Φ is the cdf of the standard Normal distribution; hence the nominal false rejection rates deviates from the true false rejection rate unless $\phi = 0$.

We have carried out a simulation experiment to see how the theory works in practice. We simulated realizations from five AR(1) models, all with sample size $N = 128$. For each simulated $\mathbf{X}_N$, 500 $\mathbf{Y}_N$'s were generated to compute the empirical fraction of $T_N(\mathbf{Y}_N)$ that are less than $T_N(\mathbf{X}_N)$, and the null hypothesis of linearity is rejected if the fraction is less than 5%. For each AR(1) generating mechanism, the experiment is replicated 500 times. Table 2 reports the empirical false rejection rates and their asymptotic counterparts for AR(1) processes with $\phi = -.5, .0, .5$ and $.9$. It shows good agreement between the theory and the empirical results. In particular, the method of surrogate data fails badly when the AR(1) coefficient is close to 1 in magnitude.

Table 1 Empirical and asymptotic false rejection rates for $T_N = \frac{X_1 - X_N}{\sqrt{2}\hat{\sigma}_N}$; nominal false rejection rate is 5%.

ϕ	-.5	.0	.5	.9
empirical	3.4	5.0	11.6	29.2
asymptotic	2.2	5.0	12.2	30.1

In practice, the above test statistic is not likely to be employed because it essentially only makes use of the information in X_1 and X_N. It seems desirable that the test statistic be asymptotically independent of any finite set of X's. For our purposes, we assume the following less restrictive assumption:

Assumption (A1): For any positive integer K, the test statistic T_N is asymptotically independent of $(X_1, X_2, \ldots, X_K, X_{N-K+1}, X_{N-K}, \cdots, X_N)^T$.

We shall prove that under assumption (A1) and another regularity condition, the asymptotic distribution of T_N is unaltered as the underlying stationary Gaussian AR(1) process is changed into a sequence of approximating stationary Gaussian circular processes. It then follows from the discussion in Section 4 that the method of surrogate data is asymptotically valid for the test statistic T_N.

Without loss of generality, we assume that for the underlying AR(1) process, $\mu = 0$ and $\sigma_a = 1$ so that $X_1 = \phi X_{t-1} + a_t, t = 1, 2, \ldots,$ where (a_t) is iid $N(0,1)$. The stationary distribution of X_t is then $N(0, 1/(1-\phi^2))$. The joint pdf of $\mathbf{X}_N$ is

$$f_{\text{AR}}(\mathbf{X}_N) \quad = \quad \frac{\sqrt{1-\phi^2}}{(2\pi)^{N/2}} \exp(-\{(1-\phi^2)X_1^2 + \sum_{t=2}^{N}(X_t - \phi X_{t-1})^2\}/2). \quad (5.1)$$

In view of the results in Section 4, we want to approximate the Gaussian pdf of $\mathbf{X}_N$ by another Gaussian pdf with a circular covariance matrix. (The approximating pdf will change with the sample size.) Note that the latter pdf may be taken as proportional to $\exp(\psi(\mathbf{X}_N))$ where ψ is a circular function, that is, $\psi(x_1, x_2, \ldots, x_N) = \psi(x_2, x_3, \ldots, x_N, x_1)$ for all $x_1, \ldots, x_N$. A simple circular approximation of $-\{(1-\phi^2)X_1^2 + \sum_{t=2}^{N}(X_t - \phi X_{t-1})^2\}/2$ is $-\sum_{t=1}^{N}(X_t - \phi X_{t-1 \text{ (mod } N)})^2/2$. Hence, we approximate the AR(1) pdf of $\mathbf{X}_N$ by the following circular Gaussian pdf defined by the formula

$$f_{\text{cyc}}(\mathbf{X}_N) \quad = \quad \frac{\sqrt{|\Sigma_N|}}{(2\pi)^{N/2}} \exp(-\sum_{t=1}^{N}(X_t - \phi X_{t-1 \text{ (mod } N)})^2/2), \qquad (5.2)$$

where

$$\Sigma_N \quad = \quad \begin{pmatrix} 1+\phi^2 & -\phi & 0 & 0 & \cdots & 0 & -\phi \\ -\phi & 1+\phi^2 & -\phi & 0 & \cdots & 0 & 0 \\ \vdots & \vdots & \vdots & \vdots & \vdots & \vdots & \vdots \\ 0 & 0 & 0 & 0 & \cdots & 1+\phi^2 & -\phi \\ -\phi & 0 & 0 & 0 & \cdots & -\phi & 1+\phi^2 \end{pmatrix} \qquad (5.3)$$

is a circular matrix whose first row equals $(1+\phi^2, -\phi, 0, 0, \ldots, -\phi)^T$. Note that $\mathbf{X}_N^T \Sigma_N \mathbf{X}_N = \exp(-\sum_{t=1}^N (X_t - \phi X_{t-1 \pmod N})^2/2)$. To differentiate the two data generating mechanisms, we say that H_{0N} holds if $\mathbf{X}_N$ is generated by the AR(1) process, and that H_{1N} holds if it is generated by the approximating Gaussian circular pdf as defined in (5.2). Roughly speaking, we aim to show that under suitable conditions, H_{0N} and H_{1N} are asymptotically close to each other relative to T_N. If this is true and since the surrogate data method is valid (for all test statistics) under H_{1N}, its asymptotic validity for calibrating T_N can be transferred to the case that H_{0N} is the data generating mechanism. This program has not been completely carried out, although some useful results will be given below suggesting that this approach might be promising. In the following result, we assume that after suitable normalization if necessary, T_N converges weakly to some non-degenerate random variable, say T, which need not be normally distributed. (Weak convergence of T_N to T heuristically means that T_N is asymptotically distributed as T; more formally, it means that for all t with $\text{Prob}(T = t) = 0$, $\text{Prob}(T_N \le t) \to \text{Prob}(T \le t)$ as $N \to \infty$.) This assumption is likely to be satisfied in practice. For example the Grassberger-Procaccia estimator of the correlation dimension is asymptotically normally distributed; see Cutler [1991]. We note that the distribution of T_N depends on the underlying data mechanism. It can be shown that for the test statistic considered in Example 5.1, its asymptotic distribution under H_{0N} is $N(0,1)$ but it becomes $N(0, 1-\phi)$ under H_{1N}. This explains why the method of surrogate data fails for that particular test statistic. Henceforth in this section, all pdf's, expectations or probabilities that are computed under H_{1N} are subscripted with the symbol cyc; otherwise all pdf's, expectations or probabilities are computed under H_{0N}, although sometimes a subscript AR will be added to emphasize that H_{0N} is the generating mechanism.

Theorem 5.2 *Assume (1) (A1) holds and (2) as $N \to \infty$, T_N converges weakly to some non-degenerate random variable denoted by T under H_{0N}. Then, as $N \to \infty$, T_N also converges weakly to T under H_{1N}.*

Thus, if the conditions of the above theorem are satisfied, for large sample size, the distribution of T_N remains approximately the same when the underlying stationary Gaussian AR(1) process is changed into the approximating stationary Gaussian circular process. Suppose that the limiting distribution T is continuous. Then $\forall t$,

$$\text{Prob}_{\text{AR}}(T_N \le t) \approx \text{Prob}(T \le t) \approx \text{Prob}_{\text{cyc}}(T_N \le t) = E_{\text{cyc}}(E_{\text{cyc}}(I(T_N \le t)|S_N)). \tag{5.4}$$

Thus, T_N may be regarded as generated by the approximating stationary Gaussian circular process. The discussion in Section 4 then suggests that the method of surrogate data is asymptotically valid for T_N. However, it doesn't follow immediately from this argument that the true false rejection rate of T_N is approximately equal to the nominal false rejection rate. In fact, for the latter to be valid, we need to show that $E_{\text{AR}}(I(T_N \le t)|S_N) \approx E_{\text{cyc}}(I(T_N \le t)|S_N)$, which unfortunately does not follow immediately from (5.4). Further research in this area is desirable. We notice that it suffices that T_N can be theoretically normalized so that condition (2) in the preceding theorem holds. There is no need to actually carry out the normalization in the implementation of the method of surrogate data because quantities like p-values or the t-ratio of the observed value from the mean, based on the surrogate data, are invariant to linear scaling.

Proof We shall prove that $\{H_{1N}\}$ is contiguous to $\{H_{0N}\}$. For an introduction to the concept of contiguity and some of its uses, see Hajek and Sidak [1965] and Le Cam [1986]. $\{H_{1N}\}$ is said to be contiguous to $\{H_{0N}\}$ if each sequence of random variables $\{g_N\}$ converging to 0 in probability under H_{0N} also does so under $\{H_{1N}\}$. The likelihood ratio of H_{1N} to H_{0N} equals

$$\begin{aligned} \ell_N(\mathbf{X}_N) &= \frac{f_{\text{cyc}}(\mathbf{X}_N)}{f_{\text{AR}}(\mathbf{X}_N)} \\ &= \frac{\sqrt{|\Sigma_N|}}{1-\phi^2}\exp(\{-\phi^2 X_1 - \phi^2 X_N^2 + 2\phi X_1 X_N\}/2) \qquad (5.5) \end{aligned}$$

which depends only on X_1 and X_N. It follows from Le Cam's first lemma (pp. 203–4 in Hajek and Sidak [1965]) that contiguity of $\{H_{1N}\}$ to $\{H_{0N}\}$ holds if under H_{0N}, ℓ_N converges weakly to some random variable, say ℓ, whose mean equals 1. The following two lemmas are instrumental for proving the theorem.

Lemma 5.3 *Assume (1) ℓ_N and T_N are asymptotically independent and (2) under H_{0N}, $(\ell_N, T_N)^T$ converges weakly to $(\ell, T)^T$ with $E(\ell) = 1$. Then under H_{1N}, T_N converges weakly to T.*

Proof The asymptotic independence of T_N and ℓ_N implies that T is independent of ℓ. Let γ be a constant to be determined later. Let t be a continuity point of T, i.e., $\text{Prob}(T = t) = 0$. Also for any function g, define $(g \wedge \gamma)(x) = \min(g(x), \gamma)$. Now,

$$\begin{aligned} \text{Prob}_{\text{cyc}}(T_N \le t) &= E(\ell_N I(T_N \le t)) \\ &= E(\ell_N \wedge \gamma I(T_N \le t)) + R_N \end{aligned}$$

where I denotes the indicator function and $R_N = E(\ell_N I(T_N \le t; \ell_N > \gamma))$ and hence

$$\begin{aligned} |R_N| &\le E(\ell_N I(\ell_N > \gamma)) \\ &= E(\ell_N) - E(\ell_N \wedge \gamma) \\ &= 1 - E(\ell_N \wedge \gamma) \\ &\to 1 - E(\ell \wedge \gamma). \end{aligned}$$

Also,

$$\begin{aligned} E(\ell_N I(T_N \le t; \ell_N \le \gamma)) &\to E(\ell I(T \le t; \ell \le \gamma)) \\ &= E(I(T \le t))E(\ell \wedge \gamma). \end{aligned}$$

Because $E(\ell) = 1$, by letting $\gamma \to \infty$, we have $\text{Prob}_{\text{cyc}}(T_N \le t) \to \text{Prob}(T \le t)$; hence the conclusion of the lemma. □

Lemma 5.4 $|\Sigma_N| = 1 + o(1)$, as $N \to \infty$.

Proof Because Σ_N is a circular matrix, its eigenvalues are equal to $(1+\phi^2) - \phi\exp(ik\omega_1) - \phi\exp(ik\omega_{N-1}), 1 \le k \le N$. (See the discussion in Section 4.) Hence, $\log|\Sigma_N| = \frac{1}{2}\sum_{k=1}^N \log|1 - \phi\exp(i\omega_k)|^2$. Let $g(\omega) = \frac{1}{2\pi}\log|1-\phi\exp(i\omega)|^2$. Then,

$$\log|\Sigma_N| = \log((2\pi)^{N/2}) + \frac{N}{4\pi}\sum_{k=1}^{N}\frac{2\pi\log g(\omega_k)}{N}. \qquad (5.6)$$

But Taylor expansion (p. 327 in Hardy [1952]) implies that

$$\int_{\omega_k}^{\omega_{k+1}} \log g(\omega)d\omega = \log g(\omega_k)(\omega_{k+1}-\omega_k) + D\log g(\omega_k)(\omega_{k+1}-\omega_k)^2/2 + \\ + \int_{\omega_k}^{\omega_{k+1}} (D^2 \log g(\omega))(\omega_{k+1}-\omega)^2/6d\omega$$

where D denotes differentiation w.r.t. ω. Note that (1) $g(\omega)$ is an even function about $\omega = \pi$ and (2) $D\log g(0) = D\log g(2\pi) = 0$. Hence, $\sum_{k=1}^N D\log g(\omega_k) = 0$. Also, it can be verified that $D^2 \log g(\omega)$ is a bounded function. Therefore, on noting that $g(\omega_N) = g(\omega_0)$, we have $\int_0^{2\pi} \log g(\omega)d\omega = \frac{2\pi}{N}\sum_{k=1}^N \log g(\omega_k) + O(\frac{1}{N^2})$. Consequently, $\log|\Sigma_N| = \frac{N}{2}\log 2\pi + \frac{N}{4\pi}\int_0^{2\pi} \log g(\omega)d\omega + O(\frac{1}{N})$, and hence $|\Sigma_N| = (2\pi)^{N/2}\exp(\frac{N}{4\pi}\int_0^{2\pi}\log g(\omega)d\omega + O(\frac{1}{N}))$. But $g(\omega)$ is the spectral density function of the MA(1) process: $Z_t = a_t - \phi a_{t-1}$ where $\{a_t\}$ is iid $N(0,1)$. By formula (10.1.41) in Priestley [1981], $2\pi\exp(\frac{1}{2\pi}\int_0^{2\pi}\log g(\omega)d\omega) = 1$, the one-step ahead prediction variance based on all the infinite past data. Therefore, $|\Sigma_N| = 1 + o(1)$. □

It follows from Lemma 5.4 that

$$\ell_N = \frac{1}{\sqrt{1-\phi^2}}\exp(\{-\phi^2 X_1^2 - \phi^2 X_N^2 + 2\phi X_1 X_N\}/2)(1+o(1)). \quad (5.7)$$

Under the stationary AR(1) process, X_1 and X_N are asymptotically iid $N(0, 1/(1-\phi^2))$. Therefore, ℓ_N converges weakly to $\ell = \frac{1}{\sqrt{1-\phi^2}}\exp(\{-\phi^2 Z_1^2 - \phi^2 Z_2^2 + 2\phi Z_1 Z_2\}/2)$ where Z_1 and Z_2 are iid $N(0, 1/(1-\phi^2))$. Assuming that $E(\ell) = 1$, then all the conditions of Lemma 5.3 are satisfied, and hence the conclusion of the theorem. It remains to verify that $E(\ell) = 1$. This essentially follows from the classical formula for the moment generating function of a quadratic form of a multivariate Normal variable: Let $\mathbf{Z}$ be a multivariate Normal random vector with zero mean and positive definite covariance matrix Γ. Then for any square matrix A, $E\exp(t\mathbf{Z}^T A\mathbf{Z})) = |I - 2t\Gamma^{1/2}A\Gamma^{1/2}|^{-1/2}$, where I denotes the identity matrix. □

A Appendix

Proof of Claim 2.1: From (2.5), we have

$$\frac{\partial X_t}{\partial \bar{X}} = 1,$$

$$\frac{\partial X_t}{\partial\sqrt{I(\omega_j)}} = 2\sqrt{\frac{2\pi}{N}}\cos(\omega_t j + \theta_j), \quad j = 1, \ldots, (N-1)/2,$$

$$\frac{\partial X_t}{\partial\theta_j} = -2\sqrt{\frac{2\pi}{N}}\sqrt{I(\omega_j)}\sin(\omega_t j + \theta_j), \quad j = 1, \ldots, (N-1)/2.$$

It can be verified that $\frac{\partial \mathbf{X}_N}{\partial \eta}$ is an orthogonal matrix. This is because, for example,

$$\begin{aligned}\sum_{t=1}^{N} \frac{\partial X_t}{\partial \bar{X}} \frac{\partial X_t}{\partial \sqrt{I(\omega_j)}} &= 2\sqrt{\frac{2\pi}{N}} \sum_{t=1}^{N} \cos(\omega_t j + \theta_j) \\ &= 2\sqrt{\frac{2\pi}{N}} \mathrm{Re}(\sum \exp(i\{\omega_t j + \theta_j\}) \\ &= 2\sqrt{\frac{2\pi}{N}} \mathrm{Re}\left(\frac{\exp(i\{\omega_{N+1} j + \theta_j\}) - \exp(i\{\omega_1 j + \theta_j\})}{\exp(i\omega_1 j) - 1}\right) \\ &= 0,\end{aligned}$$

where $\mathrm{Re}(\cdot)$ denotes the real part of the complex number in parentheses. It can be similarly shown that any two distinct column vectors of $\frac{\partial \mathbf{X}_N}{\partial \eta}$ are orthogonal to each other. Moreover, the (Euclidean) norm of $(\frac{\partial X_1}{\partial \bar{X}}, \ldots, \frac{\partial X_N}{\partial \bar{X}})^T$ equals $\sqrt{N}$; those of $(\frac{\partial X_1}{\partial \sqrt{I(\omega_j)}}, \ldots, \frac{\partial X_N}{\partial \sqrt{I(\omega_j)}})^T$ equal $\sqrt{4\pi}$ because $\sum_{t=1}^{N} \cos^2(\omega_t j + \theta_j) = \sum_{t=1}^{N}(1 + \cos(\omega_t j + \theta_j))/2 = N/2$. It can be similarly shown that the norms of $(\frac{\partial X_1}{\partial \theta_j}, \ldots, \frac{\partial X_N}{\partial \theta_j})^T$ equal $\sqrt{4I(\omega_j)\pi}$. Because the determinant of an orthogonal matrix is the product of the norms of the column vectors of the matrix, this completes the proof of the claim.

Derivation of (4.3): Note that $\log(|B_h|) = \sum_{j=1}^{N} \log \tilde{h}_N(\omega_j)$. Also, we could simplify the quadratic form

$$\begin{aligned}(\mathbf{X} - \mu\mathbf{1})^T B_h^{-1}(\mathbf{X} - \mu\mathbf{1}) &= \mathbf{X} B_h^{-1}\mathbf{X} - 2\mu\mathbf{1}^T B_h^{-1}\mathbf{X} + \mu^2\mathbf{1}^T B_h^{-1}\mathbf{1} \\ &= \mathbf{X}P^*\Lambda^{-1}P\mathbf{X} - 2\mu\mathbf{1}^T P^*\Lambda^{-1}P\mathbf{X} + \mu^2\mathbf{1}^T P^*\Lambda^{-1}P\mathbf{1} \\ &= \sum_{j=1}^{N} \frac{2\pi I_N(\mathbf{X};\omega_j)}{\tilde{h}_N(\omega_j)} - 2\bar{X}\frac{N\mu}{\tilde{h}_N(0)} + \frac{N\mu^2}{\tilde{h}_N(0)}. \qquad \text{(A.1)}\end{aligned}$$

(A.1) follows because (1) $\zeta_N = (\zeta(\omega_1), \zeta(\omega_2), \ldots, \zeta(\omega_N))^T = \frac{1}{\sqrt{2\pi}} P_N \mathbf{X}_N$ and $I(\omega) = |\zeta_{\mathbf{X}_N}(\omega)|^2$ and (2) $\mathbf{1}$ is orthogonal to all rows of P_N except the last row which is proportional to $\mathbf{1}$; see Section 2. Then, (4.3) follows readily from (4.1) and the preceding formulas.

References

Cox, D. R. and Hinkley D. V. [1974] *Theoretical Statistics*, Chapman and Hall, London.

Cutler, C. D. [1991] *Some results on the behavior and estimation of the fractal dimensions of distributions on attractors,* J. Statist. Phys., **62**, 651–708. Errata and Addenda (1991), **65**, 417–418.

Dubrovin, B. A., Formenko, A. T. and Novikov, S. P. [1984] *Modern Geometry — Methods and Applications*, Springer-Verlag, New York.

Dzhaparidze, K. [1986] *Parameter Estimation and Hypothesis Testing in Spectral Analysis of Stationary Time Series*, Springer-Verlag, New York.

Hajek, J. and Sidak, Z. [1967] *Theory of Rank Tests*, Academic Press, New York.

Hannan, E. J. [1962] *Time Series Analysis*, Muthen, London.

Hannan, E. J. [1970] *Multiple Time Series*, Wiley.

Hannan, E. J. [1979] *The statistical theory of linear systems*, Developments in Statistics, (Krishnaiah, P. R., ed.), Volume 2, Academic Press, 83-122.

Hardy, G. H. [1952] *A Course of Pure Mathematics*, Tenth edition, The English Language Book Society & Cambridge University Press.

Kennel, M. B. and Isabelle, S. [1992] *Method to distinguish chaos from colored noise and to determine embedding parameters*, Phys. Rev. A, **46**, 3111–3118.

Le Cam, L. [1986] *Asymptotic Methods in Statistical Decision Theory*, Springer-Verlag.
Lehmann, E. L. [1983] *Theory of Point Estimation*, Wiley.
Press, W. H., Flannery, B. P., Teukolsky, S. A. and Vetterling, W. T. [1986]. *Numerical Recipes*, Cambridge University Press.
Priestley, M. B. [1981] *Spectral Analysis and Time Series*, Academic Press.
Sen, P. K. [1985] *Permutational central limit theorems*, in Encyclopedia of Statistical Sciences (Kotz, S. & Johnson, N. L., eds.), Wiley.
Shiryayev, A. N. [1984] *Probability*, Springer-Verlag, New York.
Theiler, J., Eubank, S., Longtin, A., Galdrikian, B. and Farmer, J. D. [1992] *Testing for nonlinearity in time series: the method of surrogate data*, Physica D, **58**, 77–94.
Whittle, P. [1952] *Estimation and information in time series*, Skand. Aktuar., **35**, 48–60.
Yao, Q. and Tong, H. [1996] *A bootstrap detection for operational determinism*, Technical Report, Institute of Mathematics and Statistics, University of Kent at Canterbury.

Fields Institute Communications
Volume 11, 1997

Using "Surrogate Surrogate Data" to Calibrate the Actual Rate of False Positives in Tests for Nonlinearity in Time Series

James Theiler
Astrophysics and Radiation Measurements Group, NIS-2
Los Alamos National Laboratory, MS-D436
Los Alamos, NM, USA 87545

Dean Prichard
Computer Research and Applications Group, CIC-3
Los Alamos National Laboratory, MS-B213
Los Alamos, NM, USA 87545

Abstract. A distinction can be drawn between two approaches for using surrogate data to test for nonlinearity in a time series. The first is a "typical realizations" model which can be implemented in terms of a direct autoregressive (AR) fit to the data, and the second is a "constrained realizations" model which is implemented using a Fourier transform (FT). Earlier comparisons of these two methods found that the FT surrogate data test was the more powerful of the two for the same *nominal* false-positive rate, but that the *actual* false-positive rate was much lower for the AR test. In this paper, further comparisons are made of the FT and AR tests. A kind of "double bootstrap" approach (that involves taking surrogate data of surrogate data) is suggested for calibrating hypothesis tests so that their actual rate of false positives matches a specified nominal false-positive rate.

Numerical experiments are performed using a noise-corrupted high-dimensional strange attractor to generate the data, and a global nonlinear predictor as a discriminating statistic. These experiments indicate that the FT algorithm still provides a more powerful test than the AR algorithm, when compared at the same *actual* size, though the difference is considerably less than when compared at the same *nominal* size. Further experiments comparing in-sample with out-of-sample prediction show that in-sample prediction error, despite its flaws as an indicator of actual predictability, can provide a more powerful discriminating statistic when used for hypothesis testing.

1 A Bridge Between the Natural and the Statistical Sciences

Over thirty years ago, in a dozen dense pages in the Journal of the Atmospheric Sciences, Edward N. Lorenz described an extensive set of numerical experiments

1991 *Mathematics Subject Classification.* Primary 58F13, 62G09; Secondary 62M10, 62F03.

Work by the first author is supported by the United States Department of Energy.

which explained and illustrated the notion of sensitive dependence on initial conditions from deterministic dynamical equations (Lorenz [1963]). That same year, in a single brief paragraph in the Journal of the Royal Statistical Society B, George A. Barnard outlined a straightforward Monte-Carlo approach for testing simple null hypotheses (Barnard [1963]). Barnard wrote with optimism that "...provided one has access to a reasonable amount of time on a reasonably powerful computer, an exact test of significance is something one never need be without." A circa-1963 computer would surely test the patience of any modern user desiring "reasonably powerful" computation, but for Lorenz and Barnard, the natural and the statistical scientist, the power and utility of electronic computation was evident.

To be sure, the computer provides a kind of cross-disciplinary *lingua franca*, but the common enemy that brings together statisticians and physicists is not MS-DOS, but the question: Given a time series of data, what is the nature of the system that generated it?

As Lorenz's work became more widely known,[1] and the notion that deterministic few-degree-of-freedom systems could exhibit irregular aperiodic behavior became more widely appreciated, experimentalists began to ask a more specific form of this same question: is *my* irregular aperiodic system chaotic?

The question has proved to be a difficult, and arguably ill-posed one. Data sets from the natural world (including that stripped-down version of nature found in the laboratory) are neither purely chaotic nor totally random. By and large, they aren't even stationary. The disentangling of the deterministic and stochastic aspects of the dynamical system that generated a given data set is the ultimate goal of the time series analyst. To make progress on this problem requires a fusion of statistical inference and nonlinear dynamics; and a reasonably powerful computer.

This paper will discuss a very limited facet of this question, but our motivation is this grand promise that with enough good data, enough computational power, and enough intellectual elbow grease, we can achieve insight into the processes and phenomena whose collateral effects bombard our senses (and our sensors) daily.

1.1 Chaos and Nonlinearity. Directly testing for chaos in time series is an enterprise with a long and dubious tradition. The direct approach is to test short noisy data sets for signatures of the idealized mathematical properties that are exhibited by infinite trajectories of purely deterministic nonlinear dynamical systems. These properties include sensitive dependence on initial conditions, fractal attractors, and the existence of functions which map the past precisely into the future. Algorithms have been developed for *characterizing* time series that are known to be chaotic, based on Lyapunov exponents: Eckmann et al. [1985], Sano and Sawada [1985], fractal dimension: Russell [1980], Grassberger and Procaccia [1983], Theiler [1990], Cutler [1991], and short-term predictability: Farmer and Sidorowich [1987], Casdagli [1989], but the suitability of these algorithms for *distinguishing* chaos from other alternatives has been difficult to pin down.

These difficulties have breathed new relevance into the somewhat longer (and considerably less dubious) tradition of testing time series for evidence of nonlinearity. Though nonlinearity does not imply chaos, it is a prerequisite. Furthermore, it is a convenient landmark in the vast and largely uncharted territory between deterministic chaos and total randomness.

[1] Interestingly, neither Lorenz's nor Barnard's work received much attention at the time they were published, yet both have proven to be fundamental to modern nonlinear time series analysis.

Classical tests for nonlinearity, such as those proposed by Subba Rao and Gabr [1980], Hinich [1982], McLeod and Li [1983], Keenan [1985], and Tsay [1986], apply the methodology of statistical hypothesis testing, which provides a general formalism for addressing questions of inference from finite data sets. One posits a *null hypothesis*, which is a kind of default explanation for the data, and then attempts to determine whether this hypothesis is consistent with the data. If it is not, then it is rejected, and one moves on to alternative explanations for the data. The ultimate goal is to find the *least* interesting explanation that cannot be invalidated by the data.

In our view, chaos is a very interesting explanation for data, and for that reason alone, we are inclined to be skeptical of chaos "sightings" until the more pedestrian possibilities have been rejected. Linearly correlated noise is just one of those possibilities.

1.2 Surrogate Data. The term "surrogate data" was coined by the authors of Theiler et al. [1992], though the basic idea appeared in a number of earlier and contemporaneous publications, not the least of which was Barnard [1963]. But see also Osborne et al. [1986], Grassberger [1986], Scheinkman and LeBaron [1989], Elgar and Mayer-Kress [1989], Kaplan and Cohen [1990], Gantert et al. [1992], and Kennel and Isabelle [1992].

A surrogate data test involves the following steps:

1. Compute some statistic for the original data set, and call its value Q_o.
2. Based on the original data set, generate N_S surrogate data sets.
3. Compute the same statistic for each of the surrogate data sets: $Q_1, \dots, Q_{N_S}$.
4. Consider the full list of original and surrogate statistics $\{Q_o, Q_1, \dots, Q_{N_S}\}$, and rank them in numerical order, with r_i denoting the rank of Q_i. The smallest has rank 1 and the largest has rank $N_S + 1$.

Of particular interest is the rank r_o of the original statistic. If the null hypothesis is true, *and if the surrogate data generator is good*, then the original data set should be indistinguishable from its surrogates, and r_o can equally likely be any value from 1 to $N_S + 1$. Thus, a one-sided test that rejects the null hypothesis when the rank r_o is less than or equal to some R_o should produce false positives a fraction $\alpha = R_o/(N_S + 1)$ of the time. The probability of producing a false positive is called the "size" of a test. The "nominal" size is the value $R_o/(N_S + 1)$, and the "actual" size is the rate of false positives one actually obtains. The "power" of a test is the rate of "true positives", that is, the probability of (correctly) rejecting the null when the null is not true. The power obviously depends on the nature of the non-null process that generated the data.

Sometimes, one makes an implicit assumption of normality in the distribution of the Q's, and reports a "number of sigmas" S given by

$$S = \frac{Q_o - \mu_Q}{\sigma_Q} \tag{1.1}$$

where μ_Q and σ_Q are the mean and standard deviation, respectively, of the list $Q_1, \dots, Q_{N_S}$. One advantage of this approach is that it distinguishes a "five sigma" from a "fifty sigma" effect, but one has to be beware that the extra assumption also provides another opportunity for error. Whenever you analyze short data sets with highly nonlinear algorithms, nongaussian outliers are inevitable.

1.3 Generating Surrogate Data. The above recipe does not specify *how* the surrogate data are to be generated. Informally, surrogate data should mimic the original data set as closely as possible while remaining consistent with the null hypothesis. We will not attempt to make this statement more precise, because formally, all that is required is that the algorithm eventually produces the right answer; that is, the *nominal* size $\alpha = R_o/(N_S + 1)$ should correspond to the *actual* rate of false positives (an exact test). One might be a little less demanding, and ask only that the actual false positive rate not be greater than the nominal size (a conservative test), but as Good [1994] and others have pointed out, and as we will demonstrate below, this leads to less powerful tests.

When the null hypothesis is *simple*, which means that the null consists of a single unique process (for example, that the data are generated by a zero-mean unit-variance gaussian) then Barnard's prescription applies, and the surrogate data are generated as realizations of that process that defines the null hypothesis. When the hypothesis is *composite*, which means that the null consists of a class of processes (for example, that the data are generated by a gaussian of unspecified mean and variance), then things are not as easy. In a previously published paper (Theiler and Prichard [1996]), we discussed two approaches for generating surrogate data, which we called "typical realizations" and "constrained realizations." This distinction and its consequences are also addressed (with considerably more rigor) by Chan [1996]. Our emphasis was on tests for nonlinearity, and our comparisons were between autoregressive (AR) and discrete Fourier transform (FT) based methods for generating surrogate data.

The AR based tests involve fitting an autoregressive model to the data[2] and then generating realizations of this model as the surrogate data sets. The FT approach involves computing the discrete Fourier transform of the data, randomizing the phases in Fourier space, and then transforming back into the time domain. By construction, the FT surrogates have precisely the same power spectrum as the original data.

In Theiler and Prichard [1996], we found that the FT based surrogate data tests were more powerful than AR tests at the same nominal false positive rate, and that the AR tests had fewer false positives than nominally expected. In other words, the AR tests were less powerful and less accurate. But it is really unfair to complain about this inaccuracy, since the test produces a "better than spec" false-positives rate. One might ask[3] whether the AR tests' low power is merely a consequence of their too-low false positives rate. In particular, if one were to re-calibrate the AR tests so that they exhibited the nominal false positive rate, would the consequent enhancement in power (loosely speaking, the "true positive rate") match the power of the FT tests? This is the question that the next section will address.

2 Numerical Illustration

We will begin this section with a brief diatribe on the importance of avoiding (that's right – avoiding) "real" data. Too many studies, we feel, have been corrupted by the dogma that a methodology is not tested unless it is tested on real data. Our view is that a methodology should *not* be applied to real data, should not be permitted anywhere near real data, until it has been extensively tested on artificial data whose properties are known in advance.

---

[2]The order is generally determined by an Akaike-like criterion (Akaike [1974], Schwarz [1978]), though the case against overfitting does not necessarily apply to the hypothesis testing problem. We will have more to say about this in Section 4 of this article.

[3]Indeed, a number of people have asked just this question, and it was Peter Hall's asking in particular that clarified the issue in our minds, and led to this paper.

This is not to deny the crucial role that real data plays in the *development* of algorithms for nonlinear time series analysis. We all agree that the ultimate purpose of all this machinery is its application to real data, and we all agree that a thorough familiarity with real data — with all of its noisy glories and nonstationary disappointments — is an essential prerequisite for the interpretation of the "results" that data analysis provides. Real data, to paraphrase the popular bumper sticker, happens.

But it won't happen in this article. Our illustration of the surrogate data method will begin with an "original" data set that is generated on the computer according to the following prescription. We will add K independent realizations of the low-dimensional chaotic Hénon process (Hénon [1976])

$$h_t = \frac{1}{\sqrt{K}} \sum_{i=1}^{K} h_t^{(i)} \tag{2.1}$$

where the prefactor of $1/\sqrt{K}$ ensures that the variance of h_t will be the same as for the individual realizations. The power spectrum of the multi-Hénon time series is identical to that of a single Hénon time series, but the fractal dimension is a factor of K larger. We then add in-band noise by taking the sum

$$x_t = \sqrt{1-\theta}\, h_t + \sqrt{\theta}\, s_t \tag{2.2}$$

with h_t the noise-free multi-Hénon data set, and s_t an FT-based surrogate time series of the Hénon data set. Again, the power spectrum is preserved, and $0 \leq \theta \leq 1$ defines the fraction of the total variance that is noise. The square roots ensure that the total variance of x_t will be independent of θ. Note that the signal-to-noise ratio in this case is

$$\mathrm{SNR} = \frac{1-\theta}{\theta}. \tag{2.3}$$

Using $K = 5$, $\theta = 0.3$, and $N = 256$ data points, we will generate a single realization and call that our "original" data set. The signal-to-noise ratio for this set is 2.3, or 3.7dB. We use as our discriminating statistic the root-mean-square in-sample prediction error (normalized by the variance of the time series) of a global quadratic predictor that uses a time delay embedding with delay time $\tau = 1$ and embedding dimension $m = 2$. Applied to our original data set, we find $Q_o = 0.964724$.

For a one-sided test at the $\alpha = 0.05$ level, we can compute as few as $N_S = 19$ surrogates, and require for rejecting the null that the value of the discriminating statistic be smaller for the original data than for all of the surrogates. In Figure 1, we generate 19 FT surrogates and 19 AR surrogates, and compute the discriminating statistic for each surrogate. What we see in this case is that the FT correctly rejects the null hypothesis, while the AR incorrectly fails to reject.

In Figure 2, we used the same original data set, but many more ($N_S = 4999$) surrogates. Again, the FT rejects the null hypothesis, and the AR fails to reject at the $\alpha = 0.05$ level. In fact, this result is obtained for any level in the range $0.0066 < \alpha < 0.114$. The smallest nominal size at which a rejection of the null hypothesis would be significant is called the p-value. Note that if p is the p-value of a test with respect to a set of data, then the probability of rejecting the null using the test with 19 surrogates is given by $(1-p)^{19}$. Thus, for this data set, based on these more accurately estimated p-values, we can say that comparison with 19

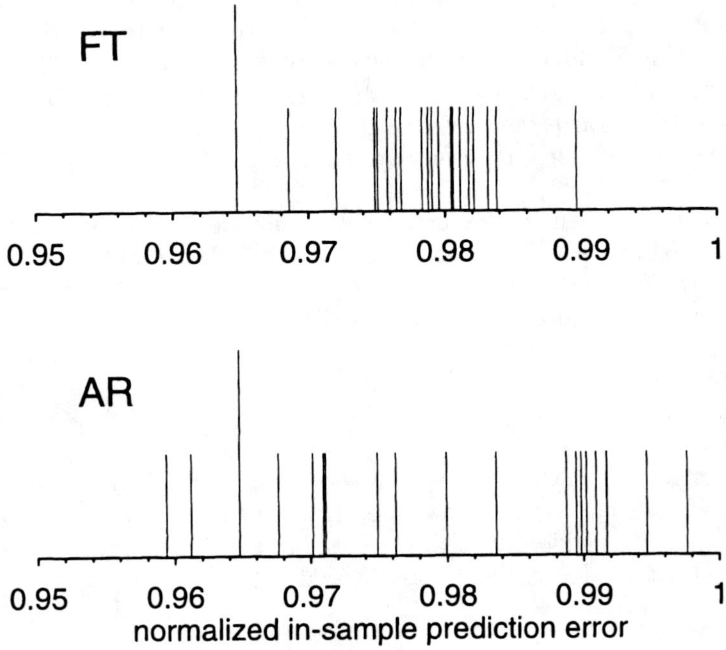

Figure 1 The discriminating statistic Q_o for an original data set (shown as the tall line) is compared to the discriminating statistics Q_i for each of $N_S = 19$ surrogates (short lines). FT surrogates are shown in the upper panel, and AR surrogates in the lower panel. The normalized prediction error for the data set generated in this realization is 0.964724. For the surrogates, on average, the in-sample error is around 0.98. This difference is arguably small, but in the case of the FT surrogates, is clearly significant. In this run, the FT rejects the null while the AR does not.

FT-based surrogates will reject the null 88% of the time, while comparing to 19 AR-based surrogates will reject only 10% of the time.

This also points out the value, mentioned by Hope [1968] and emphasized by Marriott [1979] and Hall and Titterington [1989], of using more than the minimum $N_S = 1/\alpha - 1$ surrogates to test at the level α. For instance, if we were to use $N_S =$ 39 surrogates, and reject if Q_o were smaller than the second smallest surrogate, then the probability of rejection would be $(1-p)^{39} + 39p(1-p)^{38}$. So for this data set, we'd expect the FT method to reject 97% of the time, and the AR method to reject only 5% of the time. Using $N_S = 4999$, the FT method will virtually always reject the null for this data, and the AR method will virtually always fail to reject.[4]

The above calculations were done with a single data set. One advantage of using data that does not come from a "real" experiment is that you can always generate more of it. In Figure 4, we plot the results from 1000 trials, each of which uses as its original data set a new realization of the same process that generated our first original data set. The experiment is slightly different from the above in that we used $N_S = 99$ instead of $N_S = 19$ surrogates, but the same overall result is obtained: at the $\alpha = 0.05$ level, most of the FT tests rejected the null, while fewer than 2% of the AR tests did. This confirms our statement that the FT test is more powerful than the AR test, *when compared at the same nominal size.*

[4]In fact, from Equation (2) in Marriott [1979], these probabilities are $1 - \Phi(-37.82) = 1 - 2.7 \times 10^{-313}$ that the FT will reject and $\Phi(-14.26) = 1.9 \times 10^{-46}$ that the AR will reject.

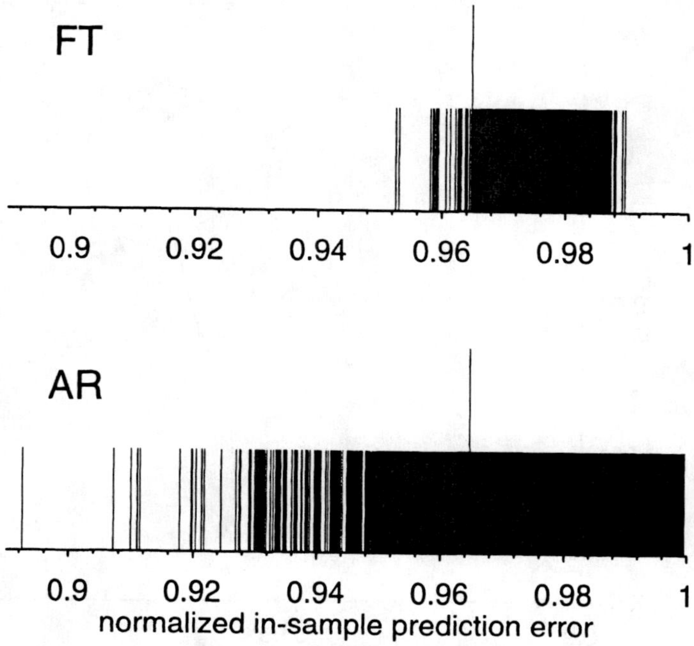

Figure 2 Same as Figure 1 but using $N_S = 4999$ surrogates. With this many surrogates, we can more precisely estimate the level of significance at which the null would be rejected. For the FT surrogates, $r_o = 33$, so the p-value is $33/5000 = 0.0066$. For the AR surrogates, we have $r_o = 570$, so $p = 0.114$.

In Figure 4(b), we estimate the *actual* size of the test by doing the same experiment as in Figure 4(a), but with original data sets that are consistent with the null hypothesis. These data sets are generated by taking surrogates of the original data sets in Figure 4(a). For these data sets, a "good" test should not reject the null, and by this measure, the AR test is very good. With 1000 trials, at the $\alpha = 0.05$ level, the AR test never rejected the null. In other words, the actual rate of false positives (the "actual size") is much less than the nominal size.

If the AR test could be recalibrated so that its actual size was equal to its nominal size, it is intuitively plausible that this enhanced proclivity to reject the null would also increase its power. Figure 4(c) indicates that, indeed, recalibration does substantially increase the AR test's power, though for this example, it is still somewhat less powerful then the equivalent FT test.

3 Surrogate Surrogate Data

With the results of the previous section in mind, we will write down an explicit strategy for calibrating surrogate data. The idea follows Hall [1986], Hall and Martin [1988], and Efron and Tibshirani [1994, see Section 18.3], and uses a kind of "double bootstrap." For us, this involves making surrogate data sets of our surrogate data sets. In particular,

1. Compute some statistic on the original data set; call its value Q_{oo}
2. Compute N_S surrogates and compute the statistic for each time series. Call these: $Q_{o1}, \ldots, Q_{oN_S}$. Let r_o be the rank of the original Q_{oo} in a sorted list that contains $Q_{oo}, Q_{o1}, \ldots, Q_{oN_s}$.

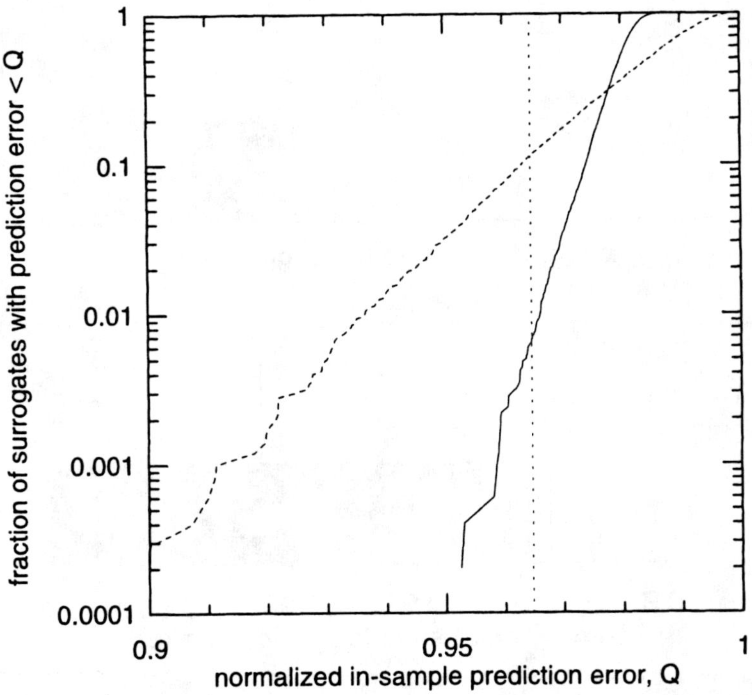

Figure 3 Same information as Figure 2, but plotted as a cumulative histogram, so that the low Q tail of the distribution is more apparent. The vertical dotted line is $Q_o = 0.964724$, the prediction error for the original time series. The dashed line is for the AR surrogates, and the solid line is for the FT surrogates. We can again see that the p-values for these respective cases are $p = 0.114$ and $p = 0.0066$.

3. Generate N_T new surrogate data sets of the original data set; these are the "fiducial" surrogate data sets. Compute the statistic for each of these and call those values: $Q_{1o}, \dots, Q_{N_T o}$.
4. For each of these N_T fiducial surrogates, generate N_S surrogate data sets (these are the surrogate surrogate data sets). Let Q_{ij} denote the value of the statistic for the j-th surrogate of the i-th fiducial. For each i, compute the rank $1 \le r_i \le N_S + 1$ of the fiducial Q_{io} in the list of surrogate values: $\{Q_{io}, Q_{i1}, \dots Q_{iN_S}\}$.
5. Finally, consider the ranks as if they were discriminating statistics, and define r_* as the rank of r_o in the sorted list of ranks that includes r_o and $r_1, \dots, r_{N_T}$. That is, $1 \le r_* \le N_T + 1$. If there are ties, take r_* as the largest possible value.
6. The reported p-value for this test is then $p = r_*/(N_T + 1)$.

Because of the coarse discretization in the rank values (they are by definition small positive integers), we can expect a fairly high incidence of ties, and for this reason, this test ends up being somewhat conservative. This can be alleviated, to some extent, by taking larger N_S.

For instance, if the uncalibrated test were exact to begin with, and the time series were *very* nonlinear (so that $r_o = 1$ could be assumed), then the null would be rejected at the $\alpha = 1/(N_T + 1)$ level only if $r_i > 1$ for all i, and this is expected to occur a fraction $f = (N_S/(N_S+1))^{N_T}$ of the time. In particular, if $N_S = N_T = 19$, then the null is rejected at the $\alpha = 0.05$ level only 38% of the time. Taking $N_S = 99$,

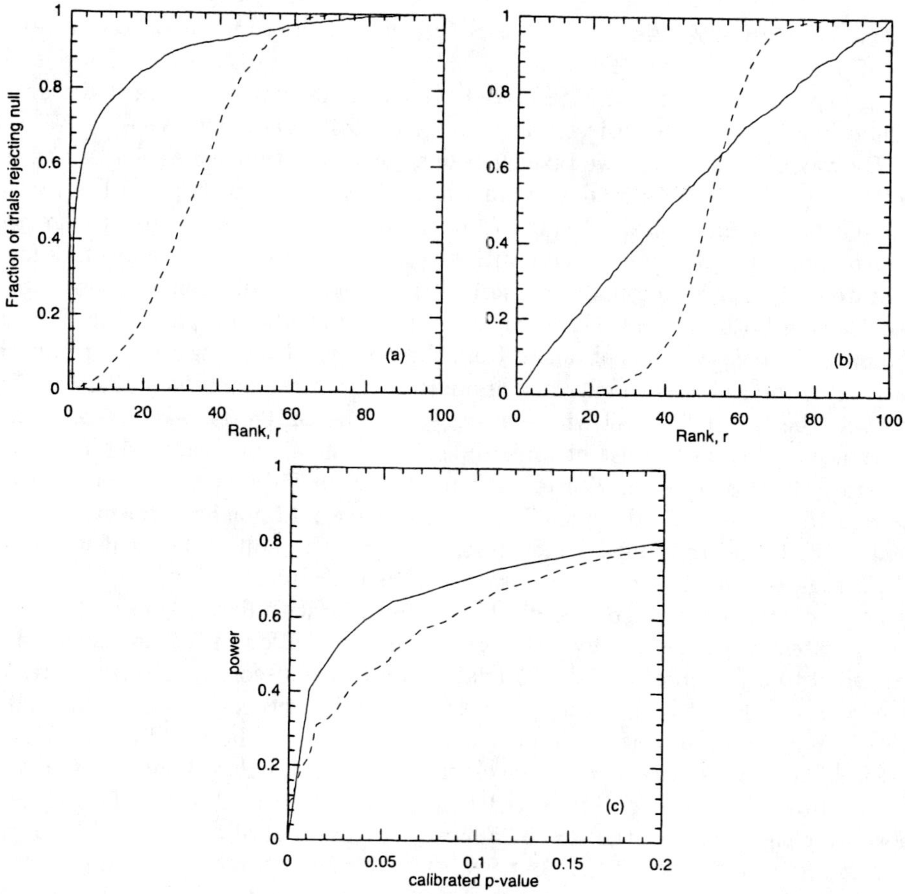

Figure 4 Comparison of the size and power of AR (dashed lines) and FT (solid lines) surrogate data schemes for testing nonlinearity in time series. These figures are based on an average of 1000 trials. In panel (a), each trial consists of a new realization of the "original" time series, and a test with $N_S = 99$ surrogates. For each trial, the rank r_o of that trial's original time series in a list of $N_S + 1 = 100$ real and surrogate Q's is computed. The plot shows the fraction of trials in which r_o was less than or equal to r as a function of r. For instance, rejection of the null at the $\alpha = 0.05$ level requires that $r_o \leq r = 5$; that occurs in 66.3% of the FT tests, and in 1.6% of the AR tests. Panel (b) is similar to (a), except that the "original" time series are consistent with the null hypothesis. In particular, each was generated by taking a surrogate data set of the data sets generated for panel (a). In this case, we expect to reject the null at the $\alpha = 0.05$ level five percent of the time. However, the AR test failed to reject the null for every one of the 1000 trials, while the FT test rejected the null 7.1% of the time. (This is significantly, though not dramatically, more often than it should have, and points to an inaccuracy in the FT algorithm – we suspect that the problem is an edge effect which is noticeable because there are only $N = 256$ data points.) For a perfectly calibrated test, we'd see a diagonal line, with a fraction $r/100$ rejections at the level $\alpha = r/100$. In panel (c), we have replotted the power in (a) against the "actual" size as estimated in (b). The AR test is seen to be much more powerful than when plotted against nominal size, but still slightly less powerful than the FT test.

this increases to 83%. As long as $N_S > 1/\alpha$, then increasing N_T also reduces the loss of power due to ties. (For example, with $N_S = 39$ and $N_T = 49$, the rejection rate is expected to be about 81%.) We have not investigated the optimal trade-off in choosing N_S and N_T subject, say, to a fixed "cost" given by $(N_S + 1)(N_T + 1)$.

By way of illustration, we take the same "original" time series that we used in Section 2 (that is, $N = 256$ data points from $K = 5$ Hénon maps with a fraction $\theta = 0.3$ of in-band noise). Using $N_S = 39$ and $N_T = 39$, we use the method of surrogate-surrogate data to compute a "calibrated" p-value. We did this for 25 separate runs (using "original" time series, generated with different random number seeds), using both FT and AR surrogates, and the results are plotted in Figure 5. The *uncalibrated* p-value is obtained from the rank r_o of the original discriminating statistic in the list of surrogate discriminating statistics; since $N_S = 39$, we would reject at the $\alpha = 0.05$ level whenever $r_o \leq 2$. The *calibrated* p-value comes from the rank r_* of r_o in the list of ranks r_k of the fiducial surrogate's discriminating statistics. In this series of 25 runs, the uncalibrated FT test would have rejected the null 15 times, while the uncalibrated AR test would not have rejected the null at all. But by using the calibrated test, we reject the null 11 times if we use the FT surrogates, and 10 times if we use the AR surrogates.

As a control experiment, we did the same thing with $\theta = 1.0$, so that the data are consistent with the null hypothesis. In this case we found the uncalibrated AR test failed to reject in any of the 25 trials, even if we raised the rejection threshold to $\alpha = 0.10$. But the calibrated AR test exhibited two rejections at $\alpha = 0.10$, one of which was also a rejection at the $\alpha = 0.05$ level. The uncalibrated FT test rejected five times at the $\alpha = 0.10$ level, one of which was also at the $\alpha = 0.05$ level. The calibrated FT test only rejected twice at the $\alpha = 0.10$ level, and neither of those were significant at the $\alpha = 0.05$ level. The number of trials here is too small to make definitive conclusions,[5] but we see that calibration certainly improves the performance of the AR test without raising too many false alarms. The effect of calibration on the FT test is slight, but possibly just large enough to pick up an inaccuracy in the FT test which we believe is due to the small sample size.

4 In-Sample and Out-of-Sample Statistics

In previous sections, we have been using in-sample prediction error as the discriminating statistic. One "problem" with in-sample prediction is that the same data that are used to train the model are used to test the model. It's like knowing beforehand what questions are going to be on the test, and studying only those.

Generally, if one is trying to assess the "predictability" of a given data set, or the "performance" of a given learning algorithm, one is careful to divide the data into two non-overlapping sets, a "training" set and a "testing" set. The training set is used to fit a model to the data, and the testing set is used to evaluate the model. The extent to which the model is able to predict the future in data that is out-of-sample (that is, data that are not in the training set) provides a measure of the predictability of the data.

When using surrogate data for testing hypotheses, however, in-sample prediction error is perfectly valid, and in many cases provides a more powerful test than out-of-sample prediction error. For example, in Figure 6 we compare the power of

[5] We refer the reader to Figure 4 in the previous section, which is a less direct test of the surrogate-surrogate algorithm, but provides more precise estimates of the various rates of false and true alarms.

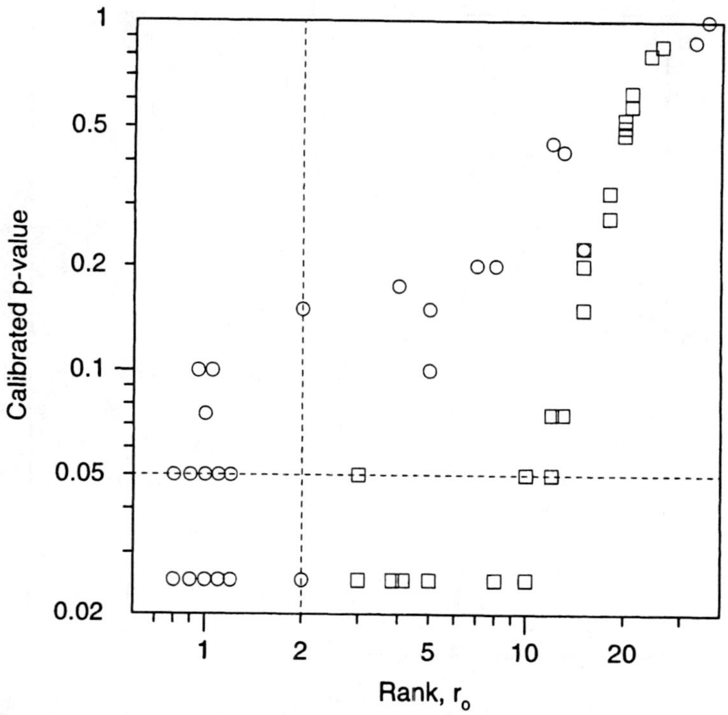

Figure 5 This scatter-plot shows the results of 25 runs of the surrogate-surrogate data method, based on 25 "original" time series. The squares correspond to AR surrogates and the circles to FT surrogates. On the horizontal axis is the rank r_o of the original statistic Q_{oo} in the sorted list of $Q_{oo}, Q_{o1}, \ldots, Q_{oN_S}$, and on the vertical axis is the calibrated p-value based on the rank of r_o in the sorted list of ranks $r_o, r_1, \ldots, r_{N_T}$. The horizontal and vertical dotted lines correspond to the $\alpha = 0.05$ levels of the calibrated and uncalibrated tests, respectively. Points which lie on or below the horizontal line indicate runs in which the null is rejected by the *calibrated* surrogate-surrogate test. Points on or to the left of the vertical line are the runs for which the null would be rejected by the ordinary (uncalibrated) surrogate data test.

FT surrogates using two discriminating statistics: one is in-sample prediction error and the other is out-of-sample. In particular, we plot the fraction of times (out of 500) that the null is rejected as a function of the amount of noise in the data. The data set used was the sum of 5 Hénon maps plus in-band noise with θ defined in Equation (2.2). The dashed line is power of the test when in-sample prediction error is used as the statistic, the solid line is the power of the test when out-of-sample prediction error is used (with equal size $N = 128$ testing and training sets). The error bars are estimated as $\sqrt{p(1-p)/N_T}$ where p is the fraction of rejections, and $N_T = 500$ is the number of trials.

There is no reason that the testing and training sets need be the same size.[6] The solid line in Figure 7 is the power of the test as a function of the training set size for a fixed $\theta = 0.3$. (The first part of the data set was used for training, and what was left was used for testing.) Notice in this case is it better to have the testing set smaller than the training set.

[6]Galdrikian and Farmer (personal communication) did some calculations of optimal partition of data into training and testing so as to maximize the evidence for nonlinearity produced by the resulting prediction error. However, in-sample prediction was not considered.

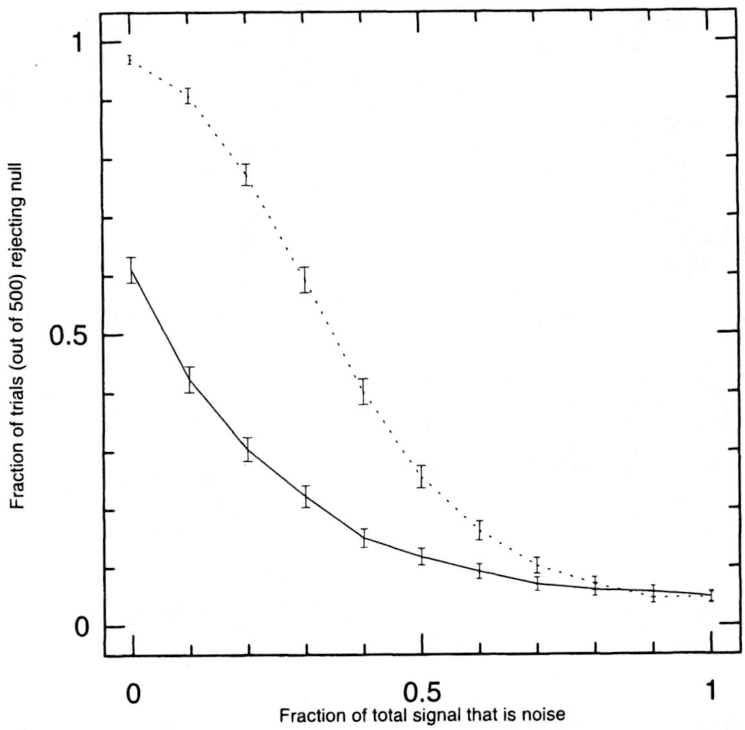

Figure 6 Comparison of the power of in-sample (dotted line) and out-of-sample (solid line) prediction error as discriminating statistics. The data set was generated as a a high-dimensional chaotic attractor with a fraction θ of added in-band noise. FT surrogates were used.

Another reason to prefer in-sample statistics to out-of-sample statistics is that since the statistic is only calculated for the testing set (while the surrogate data sets are made for the full data sets), the linear properties of the original testing set and the surrogate testing sets are not exactly the same, and this can lead to a less powerful test. (This is the same "nuisance parameter" problem that one runs into when making AR surrogates.) One way to get around this problem is to make surrogates of just the test set (and use the original training data). The dotted line in Figure 7 is power of the test using this method. In order to reduce the variance of the out-of-sample prediction error one can use k-fold cross-validation methods for estimating the true (out-of-sample) prediction error with much less variance than a single training and single testing subset. For example the short dashed line in Figure 7 is the result of k-fold cross-validation. The point at an x value of 0.5 is for 2-fold cross-validation. That is, the data set is broken into 2 equal parts. The first part is used for training and the second part for testing, giving a prediction error E_1. The testing and training sets are then switched which leads to an error E_2. The discriminating statistic is given by $Q = (E_1^2 + E_2^2)^{1/2}$. The other two points are for 4-fold and 8-fold cross validation. That is, the data set is broken into k parts, $k-1$ of these are used for training and the remaining one is used for testing, this process is repeated for the k possible testing sets, and these k errors are then combined to form the statistic $Q = (\sum_{i=1}^{k} E_k^2)^{1/2}$. The long dashed line in Figure 7 is also for k-fold cross validation but in this case surrogates were made only for the test set (rather than for the full data set).

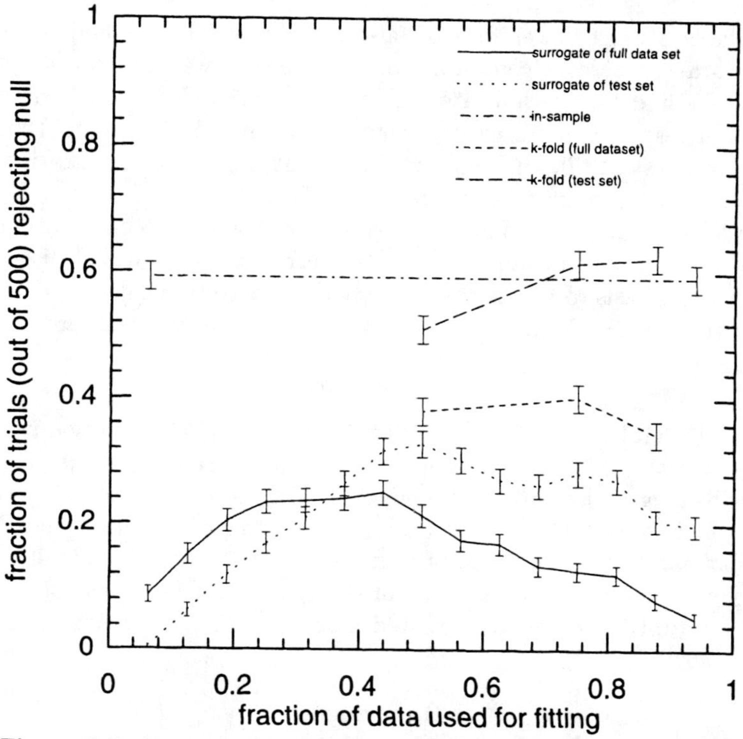

Figure 7 Comparison of the power of various out-of-sample statistics to in-sample statistics. Computing out-of-sample prediction error requires a partition of the data set into a training set and a testing set. The solid line is the power of the test as a function of the size of the training set, notice the optimal split favors a larger testing set than training set. The dotted line is the same as the solid line except that surrogates were made only for the testing set, rather than for the full data set. The short dashed line is the result using a k-fold cross validation method, and the long dashed line is also using a k-fold cross validation but in this case the surrogates were only made for the testing sets. (See text for further details.)

These results suggest that it is possible, but nontrivial, to design an out-of-sample error measure which is as powerful as the straightforward in-sample error. For example, using our global quadratic predictor with $m = 2$, there are only six free parameters, so overfitting is not as much of a problem as it would be with a more complicated predictor such as a neural network. But in the extreme and degenerate case that we overfit entirely, then there is no way to distinguish chaos from noise using in-sample prediction error (since it will be zero in both cases). If we think of power in terms of "number of sigmas", as seen in Equation (1.1), then we see that this power is increased either by increasing the "distance" in absolute units between the average Q for the real and surrogate data sets, or else by decreasing the variance of the surrogate Q's. In-sample error generally has a smaller variance than out-of-sample error, and so it will provide a more powerful discriminating statistic, unless because of overfitting the difference $Q_o - \mu_Q$ becomes too small.

5 Conclusions

Surrogate data tests based on straight autoregressive (AR) surrogates are much less powerful than the same test based on the Fourier transform (FT). But the AR tests also have a much lower rate of false positives. The method of "surrogate surrogate data" provides a computationally expensive way to calibrate surrogate data tests. We find that a calibrated AR test is substantially more powerful than a straight AR test, but in our experiments, still slightly less powerful than the FT surrogate data test. Still, the calibrated AR test is a viable alternative for data analysts who are uncomfortable with the artifacts of the FT test.[7] Calibrating the FT test has little effect, and is probably not worth the computational effort. Finally, we showed that in-sample prediction error can be a useful discriminating statistic for hypothesis testing with surrogate data, despite the deservedly tarnished reputation that in-sample prediction has as an indicator of actual predictability.

Acknowledgments

We are grateful to Colleen Cutler and Danny Kaplan, for organizing dozens of dense scientists (natural, statistical, and otherwise) in a workshop that was thought-provoking and thoroughly enjoyable. We are pleased to acknowledge the workshop participants for many useful and interesting discussions, and Peter Hall in particular, for asking the question that stimulated this study. Finally, we thank Bette Korber, Thomas Schreiber, and an anonymous referee for helpful comments on the manuscript. Had we incorporated more of their suggestions, this would be a better paper.

References

Akaike, H. [1974] *A new look at the statistical model identification*, IEEE Trans. Auto. Control, **19**, 716–723.

Casdagli, M. [1989] *Nonlinear prediction of chaotic time series*, Physica D, **35**, 335–356.

Chan, K. S. [1996] *On the validity of the method of surrogate data*, The Fields Institute Communications Series, Volume 11, American Mathematical Society, Providence, R.I., 77–97.

Cutler, C. D. [1991] *Some results on the behavior and estimation of the fractal dimensions of distributions on attractors*, J. Stat. Phys., **62**, 651–708.

Eckmann, J.-P. and Ruelle, D. [1985] *Ergodic theory of chaos and strange attractors*, Rev. Mod. Phys., **57**, 617–656.

Efron, B. and Tibshirani, R. [1993] *An Introduction to the Bootstrap* (Chapman and Hall, London).

Elgar, S. and Mayer-Kress, G. [1989] *Observations of the fractal dimension of deep- and shallow-water ocean surface gravity waves*, Physica D, **37**, 104–108.

Farmer, J. D. and Sidorowich, J. J. [1987] *Predicting chaotic time series*, Phys. Rev. Lett., **59**, 845–848.

Gantert, C., Honerkamp, J. and Timmer, J. [1992] *Analyzing the dynamics of hand tremor time series*, Biol. Cyber., **66**, 479–484.

Good, P. [1994] *Permutation Tests: A Practical Guide to Resampling Methods for Testing Hypotheses*, Springer series in statistics (Springer-Verlag).

Grassberger, P. [1986] *Do climatic attractors exist?* Nature, **323**, 609–612.

Grassberger, P. and Procaccia, I. [1983] *Characterization of strange attractors*, Phys. Rev. Lett., **50**, 346–349.

Hall, P. [1986] *On the bootstrap and confidence intervals*, Ann. Statist., **14**, 1431–1452.

Hall, P. and Martin, M. A. [1988] *On bootstrap resampling and iteration*, Biometrika, **75**, 661–671.

Hall, P. and Titterington, D. M. [1989] *The effect of simulation order on level accuracy and power of Monte-Carlo tests*, J. R. Statist. Soc. B, **51**, 459–467.

Hénon, M. [1976] *A two-dimensional mapping with a strange attractor*, Comm. Math. Phys., **50**, 69–77.

Hinich, M. J. [1982] *Testing for gaussianity and linearity of a stationary time series*, J. Time Series Anal., **3**, 169–176.

[7] But of course, AR tests have their own artifacts. For instance, when Theiler, Linsay, and Rubin [1993] looked at time series with long coherence times, they found them to be problematic for AR surrogates as well as FT surrogates.

Hope, A. C. A. [1968] *A simplified Monte-Carlo significance test procedure*, J. R. Statist. Soc. B, **30**, 582–598.

Kaplan, D. T. and Cohen, R. J. [1990] *Is fibrillation chaos?* Circulation Res., **67**, 886–892.

Keenan, D. M. [1985] *A Tukey nonadditivity-type test for time series nonlinearity*, Biometrika, **72**, 39–44.

Kennel, M. B. and Isabelle, S. [1992] *Method to distinguish possible chaos from colored noise and to determine embedding parameters*, Phys. Rev. A, **46**, 3111–3118.

Marriott, F. H. C. [1978] *Barnard's Monte-Carlo tests: how many simulations?* App. Statist., **28**, 75–77.

McLeod, A. I. and Li, W. K. [1983] *Diagnostic checking ARMA time series models using squared-residual autocorrelations*, J. Time Series Anal., **4**, 269–273.

Osborne, A. R., Kirwin, A. D., Provenzale, A. and Bergamasco, L. [1986] *A search for chaotic behavior in large and mesoscale motions in the Pacific Ocean*, Physica D, **23**, 75–83.

Russell, D. A., Hanson, J. D. and Ott, E. [1980] *Dimension of strange attractors*, Phys. Rev. Lett., **45**, 1175.

Sano, M. and Sawada, Y. [1985] *Measurement of the Lyapunov spectrum from chaotic time series*, Phys. Rev. Lett., **55**, 1082.

Scheinkman, J. A. and LeBaron, B. [1989] *Nonlinear dynamics and stock returns*, J. Business, **62**, 311–338.

Schwarz, G. [1978] *Estimating the dimension of a model*, Ann. Statist., **6**, 461–464.

Subba Rao, T. and Gabr, M. M. [1980] *A test for linearity of stationary time series*, J. Time Series Anal., **1**, 145–158.

Theiler, J. [1990] *Estimating fractal dimension*, J. Opt. Soc. Am. A, **7**, 1055–1073.

Theiler, J., Eubank, S., Longtin, A., Galdrikian, B. and Farmer, J. D. [1992] *Testing for nonlinearity in time series: the method of surrogate data*, Physica D, **58**, 77–94.

Theiler, J., Linsay, P. S. and Rubin, D. M. [1993] *Detecting nonlinearity in data with long coherence times*, in Time Series Prediction: Forecasting the Future and Understanding the Past, (Weigend, A. S. and Gershenfeld, N. A. eds.), vol. XVII of SFI Studies in the Sciences of Complexity, (Addison-Wesley), pp. 429–455.

Theiler, J. and Prichard, D. [1996] *Constrained-realization Monte-Carlo method for hypothesis testing*, Physica D, **94**, 221.

Tsay, R. S. [1986] *Nonlinearity tests for time series*, Biometrika, **73**, 461–466.

Fields Institute Communications
Volume 11, 1997

Chaos with Confidence: Asymptotics and Applications of Local Lyapunov Exponents

Barbara A. Bailey
Stephen Ellner
and
Douglas W. Nychka
Biomathematics Graduate Program
Department of Statistics
North Carolina State University
Raleigh, NC, USA 27695-8203

Abstract. In this paper we define a version of Local Lyapunov Exponents (LLEs) for discrete-time dynamical systems perturbed by noise, which are related to short-term predictability and noise amplification and thus can be used to detect regions of state space where the dynamics may be more predictable. We present a Central Limit Theorem for fluctuations of LLEs about the global LE, as the number of forward time steps used in computing LLEs is increased. We also present some large-sample statistical properties when LLEs are estimated by fitting a statistical model to a time series, that justify the construction of confidence intervals for maximum likelihood estimates.

As an application of these results we analyze time series of measles quarterly case reports, using a neural network time series model. The results show large changes in dynamic structure and predictability over the course of an epidemic.

Because LLEs provide a detailed description of a systems' nonlinearity, they are well suited to compare mechanistic models with data. Comparison of observed with model LLEs shows that current models for measles do not correctly predict the regions of state space with the highest short-term sensitivity to initial conditions. This disagreement is not apparent in more common measures used to assess these models.

1 Introduction

The global Lyapunov exponent of a dynamical system measures the average rate at which nearby trajectories diverge. A positive exponent indicates "sensitive dependence on initial conditions", hence chaotic dynamics and unpredictability, but there may still be short-term trajectory convergence and hence higher predictability in some regions of state space. Conversely, a non-chaotic system may have short-term trajectory divergence in some regions of state space. One way of quantifying these fluctuations is by calculating finite-time "local" Lyapunov exponents (LLEs).

1991 *Mathematics Subject Classification.* Primary 62M10, 58F13; Secondary 60F05.

Local Lyapunov exponents provide a more detailed description of a system's dynamics than just the global Lyapunov exponent. In many real data sets, such as case reports of measles, and field and laboratory population dynamics, the global exponent is close to zero and the system is near the "transition to chaos" (Ellner and Turchin [1995]). In these cases, it may be more useful to quantify the short-term convergence (or divergence) of trajectories and thereby determine whether or not the system exhibits "local" chaos. Local Lyapunov exponents therefore provide a useful tool for characterizing nonlinear dynamics. The analysis involves examining the short term predictability (or unpredictability) of the system and identifying the parts of the time series or regions in state space when they occur (Abarbanel, Brown, and Kennel [1992], Smith [1992], Wolff [1992]).

The paper is organized as follows. A local Lyapunov exponent is defined and estimation procedures presented in Section 2. Section 3 is a discussion of the Central Limit Theorem behavior of local Lyapunov exponents. Construction of confidence intervals based on maximum likelihood estimates is described in Section 4, including simulation results of coverage probabilities. Finally, Section 5 contains applications analyzing time series of measles cases and model output. The analysis includes the comparison of measles dynamics between Copenhagen and New York City case data and the comparison between Copenhagen case data and output from a stochastic model.

The theory is presented mostly for the case of scalar data x_t, fitted by a model involving past values $x_{t-1}, x_{t-2}, \ldots$, which is the most common in applications, but exactly parallel results hold for "state-space" models with vector data X_t fitted by a model involving only X_{t-1}. Our technical assumptions were motivated by the time-series model that we have found most useful for estimating LLEs from data, a feedforward neural network with Gaussian noise. The neural network model has the advantage that Lyapunov Exponent estimates are robust to mis-estimation of the dimension of the model (McCaffrey et al. [1992]). For measles data in particular, we found that neural net models had the highest out-of-sample forecasting accuracy of a suite of statistical and mechanistic models (Ellner et al. [1996]).

2 Definition of Local Lyapunov Exponents

We assume that the data $\{x_t\}$ are a time series generated by a nonlinear autoregressive model

$$x_{t+1} = f(x_t, x_{t-1}, \ldots, x_{t-d+1}) + \varepsilon_t \tag{2.1}$$

where $x_t \in \Re$ and $\{\varepsilon_t\}$ is a sequence of independent random variables or perturbations with $\mathrm{E}(\varepsilon_t)= 0$ and $\mathrm{Var}(\varepsilon_t)= \sigma^2$. It is important to note that the error in (2.1) is not measurement error, but dynamic noise, an inherent part of the dynamics of the system. The state-space representation of the system is

$$X_{t+1} = F(X_t) + e_t \tag{2.2}$$

where $X_t = (x_t, x_{t-1}, \ldots, x_{t-d+1})$ and $e_t = (\varepsilon_t, 0, 0, \ldots, 0)$ are in $\Re^d$.

Intuitively, the LLE is defined by making a small perturbation of X_t to X_t^*, and following forward the perturbed and unperturbed trajectories. The LLE measures the difference between the two trajectories after m time steps,

$$\lambda_m(t) = \frac{1}{m} \ln \frac{d(X_{t+m}, X_{t+m}^*)}{d(X_t, X_t^*)},$$

where d is a measure of distance.

Different versions of local Lyapunov exponents can be defined, depending on the choice of perturbation, the distance measure used, and the choice of which trajectories to follow forward (e.g., the average trajectory at X_t, the noise-free trajectory starting at X_t, or the actual sample path from times t to $t+m$, in which case $\lambda_m(t)$ is a random variable). We would not want to argue that there is a single "right" definition (though there can be "wrong" definitions, in which the local exponents do not converge to the global exponent as $m \to \infty$). Rather, the choice should depend on what description of the dynamics is of interest. We choose to perturb only the first component of X_t and follow the growth of that perturbation along the actual sample path. This corresponds to a perturbation of the system (2.1) at time t, i.e. perturbing x_t and then following the growth of that single perturbation along the actual sample path. In the state-space representation, x_t is the first component of X_t. One advantage of perturbing only the first component is that if (2.1) is regarded as a reconstruction in time-delay coordinates from one state variable x_t of a higher-dimensional system, then the local exponents for (2.1) correspond exactly to those for the true system when x_t is used as the first component of the state vector, for both finite and infinite m.

In addition, results of Yao and Tong [1994] show that this definition of the LLE is related to noise amplification by the system, and thus to predictability. Let $\sigma_m^2(x) = Var(X_{t+m}|X_t = x)$. In a scalar system with small additive noise, Yao and Tong [1994] showed

$$\sigma_m^2(x) = \sigma^2 \mu_m(x)(1 + o(1))$$

where $\sigma^2 = Var(\varepsilon(t))$ and

$$\mu_m(x) = 1 + \sum_{j=1}^{m-1} \left\{ \prod_{k=j}^{m-1} \dot{f}(f^{(k)}(x)) \right\}^2 .$$

This can be re-written as

$$\mu_m(x) = 1 + \sum_{j=1}^{m-1} \exp\left\{2(m-j)\lambda_{m-j}(f^{(j)}(x))\right\} \tag{2.3}$$

where λ_j are the LLEs along the "skeleton" system with the noise deleted. However to leading order in σ in this expansion, the skeleton LLEs agree with those along the sample path (for any fixed finite m). For a system with dynamic noise the value of μ_m sets the limits to large-sample forecasting accuracy (Tong [1995]), so Equation (2.3) shows a close tie (for small noise) between LLEs and local unpredictability.

As perturbations become small, $\lambda_m(t)$ will depend on derivatives of the map F. Then

$$\lambda_m(t) = \frac{1}{m} \ln \|J_{m+t-1} J_{m+t-2} \cdots J_t U_0\|, \tag{2.4}$$

where J_t (sometimes written $J(X_t)$) is the Jacobian matrix of F evaluated at X_t, U_0 is a unit vector and $\|\cdot\|$ is the Euclidean vector norm. Since $\lambda_m(t)$ is a function of time, the LLE depends on the trajectory and can be thought of as an "m-step ahead" local Lyapunov exponent process. If U_0 is chosen at random with respect to the uniform measure on the unit sphere, then with probability 1 as $m \to \infty$, $\lambda_m(t)$ converges to the global Lyapunov exponent, because U_0 has zero probability of falling into the subspace corresponding to subdominant exponents. In practice, however, we usually take $U_0 = (1, 0, 0, \dots, 0)$.

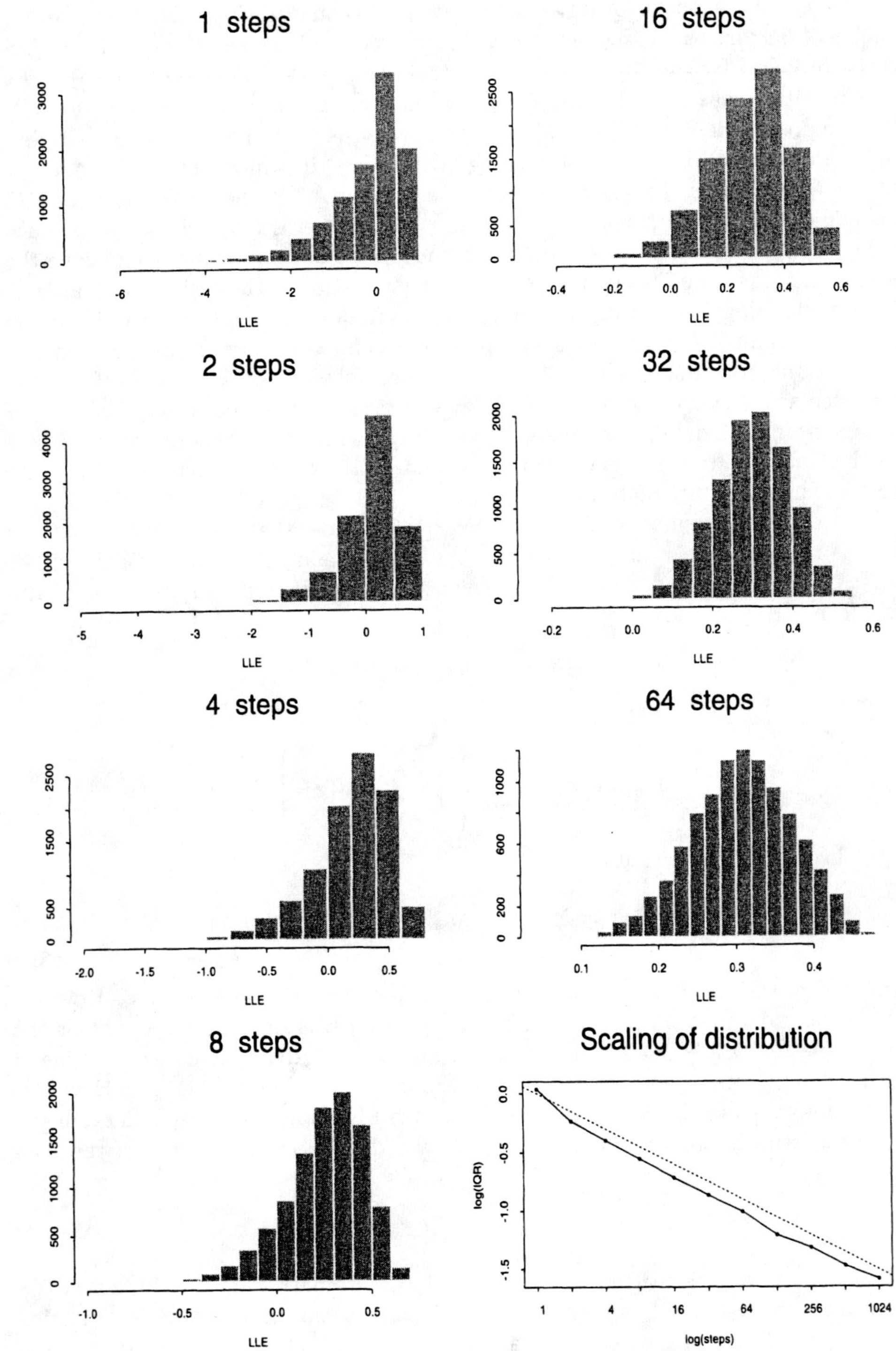

Figure 1 Histogram of LLEs of the Ikeda map. Bottom right panel shows the interquartile range of the distributions. The dashed line has slope $-1/2$ corresponding to the expected scaling of the LLEs with increasing number of steps.

3 The Limiting Distribution of Local Lyapunov Exponents

Associated with a local Lyapunov exponent is the number of forward time steps m in Equation (2.4) that are used to calculate the quantity. As m becomes large the distribution of local exponents will converge to a normal distribution. Figure 1 is a series of histograms of the LLEs from the Ikeda map which is described in Section 4. As the number of steps m increases from 1 to 64, the histograms become closer to a normal distribution. The last plot in the series is a plot of the interquartile range vs the number of steps m, on a log scale. The slope of the line is approximately one-half, indicating $1/\sqrt{m}$ scaling. In this section, a central limit theorem for local Lyapunov exponents will be stated.

The key to deriving the CLT result is the construction of a vector Markov chain that includes both the state vector and the Jacobian products. Let X_t be a Markov chain and consider the vector Markov chain W_t defined as

$$W_t = \begin{bmatrix} X_t \\ \cdots \\ U_t \end{bmatrix}$$

where X_t is the state vector and the dynamics of U_t are given by

$$U_t = \frac{J_t U_{t-1}}{\|U_{t-1}\|}.$$

Define $h(W_t) = ln(|U_t|)$. Then it follows from Equation (2.4) that

$$\lambda_m(t) = \frac{1}{m}\sum_{j=1}^{m} h(W_{t+j}).$$

Thus $\lambda_m(t)$ is the average of a functional of this process. Let λ_m denote a LLE starting at any time t. A CLT for λ_m can therefore be derived by applying standard CLT results for functionals of Markov chains (Doob [1953]). The dependence on time t has been dropped because the CLT is independent of time.

Assumption 3.1

1. *F is a bounded map.*
2. *e_t is i.i.d. and has a bounded probability density function with respect to Lebesgue measure in $\Re^d$.*
3. *Define $\mathcal{J}_l = \prod_{j=1}^{l} J(X_j)$ and let $\omega(z \mid x)$ be the probability density function of $vec(\mathcal{J}_l^\top)$ conditional on $X_0 = x$. Assume that $\omega(z \mid x)$ exists with respect to Lebesgue measure on $\Re^{d^2}$ and that it is uniformly bounded with respect to x in $L^{1+\epsilon}(\Re^{d^2})$ norm, for some $\epsilon > 0$ and $l > 0$.*
4. *$J(x)$ is uniformly bounded.*
5. *There is a single ergodic set for $\{W_t\}$.*

Theorem 3.1 (CLT) *If Assumptions 3.1 are satisfied, then*

$$\frac{\sqrt{m}\,(\lambda_m - \lambda)}{\sqrt{Var\,(\lambda_m)}} \to N(0,1) \quad as \quad m \to \infty.$$

The details of the proof of Theorem 3.1 are presented in Bailey [1996]. If there is more than one ergodic set then the limiting distribution is still normal but may depend on W_0. In the Appendix we give conditions for state space models to have a single ergodic set for W_t.

The role of nonlinearity and the noise e_t in CLT behavior is complex. If there is not much nonlinearity, i.e., the system is close to linear, then the LLEs will be nearly identical and therefore a CLT is not expected. Theiler and Smith [1995] show that if the map f is conjugate to a constant-slope map and the system is noise free, then the Lyapunov exponents scale as $1/m$ not $1/\sqrt{m}$. The meaning of Theorem 3.1 is that if the system has "enough" noise and nonlinearity, then the Lyapunov exponent does scale as $1/\sqrt{m}$ and converge to Gaussian.

4 Estimating Local Lyapunov Exponents and Construction of Confidence Intervals

The calculation of global and local Lyapunov exponents uses Equation (2.4) and requires an estimate of the Jacobian matrix J_t. In all our examples the map f in (2.1) or each component of F in a state-space model is estimated as a feed-forward neural network with a single layer of hidden units (Ellner, Nychka, and Gallant [1992]). The form of the model is

$$F(X) = \beta_0 + \sum_{i=1}^{k} \beta_i \varphi(X^T \gamma_i + \mu_i) \tag{4.1}$$

where $\varphi(u) = e^u/(1+e^u)$. The parameters β, γ and μ are found by least squares. The complexity of the model, i.e., the embedding dimension and the number of hidden units k, is chosen based on generalized cross validation (GCV). The estimated Jacobian is then the derivative of the estimated F.

The construction of confidence intervals for the estimates is based on the identification of a joint confidence region for the model parameters θ, using the likelihood ratio statistic. The approximate confidence set $\mathcal{A}_\theta$ will be all the values of θ such that

$$S(\theta) \le S(\hat{\theta}) \left[1 + \frac{p}{n-p} \mathrm{F}(p, n-p, \alpha) \right] \tag{4.2}$$

where $S(\theta)$ is the residual sum of squares and $\hat{\theta}$ is the least-squares estimate of θ. For the F distribution, p is the number of parameters in the model, n is the sample size and α is the probability level. This confidence set is based on the asymptotic normality of the nonlinear regression model parameters. Using a standard Taylor series argument, asymptotic normality of the nonlinear regression model parameters implies

$$-2\log(L(\theta)/L(\hat{\theta})) \xrightarrow{d} \chi^2(p) \tag{4.3}$$

where $L(\theta)$ is the likelihood function and $\chi^2(p)$ denotes the chi-squared distribution with p degrees of freedom. The distributional result of (4.3) implies the finite sample size result of (4.2) (Seber and Wild [1989]).

In order to insure the asymptotic normality of the nonlinear autoregression parameters and therefore justify the confidence set defined in Equation (4.2), the following assumptions are needed.

Assumption 4.1

1. *X_t is stationary and ergodic.*
2. *e_t is i.i.d. $N(0, \sigma^2)$.*
3. *$\theta \in \Theta$ and Θ is compact.*
4. *F, first, second and third partials of F exist and are continuous and uniformly bounded for all $\theta \in \Theta$.*

Theorem 4.1 *Under Assumptions 4.1, there exists a maximum likelihood (ML) estimator $\hat{\theta}_n$ of θ for which $\hat{\theta}_n \stackrel{a.s.}{\to} \theta$ and $I_n^{1/2}(\theta)(\hat{\theta}_n - \theta) \stackrel{d}{\to} N(0,1)$ as $n \to \infty$, where $I_n(\theta)$ is the conditional information matrix.*

The proof of Theorem 4.1 is presented in Bailey [1996] and it involves showing Assumptions 4.1 imply assumptions stated by Hall and Heyde [1980]. Most importantly, Theorem 4.1 implies Equation (4.3) which justifies the cutoff value used in Equation (4.2) for a nonlinear autoregressive process.

The computation of the confidence interval is a two step process. First, a neural net is fit to the data to obtain the least squares estimate for θ. The residual sum of squares from the least squares estimate is then $S(\hat{\theta})$ in Equation (4.2). The goal of the second step is to find θ's in the confidence set. The procedure is to randomly generate many parameter starting values and perform on each a partial optimization or "rough" fit. The generation of parameter sets and the partial optimization is repeated until 500 parameter sets are in the confidence set. To make sure that the parameter sets cover the entire confidence region, 100 of the parameter sets are forced to be within 20% of the cutoff (4.2). For all θ's in the confidence region, an appropriate functional φ of the parameters can be evaluated. For example, the functional $\varphi(\theta)$ could be the global Lyapunov exponent or the 10th or 90th percentile of the LLEs. The confidence interval (CI) for φ is then given by the minimum and maximum of $\{\varphi(\theta_i)\}_{i=1}^{500}$ where $\{\theta\}_{i=1}^{500}$ are the θ values found in the confidence set $\mathcal{A}_\theta$.

We conducted a simulation study to examine the actual coverage probabilities of the approximate confidence intervals. Confidence intervals for the global Lyapunov exponent and functionals of the five step ahead local Lyapunov exponents (LLE5s) were generated by the techniques described above for the Ikeda map.

The Ikeda Map is a 2-dimensional nonlinear map (Smith [1995]):

$$\begin{aligned} x_t &= 1 + \mu x_{t-1} \cos\left(0.4 - \frac{6}{x_{t-1}^2 + y_{t-1}^2 + 1}\right) \\ &\quad - \mu y_{t-1} \sin\left(0.4 - \frac{6}{x_{t-1}^2 + y_{t-1}^2 + 1}\right) \end{aligned} \qquad (4.4)$$

$$\begin{aligned} y_t &= \mu x_{t-1} \sin\left(0.4 - \frac{6}{x_{t-1}^2 + y_{t-1}^2 + 1}\right) \\ &\quad + \mu y_{t-1} \cos\left(0.4 - \frac{6}{x_{t-1}^2 + y_{t-1}^2 + 1}\right) \end{aligned} \qquad (4.5)$$

For the simulation study we used $\mu = 0.75$. Dynamic noise was added to both x_t and y_t, giving state-space dynamics of the form

$$X_t = f_1(X_{t-1}, Y_{t-1}) + e_{1,t} \qquad (4.6)$$

$$Y_t = f_2(X_{t-1}, Y_{t-1}) + e_{2,t} \qquad (4.7)$$

The noise terms $e_{i,t}$ were both given uniform distributions on (-0.1,0.1). The noise was uniform to make sure that the system stays in a locally attracting set. Figure 2(d) shows the distribution of points on the attracting set generated by this system.

A sample size of 400 was chosen for the simulation. For this sample size, 100 (x_t, y_t) realizations were generated and a neural net was fit to each realization. Since there are two variables, a neural net was fit to each data set. For each pair of neural net fits there is a corresponding confidence interval for the estimate. Confidence

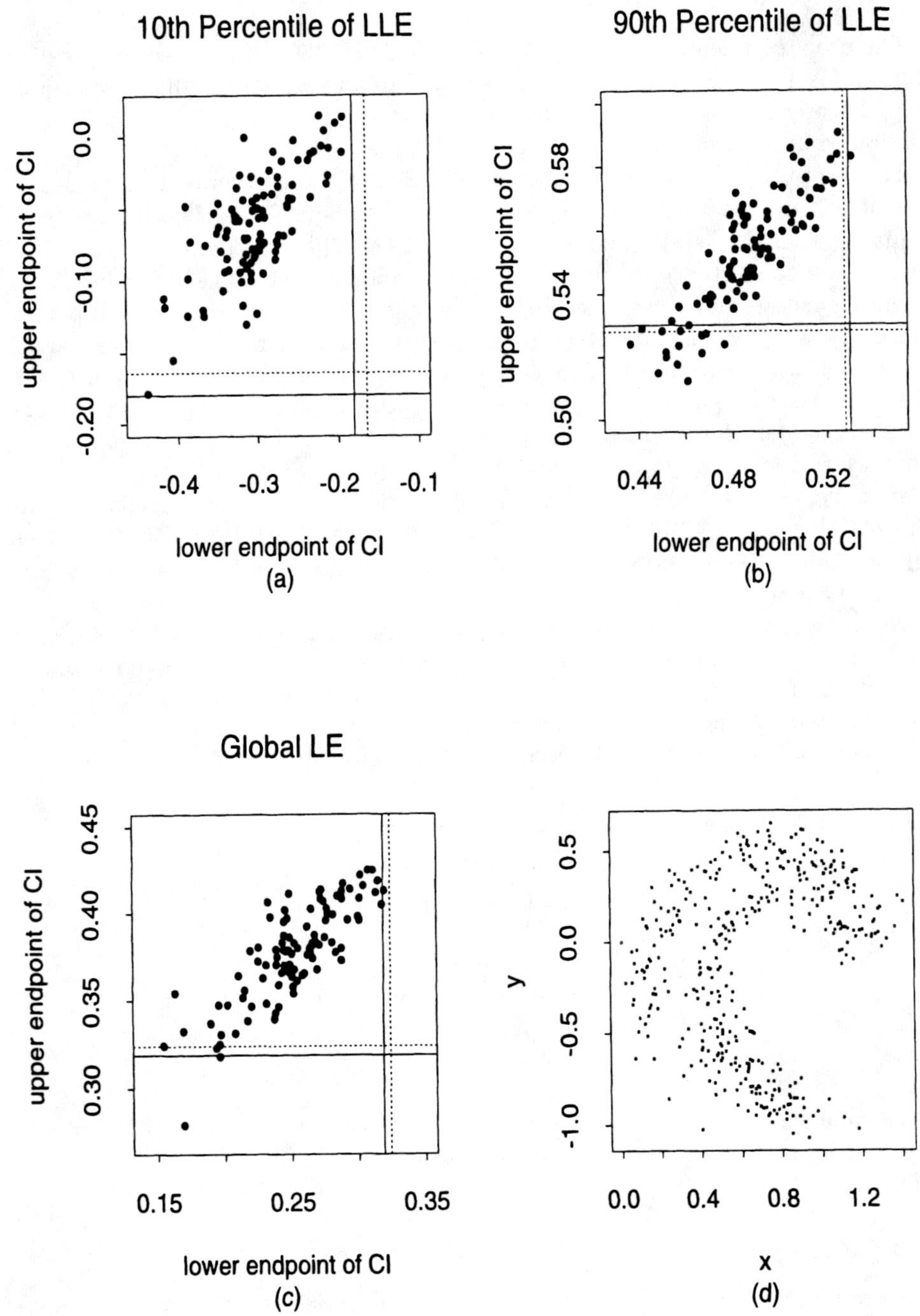

Figure 2 (a) Coverage probability plot of CI for 10th percentile of LLE5s for noisy Ikeda map. (b) Coverage probability plot of CI for 90th percentile of LLE5s for noisy Ikeda map. (c) Coverage probability plot of CI for global LE for noisy Ikeda map. (d) 400 points from a typical realization of the Ikeda map with noise uniform on (-0.1,0.1).

intervals were generated for the global Lyapunov exponent and the 10th and 90th percentile of the five step ahead local Lyapunov exponents (LLE5s).

The results of the simulation study are summarized in "coverage probability" plots (Figure 2). The axes of this plot are the lower and upper limit of the 95% confidence interval. Each plotted point (•) represents the confidence interval from one of the 100 data sets simulated. The solid lines on the plot are the true value of the quantity being estimated. The dotted lines on the plots are the expected value of the estimate, i.e. the mean of the 100 estimates corresponding to the best fit parameter values for each realization of (x_t, y_t). The intersection of the solid lines divides the plot into four quadrants. The top left quadrant is the quadrant where the lower endpoint of the confidence interval is smaller than the true value, and the upper endpoint of the confidence interval is larger than the true value. This is the quadrant where the confidence interval is covering the true value. If points are in the bottom left quadrant then the lower endpoint of the confidence interval is below the true value but the upper endpoint is also smaller than the true value. Similarly, if points are in the top right quadrant then the upper endpoint of the confidence interval is above the true value but the lower endpoint is larger than the true value. It is impossible for points to be in the bottom right quadrant. A coverage probability of 95% corresponds to 95 out of the 100 points lying in the top left quadrant.

Figure 2(a) is the coverage probability plot for the 10th percentile of LLE5s. This plot shows that all confidence intervals cover the true value. The 90th percentile of LLE5s are in Figure 2(b). The coverage is not as good for this estimate. Fifteen confidence intervals do not cover the true value. However, all but one of the fifteen are in the bottom left quadrant where the intervals are "too small", indicating that the model produces conservative estimates. Figure 2(c) is the coverage probability for the global LE. Three intervals do not cover the true value. The coverage probability plots from the Ikeda map simulation indicate that it is computationally feasible to construct confidence intervals using a reasonable sample of the likelihood surface. Also, when confidence intervals do not cover the true value they indicate that the model produces conservative estimates guarding against false positives. Other simulation studies to examine the coverage probabilities of the confidence intervals using different test systems showed similar results (Bailey [1996]).

5 Applications

The dynamics of infectious diseases with recurrent epidemics is an area where the LLE analysis can be useful. Because of the variable cycle of outbreaks, measles exhibits very complicated dynamics. A comparison of LLEs and their fluctuations in measles data from different cities may provide insight into the similarities and differences in the dynamics of disease incidence. Stochastic models have attempted to capture the dynamics of measles. Local Lyapunov exponents can be used as a "probe" into the adequacy and inadequacy of these models.

5.1 Comparison between Cities. The time series of quarterly measles cases from two cities, New York City and Copenhagen, were chosen for comparison. The model fit to the log transformed case data ($\log C_t$) involved not only lagged values of the log cases, but also a component for the number of susceptibles in the population at a given time. The susceptibles from the population are reconstructed from the reported cases data following the procedure described in Bobashev et al. [1995].

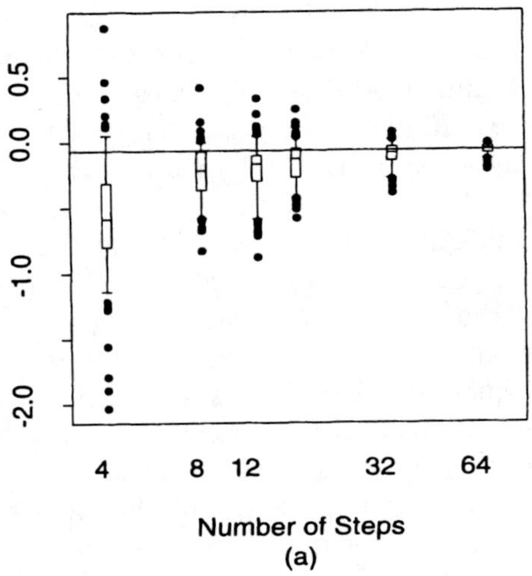

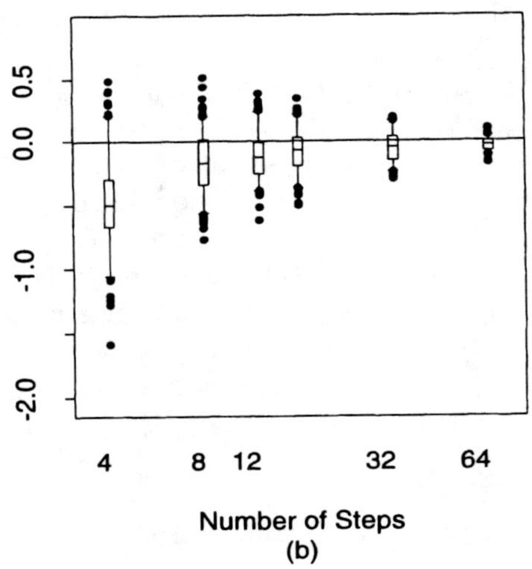

Figure 3 (a) Distribution of LLEs for New York City measles quarterly cases. (b) Distribution of LLEs for Copenhagen measles quarterly cases.

The mass balance equation for the number of susceptibles (S_t) in a population at time t is a function of the number of cases (C_t) and the birth rate (b_t). The following set of equations describe the number of cases and susceptibles at a given time.

$$\log C_{t+1} = f(\log C_t, S_t, \sin(2\pi t/4), \cos(2\pi t/4)) + e_t \tag{5.1}$$

$$S_{t+1} = S_t - C_{t+1} + b_{t+1} \tag{5.2}$$

In the first equation, f is estimated by a neural net and $\sin(2\pi t/4)$ and $\cos(2\pi t/4)$ are seasonal forcing terms with periods corresponding to quarterly data. The distribution of LLEs for both of the cities is seen in Figure 3. The solid line represents

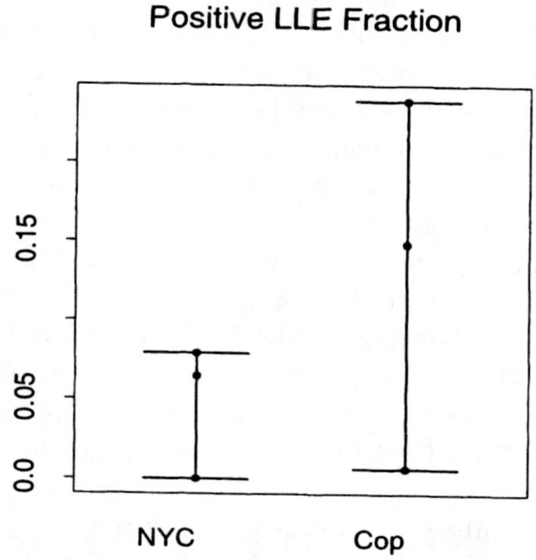

(a)

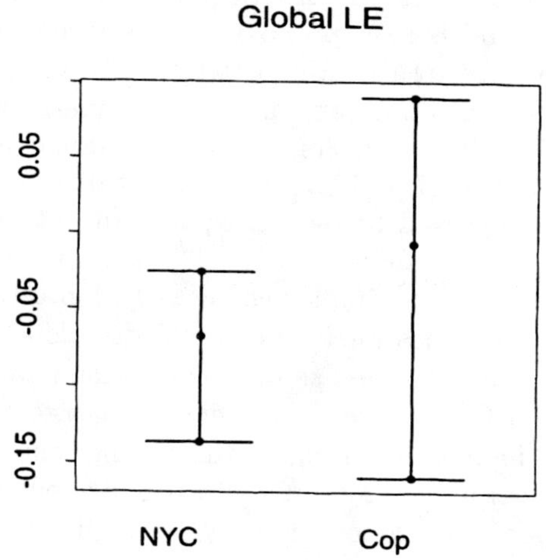

(b)

Figure 4 (a) Positive fraction of LLE4s and corresponding 95% confidence intervals for New York City and Copenhagen measles quarterly cases. (b) Global Lyapunov exponents and corresponding 95% confidence intervals for New York City and Copenhagen measles quarterly cases.

the estimated global Lyapunov exponent. The distribution of LLEs for both cities is similar.

Figure 4 shows a series of two plots to describe the dynamics of measles. Figure 4(b) is the estimated global Lyapunov exponent for each city and the corresponding 95% confidence interval. New York City has an estimated $\lambda = -0.06$ with a 95% CI (-0.13, -0.02) and Copenhagen has an estimated $\lambda = -0.006$ with a 95% CI (-0.15, 0.09). The global LE for New York City cases has a confidence interval with endpoints that are both negative, indicating that the series is not chaotic with 95% confidence. The confidence interval for the global LE for Copenhagen cases covers zero, so it is inconclusive. The time series from both cities have a λ near 0 and are thus near the transition to chaos. Investigation of the LLEs shows that both systems vary between short term stable and chaotic behavior. A four step-ahead local Lyapunov exponent (LLE4) was chosen to characterize the local behavior. The LLE4 will describe how well the system can be predicted one year ahead. Figure 4(a) show confidence intervals for the positive fraction of LLE4s. It can be seen that the confidence intervals for the positive fraction of LLE4s and consequently the negative fraction of LLE4s are both greater than zero for the Copenhagen series. This indicates that with 95% confidence there are some regions of state space where the LLE4 is positive, but there is a much larger percentile of LLE4s that is negative. The results for New York City are less conclusive. Our best estimate is that about 7% of the LLE4s are positive, but the 95% confidence interval for the positive fraction includes 0. Thus we can conclude that the Copenhagen series varies between finite time stability and chaos, (with the finite time chaos being rare), while it is possible that the New York City series has only finite time stability.

5.2 Comparison of Measles Incidence Data and Model Output. A variety of mechanistic models have proposed to describe the dynamics of measles. These range in complexity from simple deterministic models (such as the SEIR model, a system of 3 coupled ordinary differential equations) up to elaborate stochastic simulation models incorporating finite-population effects, age structure, and immigration (Anderson and May [1992], Grenfell, Bolker, and Kleczkowski [1995]). SEIR-type models have proved to be inadequate, in both deterministic and stochastic formulations (Bolker and Grenfell [1993]). One current candidate is an age-structured stochastic model, implemented as a Monte Carlo simulation with a large but finite population classified by age and disease state (Bolker and Grenfell [1993]). Here we illustrate the use of LLEs to compare this model with data on measles incidence in Copenhagen. For this comparison the model used a population of one million individuals (roughly the population of Copenhagen over the time period of the data) and was "tuned" to produce output that resembled the dynamics of the Copenhagen series. Figure 5 is a plot of the Copenhagen data (quarterly case totals) and corresponding model output.

LLEs for the two time series were estimated by fitting Equation (2.1), where f is estimated by a neural net. Figure 6 is a plot of the estimated global Lyapunov exponents and the 10th and 90th percentiles of the LLE4s with corresponding 95% confidence intervals. In Figure 6(c), the global Lyapunov exponent of the real data is $\lambda = -0.06$ with 95% CI (-0.28,-0.004) while that of the Monte Carlo model is $\lambda = -0.05$ with 95% CI (-0.16,0.03). The confidence interval for the global Lyapunov exponent of the real data is below zero indicating that the series is not chaotic. The Monte Carlo model's estimated global Lyapunov exponent is very close to that of the real data.

In Figure 6(a), the 10th percentile of the LLE4s for the real data and the model indicate that the model is capturing the smallest LLE4s. The confidence intervals for the 10th percentiles are all negative. Figure 6(b) shows that the 90th percentile confidence intervals for the data and model are positive, indicating the model is also capturing the largest LLE4s. From examining the global and the 10th and 90th percentile of the LLE4s, the Monte Carlo model seems to match the real case data very well.

Figures 7-8 are a series of three plots for the data and the model. The top plot (a) is the time series of the LLE4s. The middle plot (b) is the original time series coded to identify the times at which the LLE4 was in the top 20% of the distribution of LLE4s, i.e. indicated by a 5 overlaid on the time series. The bottom plot (c) is a scatter plot of all LLE4s plotted against the number of log cases at the time t when the LLE4s are calculated. Figures 7(b) and 7(c) show that in the real data, the largest positive LLE4s occur when there are a small number of cases, i.e. it is very hard to predict the start of an epidemic. The Monte Carlo model does not seem to capture this aspect of the dynamics very well, as seen in Figures 8(b) and 8(c). The model has the feature of unpredictability or large LLE4s when there are a large number of cases, which is not present in the real data. This can be seen in the pattern of the scatter plots of Figure 8(c).

Even though the Monte Carlo model has comparable distribution of local and global Lyapunov exponents, it does not seem to capture the positive local Lyapunov exponents or the short term unpredictability of the system when there are small numbers of infectives. The model indicates the need to investigate the critical number of infectives or susceptibles needed to start a measles outbreak and the need to look closer into the dynamics when the number of cases is small.

6 Discussion

A LLE analysis of a dynamical system provides useful information in understanding the system behavior. Measles incidence data is an example of a system where the global LE is close to zero (Ellner et al. [1996]). Even though it is possible to estimate the global LE, most likely the confidence interval will cover zero and it is not possible to make a statement about whether the system is chaotic or not. However, the examination of the behavior of LLEs can determine whether the system has short term instability. i.e. "local" chaos.

The identification of the parts of the time series that have large (or small) LLEs and consequently less (or more) predictability provides information about the heterogeneity of the system. The results of the LLE analysis of measles incidence data show large changes in dynamic structure and predictability over the course of an epidemic.

Local Lyapunov exponents are also useful as a probe into the adequacy or inadequacy of a mechanistic model. Our examination of the stochastic model for measles indicates that the model is not producing the same short term unpredictability as the data in particular regions of the state space. By "localizing" the discrepancy between model and data, these findings indicate where to focus efforts in improving the model so that it matches observed dynamics.

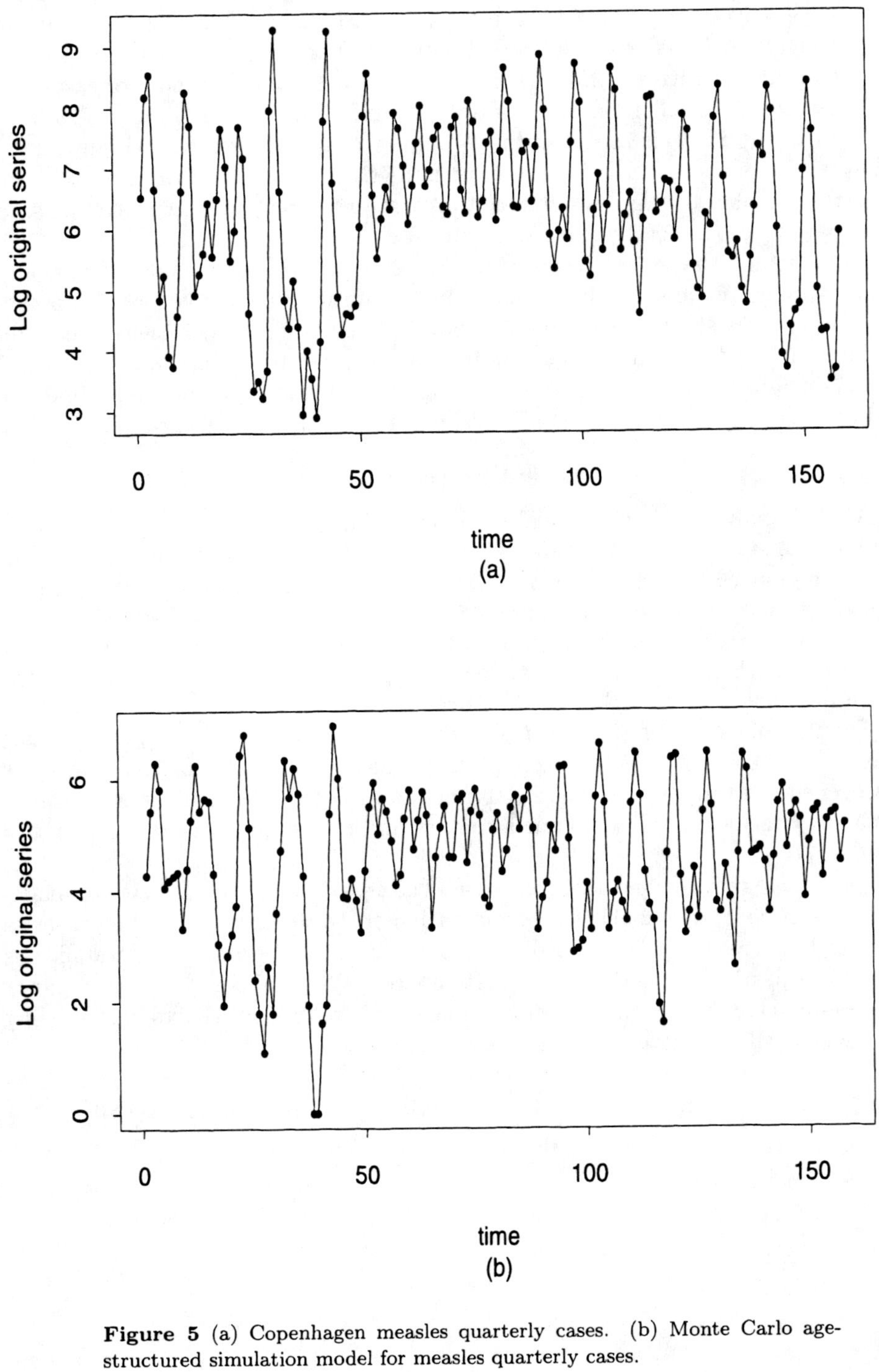

Figure 5 (a) Copenhagen measles quarterly cases. (b) Monte Carlo age-structured simulation model for measles quarterly cases.

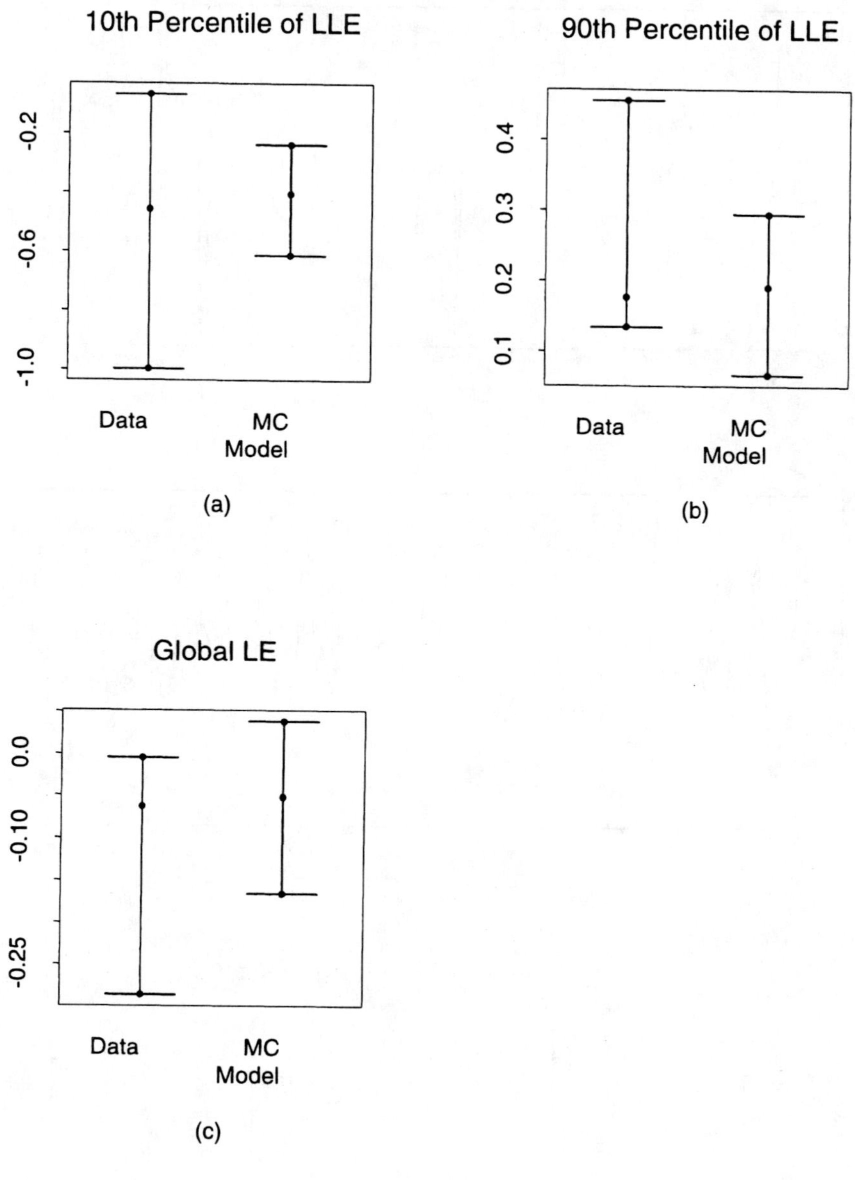

Figure 6 (a) 10th percentile of LLE4s and corresponding 95% confidence intervals for Copenhagen measles quarterly cases and Monte Carlo model output, (b) 90th percentile of LLE4s and corresponding 95% confidence intervals for Copenhagen measles quarterly cases and Monte Carlo model output, (c) Global Lyapunov exponents and corresponding 95% confidence intervals for Copenhagen measles quarterly cases and Monte Carlo model output.

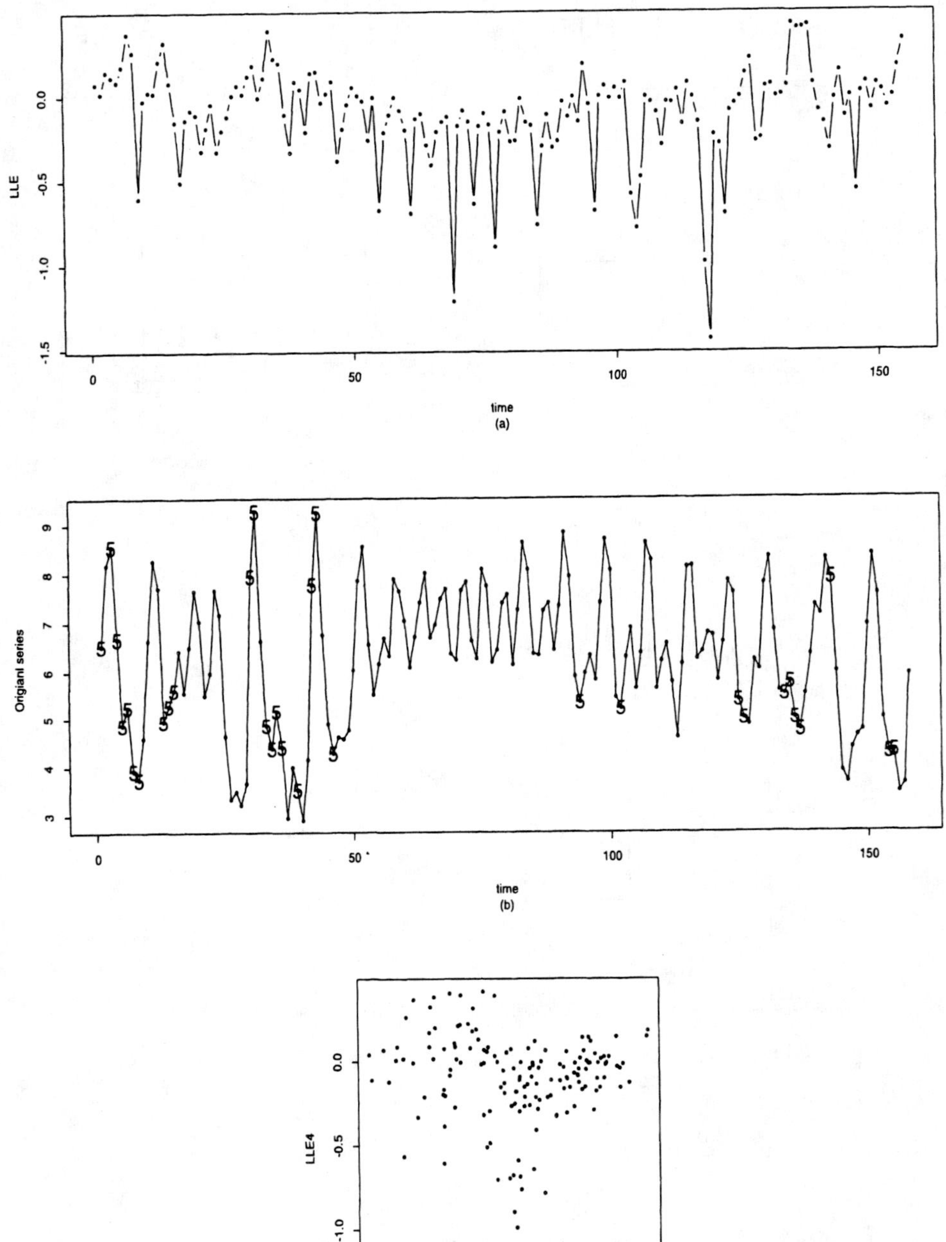

Figure 7 (a) 4-step ahead local Lyapunov exponents (LLE4s) for the Copenhagen measles quarterly cases. (b) Series coded to identify the times at which LLE4 was in top 20% of the distribution of LLE4s. (c) Log(cases) vs LLE4s for Copenhagen measles quarterly cases.

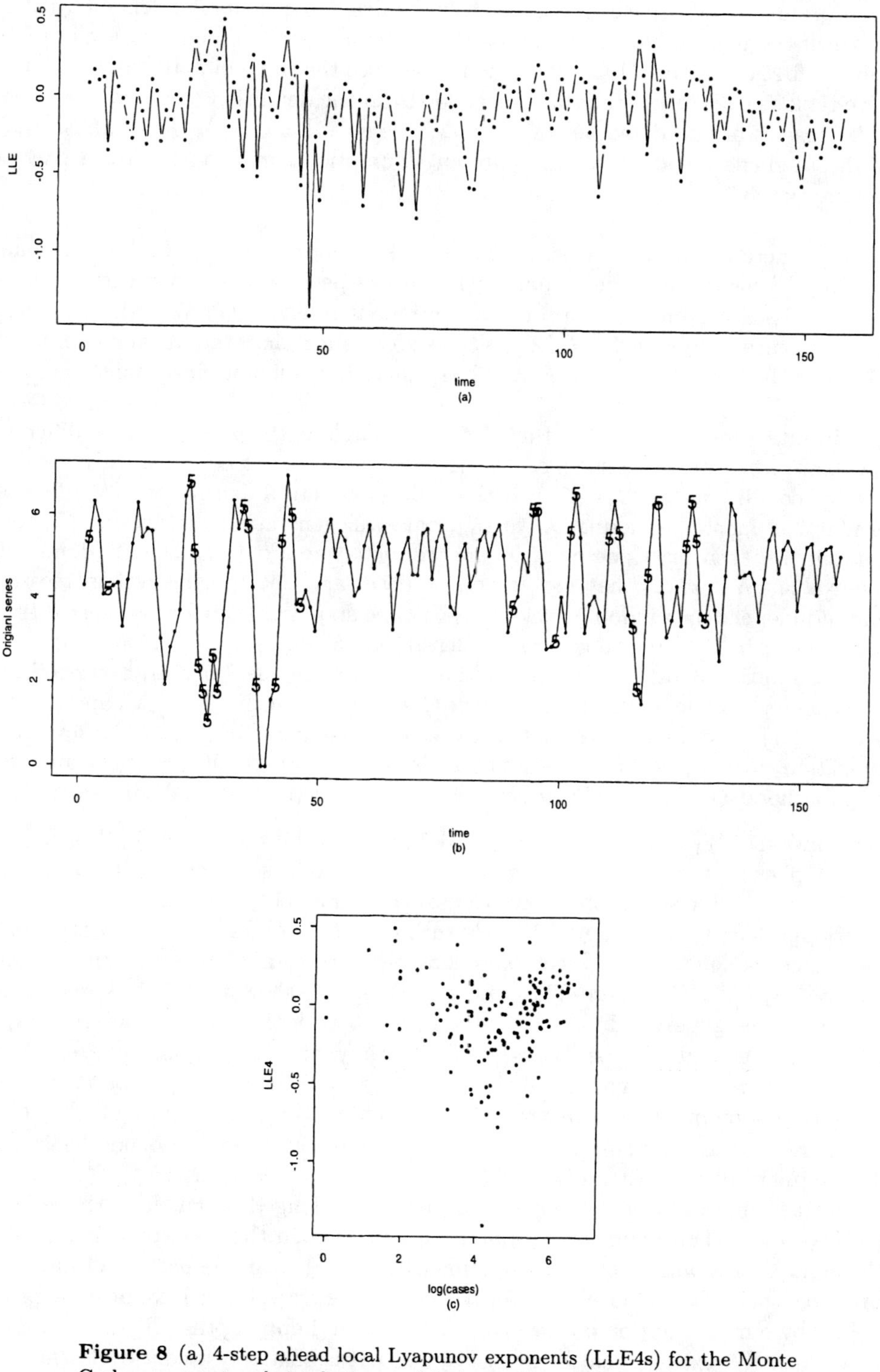

Figure 8 (a) 4-step ahead local Lyapunov exponents (LLE4s) for the Monte Carlo age-structured simulation model of measles quarterly cases. (b) Series coded to identify the times at which LLE4 was in top 20% of the distribution of LLE4s. (c) Log(cases) vs LLE4s for Monte Carlo age-structured simulation model of measles quarterly cases.

Appendix: Single Ergodic Set for State-Space Systems.

Consider a state-space system (such as the Ikeda map example) defined by (2.2) in which the noise vector e_t is i.i.d. with continuous positive density on all of $\Re^d$, and the first derivative DF is continuous. Because the value of $h(W_t)$ is unaffected if we replace U_t by $-U_t$, the state space for U_t in the definition of W_t can be taken to be the set of unordered vector pairs $\Delta = \{\{u, -u\}, u \in \Re^d\}$. To establish that W_t has a single ergodic set we require an assumption similar to the third part of Assumption 3.1:

3^*. (a) There exists $m > 0$ such that the distribution of (X_m, U_m) conditional on (X_0, U_0) has a nonsingular component with respect to Lebesgue measure whose density is positive on an open set and continuous in (X_m, U_m, X_0, U_0). (b) There is a sequence of points $\Sigma = (X_1^*, X_2^*, \cdots, X_k^*)$ such that the Jacobian product $J_k^* = DF(X_k^*)DF(X_{k-1}^*)\cdots DF(X_1^*)$ has a strictly dominant eigenvalue.

The intuitive meaning of 3^* is that different sample paths must generate different Jacobian products. The sample path $(X_1, X_2, \cdots X_{m-1})$ has a smoothly varying density conditional on (X_0, X_m); if this translates into a density on $\Re^{d\times d}$ for the product of Jacobians taking U_0 to U_m, this will generate a density for U_m and guarantee that at least one point in the product's support has a strictly dominant eigenvalue. We expect that $m \geq d$ will be necessary, because the system receives only d independent random shocks at each time step and therefore at least d time steps would be required to generate a density on $\Re^{d\times d}$.

Assumption 3^*(a) implies that W is a T-chain via the "forward accessibility" criterion for nonlinear state space models (Meyn and Tweedie [1993], Chapter 6 and Section 7.1). It will be shown that W has a reachable point W^* (for a Markov chain transition probability $P(x, A)$ a point x^* is called reachable if, for any y and any neighborhood O of x^*, $\sum_n P^n(y, O) > 0$). Therefore W is ψ-irreducible for ψ the distribution of W_1 conditional on $W_0 = W^*$ (Meyn and Tweedie [1993], Prop 6.2.1). This implies that W cannot have more than one ergodic set, since chains restricted to distinct ergodic sets cannot have the same irreducibility measure.

Suppose first that a point X^* exists such that $J^* = DF(X^*)$ has a strictly dominant eigenvalue. Let v be a corresponding eigenvector normalized to unit length, and $W^* = (X^*, U^*)$ where $U^* = \{J^*v, -J^*v\}$. To show that W^* is reachable, choose any neighborhood O of W^* and an $\varepsilon > 0$ such that O contains the ball of radius ϵ centered at W^*. Let $W_0 = (X_0, U_0)$ be any initial point, $U_0 = \{u_0, -u_0\}$. It is a generic property of matrices A that Au_0 does not lie in M, the subspace spanned by the non-dominant (generalized) eigenvectors of J^*. The support of (X_m, U_m) therefore contains a point $(\tilde{X}, \tilde{U})$ such that the vectors in $\tilde{U}$ do not lie in M. The support of the distribution of $(X_{m+1}, X_{m+2}, \cdots)$ contains $(X^*, X^*, X^*, \cdots)$ by our assumption on the error distribution, and along this sample path we have $(X, U) \rightarrow W^*$. Following this sample path long enough there exists a time τ such that (X_τ, U_τ) is within $\varepsilon/2$ of W^*. Sufficiently nearby sample paths will have the property that (X_τ, U_τ) is within ε of W^*, and these trajectories have positive probability by our assumption on the error distribution, implying that W^* is reachable. If no such X^* exists, then following sample paths that repeatedly cycle through Σ (rather than staying at X^*) shows that (X_1^*, U_1^*) is reachable, where v is an eigenvector of J_k^* normalized to unit length and $U_1^* = \{DF(X_1^*)v, -DF(X_1^*)v\}$.

References

Abarbanel, H. D. I., Brown, R., and Kennel, M. B. [1992] *Local Lyapunov exponents computed from observed data*, J. Nonlinear Sci., **2**, 343–365.

Anderson, R. M. and May, R. M. [1992] *Infectious Diseases of Humans*, Oxford University Press, Oxford.

Bailey, B. A. [1996] *Asymptotics and Applications of Local Lyapunov Exponents*, PhD Thesis, North Carolina State University, Raleigh, NC.

Bobashev, G., Ellner, S., Nychka, D. W., and Grenfell, B. [1995] *Susceptibles reconstruction from measles epidemic data*, Institute of Statistics Mimeo Series, Statistics Department, North Carolina State University, Raleigh, NC.

Bolker, B. M. and Grenfell, B. T. [1993] *Chaos and biological complexity in measles dynamics*, Proc. Roy. Soc. Lond. B, **251**, 75–81.

Doob, J. L. [1953] *Stochastic Processes*, John Wiley & Sons, Inc., New York, 221–232.

Ellner, S., Nychka, D. W., and Gallant, A. R. [1992] *LENNS, a program to estimate the dominant Lyapunov exponent of noisy nonlinear systems from time series data*, Institute of Statistics Mimeo Series, Statistics Department, North Carolina State University, Raleigh, NC.

Ellner, S. and Turchin, P. [1995] *Chaos in a noisy world: New methods and evidence from time-series analysis*, Amer. Natur., **145**, 343–375.

Ellner, S., Bailey, B. A., Bobashev, G., Gallant, A. R. Grenfell, B., and Nychka, D. W. [1996] *Noise vs. determinism in measles epidemics dynamics: estimates from nonlinear forcasting*, Manuscript.

Grenfell, B., Bolker, B., and Kleczkowski, A. [1995] *Seasonality, demography, and the dynamics of measles in developed countries* in D. Mollison (ed.) Epidemic Models: Their Structure and Relation to Data, Cambridge University Press, Cambridge.

Hall, P. and Heyde, C. C. [1980] *Martingale Limit Theory and Its Application*, Academic Press, 155–161.

McCaffrey, D., Ellner, S., Nychka, D. W., and Gallant, A. R. [1992] *Estimating the Lyapunov exponent of a chaotic system with nonlinear regression*, J. Amer. Statist. Assoc., **87**, 682–695.

Meyn, S. P. and Tweedie, R. L. [1993] *Markov Chains and Stochastic Stability*, Springer-Verlag, London.

Seber, G. A. F. and Wild, C. J. [1989] *Nonlinear Regression*, John Wiley & Sons, 228–232.

Smith, L. A. [1992] *Identification and prediction of low dimensional dynamics*, Physica D, **58**, 50–76.

Smith, L. A. [1995] *Locally optimized prediction of nonlinear systems: stochastic and deterministic*, Chaos and Forecasting (H. Tong, ed.), World Scientific, Singapore, 87–108.

Theiler, J. and Smith, L. A. [1995] *Anomalous convergence of Lyapunov exponent estimates*, Phys. Rev. E, **51**, 3738–3741.

Tong, H. [1995] *A personal overview of nonlinear time series analysis from a chaos perspective*, Scand. J. Statist., **22**, 399–445.

Wolff, R. C. L. [1992] *Local Lyapunov exponents: looking closely at chaos*, J. R. Statist. Soc. B, **54**, 353–371.

Yao, Q. and Tong, H. [1994] *Quantifying the influence of initial values on non-linear prediction*, J. R. Statist. Soc. B, **56**, 701–725.

Fields Institute Communications
Volume 11, 1997

Estimating Local Lyapunov Exponents

Zhan-Qian Lu
National Center for Atmospheric Research
P.O. Box 3000
Boulder, CO, USA 80307-3000

Richard L. Smith
Statistical Laboratory
Cambridge University, UK
and
Department of Statistics
University of North Carolina
Chapel Hill, NC, USA 27599-3260

Abstract. Lyapunov exponents are among the principal measures of the chaotic behavior of a dynamical system. However, they are difficult to estimate when the only information available is a sample of data from trajectories of the system. *Local* Lyapunov exponents are easier to estimate and, in some contexts, a more useful measure of the predictability of a system than global Lyapunov exponents. In this paper we propose a method of estimating the dynamical system and its local Lyapunov exponents using local linear and local quadratic regression. Approximations to the bias and variance of such estimators are quoted, and we illustrate the results with a number of examples.

1 Introduction

In recent years there have been many attempts to apply dynamical systems theory to understand the long-term chaotic behavior of non-linear time series. The pioneering book by Tong [1990] was followed by a number of collections of papers, such as Tong and Smith [1992], Drazin and King [1992] and Grenfell et al. [1994]; see also Isham [1993] and Jensen [1993] for reviews of this very active area.

One major theme of this research has been the attempt to estimate dynamical invariants such as the (fractal) dimension of an attractor, and entropy and information measures. This is useful for two reasons: first because they are interesting measures in their own right of the long-term behavior of a dynamical system, and second

---

1991 *Mathematics Subject Classification.* Primary 58F13, 62M20; Secondary 62G07.

This paper is based in part on the first author's PhD thesis at the University of North Carolina, Chapel Hill. The first author was supported by the NCAR Geophysical Statistics Project, sponsored by NSF grant DMS-9312686. The second author was supported in part by NSF grant DMS-9205112.

because they form the basis for tests to distinguish between chaos and randomness. Of all these measures, perhaps the most widely used is the correlation dimension, introduced by Grassberger and Procaccia [1983] and, as much as anything, popular because there is a relatively simple algorithm to estimate it, at least in its simplest form. L.A. Smith [1988], R.L. Smith [1992] and Wolff [1994] are among the authors to have studied the statistical properties of the Grassberger-Procaccia procedure and related estimators and tests. Recent work by Cutler [1993, 1994] has expanded enormously on the meaning and interpretation of the correlation dimension and other fractal dimensions, but they still lack any direct interpretation in terms of the dynamics of the system. On the other hand, *Lyapunov exponents* are much more readily interpretable: the positivity of the largest Lyapunov exponent is often taken as the definition of chaos and it provides a direct measure of the sensitivity of the system to perturbation of initial conditions. However, Lyapunov exponents have traditionally been regarded as much harder to estimate. Early attempts such as those of Wolf et. al [1985] and Eckmann et al. [1986] lacked any clear measure of statistical performance. The approach developed by McCaffrey et al. [1992] and Nychka et al. [1992] was much more statistical, being based on nonparametric regression to estimate the dynamic map followed by direct calculations on the estimated map. Nevertheless it is still hard, in their approach, to obtain analytic results for such quantities as the asymptotic bias and variance of estimators. On the other hand, *local Lyapunov exponents* – Lyapunov exponents over a finite time horizon – can be studied much more explicitly. Wolff [1992] presented one approach to local Lyapunov exponents (henceforth LLEs) in the one-dimensional case.

The main aim of the present paper is to develop the theory of LLEs, in the general multidimensional case, in conjunction with recent results on function estimation by local polynomial regression. This method of nonparametric regression has been developed a great deal in the past few years, with explicit asymptotic results being available for the bias and variance of estimators thus allowing theoretical treatment of such questions as optimal kernel bandwidth, cf. Fan [1993], Ruppert and Wand [1994].

Lu [1996] extended the work of Ruppert and Wand [1994] to consider the estimation of derivatives of multivariate functions in the standard nonparametric regression setting, in which all the observation vectors are independent, and Lu [1994] extended those results to time series with correlated observations. Developing those results, Lu [1994] showed how to obtain approximations to the bias and variance of a local Lyapunov exponent estimator based on local linear or local quadratic regression. These results are reviewed in Sections 2–4 of the present paper. In Section 5 we describe an implementation of the methodology, and in Section 6 we give some numerical and real-data examples.

2 Local Lyapunov Exponents in Stochastic Systems

The general model we consider is of the form

$$x_{t+1} = m(x_t, x_{t-1}, ..., x_{t-p+1}) + \sigma\epsilon_{t+1}, \tag{2.1}$$

where $m : \mathbb{R}^p \to \mathbb{R}$ is some nonlinear function, $\sigma \geq 0$ and $\{\epsilon_t, \ -\infty < t < \infty\}$ is a sequence of i.i.d. random variables with mean 0 and variance 1; moreover, ϵ_{t+1} is assumed independent of $\{x_s, \ s \leq t\}$. In the case $\sigma = 0$, (2.1) is often taken as a generic representation of a deterministic dynamical system in either discrete or continuous time, as a consequence of Takens' embedding theorem. See, e.g., Section 3.2.6 of Isham [1993] for one of many discussions of this. However, very few real observed time series are expected to be deterministic dynamical systems, and adding the random noise term $\sigma\epsilon_{t+1}$ is the most obvious way of introducing a random component into the model. The determination of the *embedding dimension* p is one critical decision which has itself been the focus of detailed discussion in the literature, such as Cheng and Tong [1992].

The model (2.1) may also be written in *state space form*: define X_t to be the vector $(x_{t+p-1}, x_{t+p-2}, ..., x_t)^T$ and let $\Sigma = \sigma I_p$ (I_p is the $p \times p$ identity matrix), $E_t = (\epsilon_{t+p}, 0, ..., 0)$ and

$$M(X_t) = \left(m(x_{t+p-1}, x_{t+p-2}, ..., x_t), x_{t+p-1}, ..., x_{t+1}\right)^T, \tag{2.2}$$

then

$$X_{t+1} = M(X_t) + \Sigma E_{t+1}. \tag{2.3}$$

We are interested in the question of sensitivity to initial conditions. Consider two realisations $\{X_t,\ t \geq 0\}$ and $\{X'_t,\ t \geq 0\}$ of (2.3), starting from neighboring initial values X_0, X'_0 but subject to the same noise sequence $\{E_t\}$. After ℓ steps we will have, by a first-order Taylor expansion together with the chain rule,

$$\begin{aligned} X_\ell - X'_\ell &\approx \nabla M(X_{\ell-1}) \nabla M(X_{\ell-2}) ... \nabla M(X_0)(X_0 - X'_0) \\ &= J^{(\ell)} \cdot (X_0 - X'_0) \quad \text{say} \end{aligned} \tag{2.4}$$

where $\nabla M(X)$ is the Jacobian matrix of the map M evaluated at X, and $J^{(\ell)}$ is the product of ℓ such matrices. Under very general conditions, the matrix

$$\Lambda = \lim_{\ell\to\infty} \left[\{J^{(\ell)}\}^T J^{(\ell)}\right]^{1/(2\ell)} \tag{2.5}$$

exists; moreover, the logarithms of the eigenvalues of Λ, ordered as $\lambda_1 \geq \lambda_2 \geq ... \geq \lambda_p$, are invariants of the system (not dependent on X_0 or the specific sequence of $\{E_t\}$). These are the Lyapunov exponents of the system. In particular, the largest Lyapunov exponent λ_1 is a direct measure of the exponential rate at which the trajectories $\{X_t\}$, $\{X'_t\}$ separate, and the property $\lambda_1 > 0$ is often taken as the definition of a chaotic system. The review papers of Isham [1993] and Jensen [1993] may be consulted for more details about the existence and properties of Lyapunov exponents.

This definition of the Lyapunov exponents of a stochastic system is the simplest one operationally, but it seems artificial because it introduces the rather unrealistic idea that two trajectories starting from different starting positions may nevertheless be subject to the same random shocks. In particular, this criticism is made by Tong [1995], pp. 405-6. However, a different viewpoint, also given by Tong, leads to practically the same thing. If we define $F_\ell(x) = \mathrm{E}\{X_\ell \mid X_0 = x\}$, the conditional mean of the ℓ-step prediction from a starting point x, then we may consider the effect of a small (vector) perturbation δ in x. We have

$$F_\ell(x+\delta) - F_\ell(x) = \dot{F}_\ell(x)\delta + o(||\delta||),$$

where

$$\dot{F}_\ell(x) = \mathrm{E}\left\{\prod_{j=1}^{\ell} \nabla M(X_{j-1}) \mid X_0 = x\right\}$$

(Tong [1995], p. 408). Thus the matrix $\dot{F}_\ell$ given by Tong is the same as our $J^{(\ell)}$ but with an additional "averaging over trajectories". The contrast between the two definitions is discussed by Yao and Tong [1994], who define "Lyapunov-like indices" which are not exactly the same as Lyapunov exponents but are nevertheless very similar. Our own point of view is that there are by now a number of ways of defining sensitivity to initial conditions, and ultimately one should consider all of them, but the Lyapunov exponents defined by (2.5) are a natural starting point for discussion about how one should estimate these quantities.

Nevertheless, one clear disadvantage of this definition is that, from the point of view of practical evaluation, (2.5) is difficult to handle because there is no good way to decide how large ℓ must be taken to guarantee a good approximation to Λ. In the definition of local Lyapunov exponents, we do not take the limit and work directly with

$$\Lambda^{(\ell)} = \left[\{J^{(\ell)}\}^T J^{(\ell)}\right]^{1/(2\ell)} \tag{2.6}$$

for finite ℓ; if we then denote the logarithms of the eigenvalues of $\Lambda^{(\ell)}$ as

$$\lambda_1^{(\ell)} \geq \lambda_2^{(\ell)} \geq \dots \geq \lambda_p^{(\ell)}, \tag{2.7}$$

these are called the local Lyapunov exponents of order ℓ. In contrast to the global Lyapunov exponents, they do depend on X_0 and indeed the entire trajectory $\{X_0, X_1, \dots, X_{\ell-1}\}$. However for some purposes this can be a positive feature, for example in helping to determine to what extent the short-term predictability of the system depends on the current state.

A second positive feature of local as opposed to global Lyapunov exponents is that there is a clear-cut procedure for estimating them from sample data. Indeed, if we can obtain an estimate of ∇M, then we can carry through the steps (2.4), (2.6) and (2.7) directly and so obtain estimates of the LLEs for any finite order ℓ, for any given starting point on the trajectory. The properties of the LLE estimators are then governed by those of the estimator $\widehat{\nabla M}$.

2.1 Application to the Lorenz Attractor

Figure 1 shows some plots generated from the Lorenz [1963] system. This is defined by the differential equations

$$\frac{dx(t)}{dt} = -\sigma\{x(t) - y(t)\},$$
$$\frac{dy(t)}{dt} = -x(t)y(t) + rx(t) - y(t),$$
$$\frac{dz(t)}{dt} = x(t)y(t) - bz(t),$$

with $\sigma = 10$, $b = \frac{8}{3}$, $r = 28$. A trajectory was calculated using a one-step forward Euler method with step size 0.001. The resulting system was recorded at time intervals of 0.05, and we consider LLEs corresponding to a time step of 0.1, so that $\ell = 2$ for this example. Figure 1(a) shows the time series of LLE spectra. The means of the three LLEs are 3.98, –2.11 and –15.58, which are a long way from the global Lyapunov exponents of 0.90, 0 and –14.57. There is also substantial variability in the plot, illustrating the sensitivity of the LLE to the starting position.

Figure 1(b–d) show the distribution of the largest LLE in the projected phase space (x, y), (x, z) and (y, z). We use this to classify the phase space into three

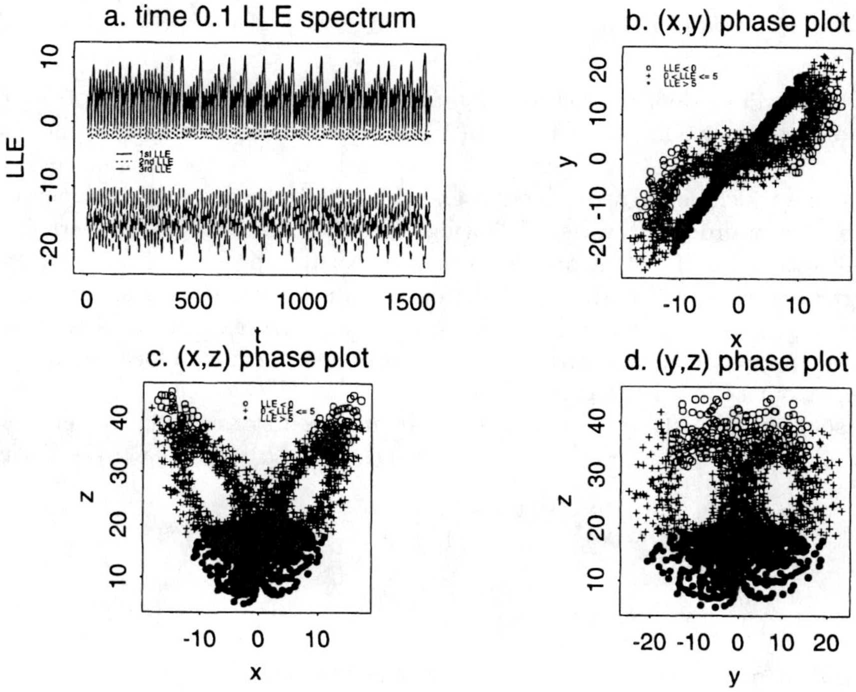

Figure 1 Local Lyapunov exponents for the Lorenz system. Plot (a) shows the two-step LLEs, $\lambda_1^{(2)}$ (top), $\lambda_2^{(2)}$ (middle) and $\lambda_3^{(2)}$ (bottom) each plotted against time t (horizontal axis). The remaining plots show three different projections of the attractor with regions of small $\lambda_1^{(2)}$ (circles), medium $\lambda_1^{(2)}$ (crosses) and high $\lambda_1^{(2)}$ (black dots) marked.

areas which are "predictable" ($\lambda_1^{(2)} < 0$, denoted by circles), "fairly unpredictable" ($0 < \lambda_1^{(2)} \leq 5$, denoted by crosses) and "very unpredictable" ($\lambda_1^{(2)} > 5$, denoted by black dots). The three regions are well separated in phase space. The predictability increases in the z direction as z increases. The most unpredictable region is in the region of low z and in the interchange of the two wings (most noticeable in the (x, y) plot). The practical implication of this result is that, for short-term forecasting, one may do better in certain regions than others. This example is, of course, deterministic, but essentially the same interpretation is made for stochastic systems.

3 Estimation by Local Polynomial Regression

Nychka et al. [1992] review a number of strategies for estimating m, concentrating on thin-plate splines and neural nets. Here we shall work with the alternative method of local polynomial regression which has been popularized in particular by the work of Fan [1993]. We shall concentrate on the case of local quadratic regression: to estimate $m(X)$ and $\nabla m(X)$ at a location $X \in \mathbb{R}^p$, based on data $X_0, ..., X_n$, choose a scalar a, a p-vector b and a $p \times p$ symmetric matrix L to minimize

$$\sum_{t=0}^{n-1} \{x_{t+1} - a - b^T(X_t - X) - (X_t - X)^T L (X_t - X)\}^2 K\left(\frac{X_t - X}{h}\right), \qquad (3.1)$$

where K is a p-dimensional kernel function and h is a scalar *bandwidth parameter.* We then take a as our point estimate of $m(X)$ and b as our estimate of the gradient $\nabla m(X)$. If we restrict $L = 0$ then (3.1) becomes local linear regression, and of course we may include higher-order powers of $X_t - X$ to get local polynomial regression of any order. However, following results in Fan [1993], Ruppert and Wand [1994] and Lu [1996], it appears to be especially appropriate to use local *linear* regression when estimating the function itself, and local *quadratic* regression for estimating first-order derivatives. Since the Lyapunov exponents depend on accurate estimation of the first-order derivatives, we concentrate on local quadratic regression as our estimation technique.

The solution of (3.1) may be expressed in matrix form as follows: let $q = (p+1)(p+2)/2$, $\mathbf{Y} = (x_1, ..., x_n)^T$, $W = \mathrm{diag}\{K((X_0 - X)/h), ..., K((X_{n-1} - X)/h)$ and $\mathbf{X}$ the $T \times q$ matrix

$$\mathbf{X} = \begin{pmatrix} 1 & (X_0 - X)^T & \mathrm{vech}^T\{(X_0 - X)(X_0 - X)^T\} \\ \vdots & \vdots & \vdots \\ 1 & (X_{n-1} - X)^T & \mathrm{vech}^T\{(X_{n-1} - X)(X_{n-1} - X)^T\} \end{pmatrix}$$

where vech^T denotes the row vector consisting of the columns on and below the diagonal of a symmetric matrix. Let $\beta = (a, b^T, \mathrm{vech}^T L)^T$. Then the problem becomes choosing β to minimize $(\mathbf{Y} - \mathbf{X}\beta)^T W (\mathbf{Y} - \mathbf{X}\beta)$ for which it is well known that the least-squares solution is

$$\widehat{\beta} = (\mathbf{X}^T W \mathbf{X})^{-1} \mathbf{X}^T W \mathbf{Y}. \qquad (3.2)$$

provided $(\mathbf{X}^T W \mathbf{X})^{-1}$ exists. From (3.2), we estimate $\widehat{m}(X) = \widehat{a}$ and $\widehat{\nabla m}(X) = \widehat{b}$ for any specified value of X. As is usual in kernel density estimation, we fix K to be some spherically symmetric kernel and concentrate our attention on the "optimal" choice of bandwidth h. Previous results by Lu [1994, 1996] have shown how to obtain bias and variance approximations for the estimates of m, ∇m and ultimately the local Lyapunov exponents derived via (2.7), as functions of the bandwidth h. In theory at least, therefore, we may specify h in such a way as to minimize the asymptotic mean squared error. We shall next summarize these results.

4 Bias and Variance Approximations

We shall consider the estimation of the ℓ-step LLE starting from a fixed value of $X \in \mathbb{R}^p$, say $X = X_{(0)}$. We assume that the first-order derivatives of m at $X_{(0)}$ and also the next $\ell - 1$ data points, $X_{(1)}, ..., X_{(\ell-1)}$, are estimated by local quadratic regression as described in Section 3. We then form an estimated matrix $\widehat{J}^{(\ell)}$ as in (2.4), and estimated LLEs $\widehat{\lambda}_1^\ell \geq ... \geq \widehat{\lambda}_1^\ell$ from (2.6) and (2.7), in each case substituting the estimates $\widehat{\nabla M}$ for ∇M. The final estimates are sensitive to the bandwidth h used in defining $\widehat{\nabla M}$. To gain some indication of suitable h values to use, we proceed by calculating approximations to the bias and variance of these estimators.

The development of bias and variance approximations proceeds in three steps. First, we consider the estimation of ∇M or equivalently ∇m (recall (2.2)). Second, we develop such results for the matrix $J^{(\ell)}$ given by (2.4). These results will in turn be applied to the estimation of the LLEs $\lambda_1^{(\ell)}, \ldots, \lambda_p^{(\ell)}$. We shall only outline the results here, concentrating on those that are directly relevant to our end objectives, and omitting all proofs. For full details the reader is referred to Lu [1994].

First, then, suppose $\nabla m(X)$ is estimated for some fixed $X \in \mathbb{R}^p$ by the procedure described in Section 3. The main assumptions we require are:

(i) m is four times continuously differentiable in a neighborhood of X,
(ii) the stationary density of X_t, denoted f, exists and is continuous, and $f(X) > 0$,
(iii) the $\{\epsilon_t\}$ are i.i.d. and $\mathrm{E}|\epsilon_t|^{2+\delta} < \infty$ for some $\delta > 0$,
(iv) the kernel K is spherically symmetric and $\int u_1^{12} K(u_1, \ldots, u_p) du_1 \ldots du_p < \infty$.

Define

$$\mu_k = \int u_1^k K(u_1, \ldots, u_p) du_1 \ldots du_p, \quad \nu_k = \int u_1^k K^2(u_1, \ldots, u_p) du_1 \ldots du_p$$

and let $b(X)$ denote the p-vector whose ith entry is

$$\mu_4 \frac{\partial^3 m}{\partial x_i^3} + 3\mu_2^2 \sum_{j \neq i} \frac{\partial^3 m}{\partial x_i \partial x_j^2}.$$

Also let A be the $p \times p$ matrix

$$A = \frac{\sigma^2 \nu_2}{\mu_2^2 f(X)} I_p.$$

Under these conditions, if $n \to \infty$, $h \to 0$ in such a way that $nh^{p+2} \to \infty$, we have that

$$n^{1/2} h^{1+p/2} \left\{ \widehat{\nabla m}(X) - \nabla m(X) - \frac{h^2}{6\mu_2} b(X) \right\} \xrightarrow{d} \mathcal{N}(0, A) \tag{4.1}$$

where $\mathcal{N}(\mu, \Sigma)$ denotes a multivariate normal distribution with mean μ and covariance matrix Σ. In terms of bias and variance considerations, (4.1) is saying that

$$\text{Bias of } \widehat{\nabla m}(X) \approx \frac{h^2}{6\mu_2} b(X),$$
$$\text{Variance of } \widehat{\nabla m}(X) \approx n^{-1} h^{-p-2} A.$$

Lu [1994] also shows that if the conditions are satisfied at k distinct points X, say $X_{(0)}, \ldots, X_{(k-1)}$, then the estimates $\widehat{\nabla m}(X_{(0)}), \ldots, \widehat{\nabla m}(X_{(k-1)})$ are asymptotically independent. In the following discussion, we are particularly interested in the case where $X_{(k)}$, $k = 0, 1, \ldots, \ell - 1$, form a time series, as outlined in the introduction to this section.

It immediately follows from (4.1) and the definition of $M(X)$ that we can write

$$n^{1/2} h^{1+p/2} \{ \widehat{\nabla M}(X_{(k)}) - \nabla M(X_{(k)}) - h^2 B_k \} \xrightarrow{d} W_k \tag{4.2}$$

where

$$B_k = \frac{1}{6\mu_2} \begin{pmatrix} b(X_{(k)})^T \\ \mathbf{0} \end{pmatrix}, \quad W_k = \frac{\sigma\sqrt{\nu_2}}{\mu_2 \sqrt{f(X_{(k)})}} \begin{pmatrix} Z_k^T \\ \mathbf{0} \end{pmatrix}.$$

Here Z_k is a p-vector of i.i.d. standard normal random variables and $\mathbf{0}$ is a $(p-1)\times p$ matrix of zeros. Moreover, $\{W_k\}$ corresponding to distinct fixed values of $\{X_{(k)}\}$ are independent.

Now we consider the second step, in which these results are extended to the estimation of $J^{(\ell)} = G_{\ell-1}...G_0$ say, where $G_k = \nabla M(X_{(k)})$.

Taking a first-order expansion using the chain rule, we have

$$\widehat{J}^{(\ell)} - J^{(\ell)} \approx \sum_{k=0}^{\ell-1} G_{\ell-1}...G_{k+1}(\widehat{G}_k - G_k)G_{k-1}...G_0.$$

Define

$$\begin{aligned} B^{(\ell)} &= \sum_{k=0}^{\ell-1} G_\ell...G_{k+1}B_kG_{k-1}...G_0, \\ W^{(\ell)} &= \sum_{k=0}^{\ell-1} G_\ell...G_{k+1}W_kG_{k-1}...G_0. \end{aligned} \tag{4.3}$$

Then it can be shown that

$$n^{1/2}h^{1+p/2}\{\widehat{J}^{(\ell)} - J^{(\ell)} - h^2B^{(\ell)} - O(h^4)\} \xrightarrow{d} W^{(\ell)}. \tag{4.4}$$

Noting that only the first rows of the matrices B_k and W_k are non-zero, the expressions for $B^{(\ell)}$ and $W^{(\ell)}$ in (4.3) can be simplified, as follows. Let $g_k^{(\ell)}$ denote the first column of the matrix $G_\ell...G_{k+1}$ and let $H_k = G_{k-1}...G_0$. Then

$$B^{(\ell)} = \frac{1}{6\mu_2}\sum_{k=0}^{\ell-1} g_k^{(\ell)} b(X_{(k)})^T H_k, \qquad W^{(\ell)} = \frac{\sigma\sqrt{\nu_2}}{\mu_2}\sum_{k=0}^{\ell-1}\frac{g_k^{(\ell)} Z_k^T H_k}{\sqrt{f(X_{(k)})}}. \tag{4.5}$$

The third step is to extend this to estimating the eigenvalues of $\{J^{(\ell)}\}^T J^{(\ell)}$. In fact it is easier to work with the *singular values* $\delta_1^{(\ell)} \geq \delta_2^{(\ell)} \geq \delta_p^{(\ell)}$ of $J^{(\ell)}$; the singular values of $J^{(\ell)}$ are the square roots of the eigenvalues of $\{J^{(\ell)}\}^T J^{(\ell)}$ so the LLEs will be given by

$$\lambda_j^{(\ell)} = \frac{1}{\ell}\log\delta_j^{(\ell)}, \quad j = 1, ..., p.$$

Our estimation procedure will be to let $\{\widehat{\delta}_j^{(\ell)},\ j = 1, ..., p\}$ denote the singular values of $\widehat{J}^{(\ell)}$ and then to define

$$\widehat{\lambda}_j^{(\ell)} = \frac{1}{\ell}\log\widehat{\delta}_j^{(\ell)}, \quad j = 1, ..., p. \tag{4.6}$$

We now need to consider the asymptotic distribution of $\{\widehat{\delta}_j^{(\ell)},\ j = 1, ..., p\}$. Lu [1994] gives the following result. Suppose X_n, B_n and W are random $p \times p$ matrices, A a fixed $p \times p$ matrix and c_n a scalar for each n. Suppose as $n \to \infty$, we have that $c_n \to \infty$, $B_n \xrightarrow{P} 0$ and $c_n(X_n - A - B_n) \xrightarrow{d} W$. For any matrix M let $\delta(M) = (\delta_1(M), ..., \delta_p(M))^T$ denote the singular values of M arranged in decreasing order. Suppose the singular values of A are all distinct and positive. We can write $A = U\Delta V^T$ where U and V are orthogonal matrices and Δ is a diagonal matrix with diagonal entries $\delta_1(A), ..., \delta_p(A)$. For any square matrix M let $\mathrm{dgv}(M)$ denote the column vector consisting of the diagonal entries of M. Then

$$c_n\left\{\delta(X_n) - \delta(A) - \mathrm{dgv}(U^TB_nV) - o_p(B_n)\right\} \xrightarrow{d} \mathrm{dgv}(U^TWV). \tag{4.7}$$

A recent paper by Eaton and Tyler [1994] has reviewed results of this nature including the case in which the singular values of A are not all distinct; this is however much harder to deal with, and we do not consider it here.

We can now see that, provided the singular values of $J^{(\ell)}$ are distinct, a combination of (4.4), (4.5) and (4.7), together with a simple delta function argument to handle the transformation (4.6), gives the result we are looking for. The final result is in the following:

Theorem 4.1 *Suppose $J^{(\ell)}$ has p distinct non-zero singular values and write its singular value decomposition in the form $J^{(\ell)} = U^{(\ell)}\Delta^{(\ell)}(V^{(\ell)})^T$ where $U_1^{(\ell)},...,U_p^{(\ell)}$ are the columns of $U^{(\ell)}$ and $V_1^{(\ell)},...,V_p^{(\ell)}$ are the columns of $V^{(\ell)}$. Then*

$$n^{1/2}h^{1+p/2}\left\{\begin{pmatrix}\widehat{\lambda}_1^{(\ell)}\\ \vdots\\ \widehat{\lambda}_p^{(\ell)}\end{pmatrix}-\begin{pmatrix}\lambda_1^{(\ell)}\\ \vdots\\ \lambda_p^{(\ell)}\end{pmatrix}\right.$$
$$\left.-\frac{h^2}{6\mu_2\ell}\begin{pmatrix}(\delta_1^{(\ell)})^{-1}\sum_{k=0}^{\ell-1}(U_1^{(\ell)})^T g_k^{(\ell)} b(X_{(k)})^T H_k V_1^{(\ell)}\\ \vdots\\ (\delta_p^{(\ell)})^{-1}\sum_{k=0}^{\ell-1}(U_p^{(\ell)})^T g_k^{(\ell)} b(X_{(k)})^T H_k V_p^{(\ell)}\end{pmatrix}-o(h^2)\right\}$$
$$\xrightarrow{d}\mathcal{N}(0,\Sigma^{(\ell)}). \tag{4.8}$$

Here the covariance matrix $\Sigma^{(\ell)}$ has entries $\sigma_{ij}^{(\ell)}$ where

$$\sigma_{ij}^{(\ell)}=\frac{\sigma^2\nu_2}{\ell^2\mu_2^2\delta_i^{(\ell)}\delta_j^{(\ell)}}\sum_{k=0}^{\ell-1}\frac{1}{f(X_{(k)})}\left\{(U_i^{(\ell)})^T g_k^{(\ell)}\right\}\left\{(U_j^{(\ell)})^T g_k^{(\ell)}\right\}(V_i^{(\ell)})^T H_k^T H_k V_j^{(\ell)}. \tag{4.9}$$

The practical interpretation of this result is that the bias of $\widehat{\lambda}_j^{(\ell)}$ is approximately

$$\frac{h^2}{6\mu_2\ell}(\delta_j^{(\ell)})^{-1}\sum_{k=0}^{\ell-1}(U_j^{(\ell)})^T g_k^{(\ell)} b(X_{(k)})^T H_k V_j^{(\ell)} \tag{4.10}$$

while its variance is approximately $n^{-1}h^{-2-p}\sigma_{jj}^{(\ell)}$.

This leads to an expression for the mean squared error (MSE=bias^2+variance) and we can choose h to minimize this, at least from a theoretical point of view, and we can also calculate the resulting MSE.

Lu [1994, 1996] also gave expressions for the bias and variance of $\widehat{\nabla M}$, and consequently the LLEs, in the case where estimation is via local linear regression instead of local quadratic regression. For estimation of M itself, rather than ∇M, we use local linear regression. In this case explicit expressions for the bias and variance of local linear regression have been given by Ruppert and Wand [1994].

5 Implementation

The kernel estimator was computed by solving the weighted least squares equation (3.2) using standard software for the solution of least squares problems. As part of the solution, this gave us the minimized weighted sum of squares

$$S=(\mathbf{Y}-\mathbf{X}\widehat{\beta})^T W(\mathbf{Y}-\mathbf{X}\widehat{\beta}). \tag{5.1}$$

For local linear regression, the method is the same, but the "vech" columns in the definition of $\mathbf{X}$ and the corresponding components of β are omitted. The kernel used here was $K(x) = C_p(1-|x|^2)$ where $|\cdot|$ denotes L_2 norm $(|(x_1,...,x_p)|^2 = \sum_1^p x_j^2)$ and C_p is chosen so that the kernel integrates to 1 (this criterion leads to $C_p = \pi^{-\frac{p}{2}}\Gamma(2+\frac{p}{2}))$.

The bias term, which depends on higher-order derivatives of m, is not directly computable except as a theoretical calculation in cases where m is known (as in the cosine map example, Section 6 below). One possible strategy is to extend the method of local polynomial regression to estimate higher-order derivatives and to use them to derive sample-based estimators of the bias. This is known to be somewhat problematic but we shall use this method in two examples in Section 6.

The variance term, however, is more directly computable from (4.9). The main difficulty with this formula is the need for a separate estimate of f, the density of data points in the assumed stationary distribution. It is possible to avoid estimating this by an alternative strategy, which we now outline.

Consider $\widehat{\beta}$ defined by (3.2). The data vector $\mathbf{Y}$ has covariance matrix $\sigma^2 I$, so the covariance matrix of $\widehat{\beta}$ is

$$\sigma^2(\mathbf{X}^T W\,\mathbf{X})^{-1}\,\mathbf{X}^T W^2\,\mathbf{X}(\mathbf{X}^T W\,\mathbf{X})^{-1}. \tag{5.2}$$

We need to estimate σ^2. However after defining S by (5.1), we can easily see that $\mathrm{E}\{S\} = K\sigma^2$, where

$$K = \operatorname{tr}\left\{W - W\,\mathbf{X}(\mathbf{X}^T W\,\mathbf{X})^{-1}\,\mathbf{X}^T W\right\}. \tag{5.3}$$

Consequently S/K is an unbiased estimator of σ^2, and by combining this with (5.2) we obtain an unbiased sample estimate of the covariance matrix of $\widehat{\beta}$, for any particular evaluation of (3.2).

Finally, by extracting the components of β which correspond to elements of ∇M, and passing these through the same operations as led to (4.9), we obtain sample-based estimators of the variances of the estimated LLEs, which avoids the direct evaluation of f.

6 Numerical Examples

Our first example is totally artificial, but is intended to illustrate one method for selecting the bandwidth h in a situation where we can calculate the "correct" answer. Then we briefly discuss two well-known data sets.

The artificial example is the cosine map, which is closely related to the better-known Hénon map. In its deterministic form, it consists of iterating the second-order nonlinear difference equation $x_{t+1} = \cos(2.8x_t) + 0.3x_{t-1}$ (Nychka et al. [1992]), and discarding initial transients. Figure 2(a) shows a plot of x_{t+1} against x_t, while Figure 2(b) shows the same plot with additive noise (as in (2.1)) with $\sigma = 0.2$. Our analysis is based on the data in Figure 2(b).

We first consider estimation of the function $m(\cdot)$, and then go on to consider the LLEs. Moreover, we do separate analyses of the full data set which consists of 10,000 data points, and a much shorter data set consisting only of the first 400 data points. This is intended to bring out the difference between large and relatively small sample sizes. Thus there are four analyses in all, and one of the features which emerges is the need for quite different bandwidths in the different cases.

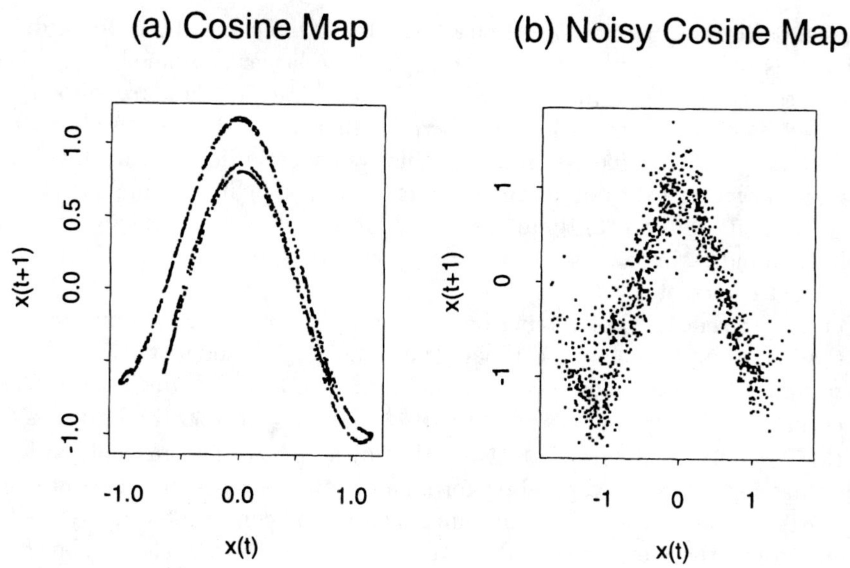

Figure 2 Data plots with and without noise

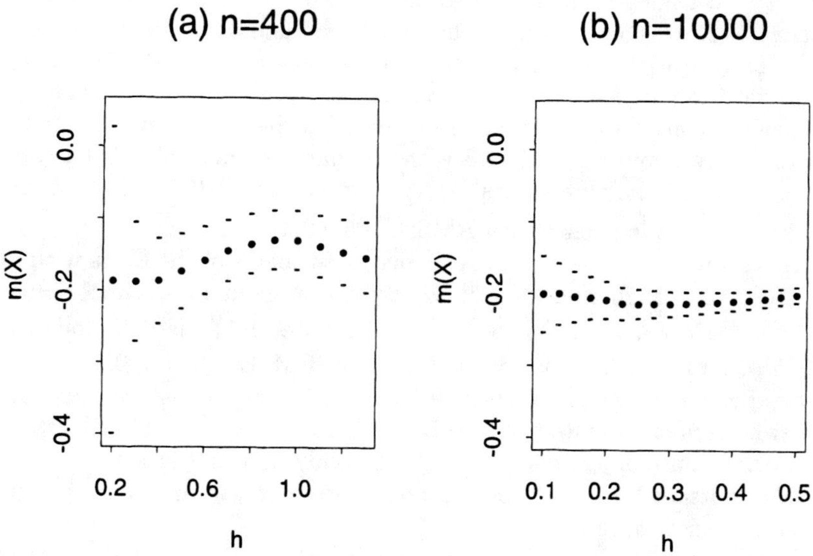

Figure 3 Estimated function values and 95% CIs

The methods are "local" in the sense that they are designed to fit the function and its LLEs in the neighborhood of a specific point, and for the present analyses we work from a single starting point, arbitrarily selected to be $x_0 = 0.74327$, $x_1 = -0.57342$. Figure 3(a) shows a series of estimates of $m(x_1, x_0)$, computed by local linear regression, for several different values of the bandwidth h, for the data set with sample size $n = 400$. Also shown are the upper and lower boundaries of the approximately 95% confidence intervals, using standard errors computed by the method described in Section 5. Figure 3(b) shows the same thing, for a different set of values of h, computed for $n = 10,000$. The larger sample size in this case allows us to use smaller values of h while still reducing the standard errors.

In each case, the standard errors increase sharply at the left hand end of the displayed range of h values, suggesting that variance is dominating bias. For h too large, however, bias dominates variance. Although it is a highly subjective judgement to pick out an "optimal" value of h from such plots, they can nevertheless be helpful in indicating a suitable range of values and the sensitivity of the estimate to the exact choice of h.

For the present example, being based on a known m, we can compute the bias term exactly, using formulae in Ruppert and Wand [1994], and so work out the exact h to minimize the MSE. For $n = 250$, the best h is 0.364, for which the root mean squared error (RMSE) is $0.038 = 0.28 \times n^{-\frac{1}{3}}$. For $n = 10,000$ the best h is 0.221, and RMSE$= 0.014 = 0.30 \times n^{-\frac{1}{3}}$. Note that although we use an analytic formula for the bias here, the corresponding formula for the variance still depends on the unknown stationary density f, and since this is not computable analytically, we continue to use the method described in Section 5 for the variances. Nevertheless the results between the two sample sizes are very consistent as reflected by the estimated constant of proportionality in the theoretical $n^{-\frac{1}{3}}$ relationship for the RMSE.

Now let us consider estimation of LLEs, for which we take $\ell = 5$. We now use local quadratic regression, and plot both the LLEs and their confidence intervals for a range of h, in Figure 4 for $n = 400$ and Figure 5 for $n = 10,000$. In Figure 4 it is rather hard to pick out an optimal value of h, since all the confidence intervals are rather wide relative to the range of estimates of the LLEs, but we might guess that the LLEs are about 0.2 and –1.3 with standard errors of about 0.08 and 0.25 respectively. In fact, the true values for $\lambda_1^{(\ell)}$ and $\lambda_2^{(\ell)}$ are 0.356 and –1.56, and the best bandwidth for $\lambda_1^{(\ell)}$ is 0.53 with RMSE .105$= 0.47 \times n^{-\frac{1}{4}}$.

Figure 5 shows a very consistent sequence of estimates of the LLEs with sample size $n = 10,000$, though even here it is hard to pick out an optimal value of h. Arbitrarily we select $h = 0.4$ for which the estimate of $\lambda_1^{(\ell)}$ is 0.31 and standard error 0.025. The best h in this case is 0.35 with RMSE=0.046$= 0.46 \times n^{-\frac{1}{4}}$. As might be expected, quoting the standard error without taking bias into account does somewhat underestimate the true RMSE.

We do not consider estimation of the second LLE in the same detail since this is less important for any "chaos" interpretation and in any case the errors of estimation are much larger.

The qualitative messages from this example are, first, that the optimal bandwidth depends on the problem being considered, being larger for smaller sample sizes, and larger for LLE estimation than for estimation of the m function itself. Second, visual inspection of the plots gives reasonable guidance to the choice of bandwidth and the error of the resulting estimates, though the error estimate is always something of an underestimate because it is based only on variance and not at all on bias.

Now we turn to two well-known real data examples, both given by Tong [1990]. Figure 6 is based on $\log_{10}$ counts of Canadian lynxes, 1821–1934; previous analyses are due to Tong [1990], Cheng and Tong [1992], amongst others. The figure shows a time series plot, phase plot, two-year LLE spectrum plot and associated confidence

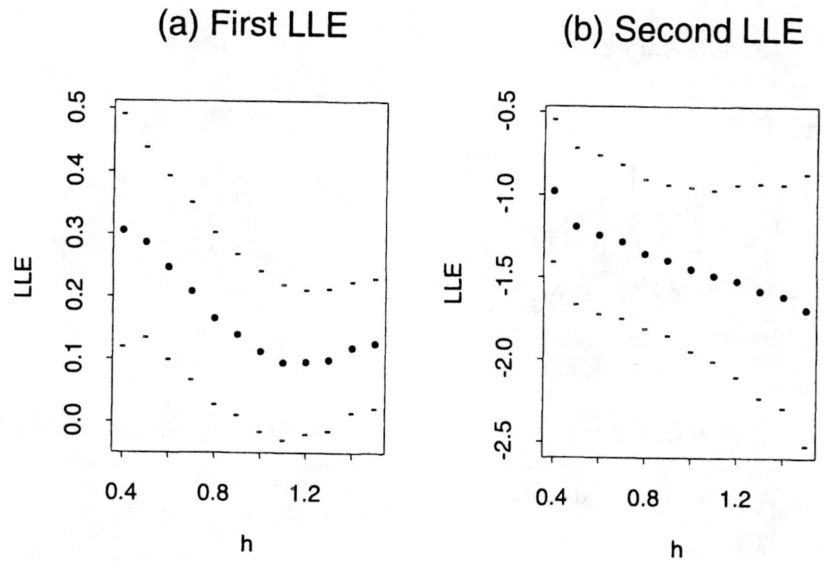

Figure 4 LLE calculations for n=400

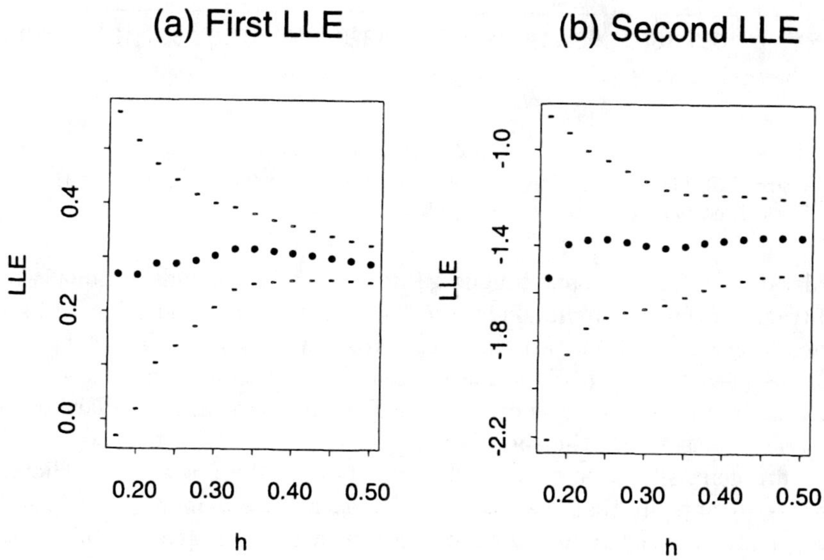

Figure 5 LLE calculations for n=10000

intervals. The LLE computations were based on embedding dimension $p = 2$. Also superimposed on the time series and phase plots are an indicator of the value of the largest LLE (large values denoted by a +, small values by o). This shows a clear subdivision of the phase space into less predictable and more predictable regions, much as we saw earlier for the theoretical Lorenz attractor. Figure 6(c) gives an indication of how both LLEs vary along the length of the series. Here the "bias-corrected point estimate" is computed by using a local cubic regression to estimate

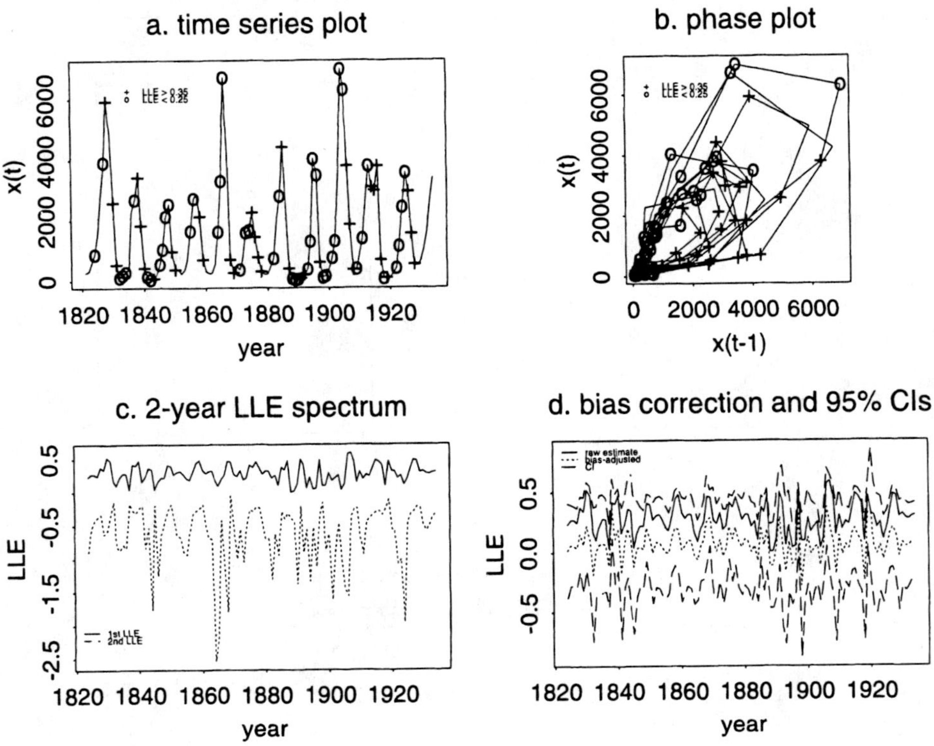

Figure 6 Annual lynx numbers. o denotes a point for which the 2-year largest LLE for $\log_{10}(\text{lynx})$ is < 0.25; + denotes a point for which it is > 0.35. The estimation parameters are $p = 2$, $h = 1.4$, $h_1 = 2$.

third-order derivatives of m, and plugging into the theoretical bias formula, which is then subtracted from the prediction interval. The upper and lower boundaries of the confidence intervals are then two standard errors either side of this. A bandwidth $h = 1.4$ was used for the LLE estimation and $h_1 = 2$ as an input to local cubic regression in computing the bias correction. The large bias and standard error are not surprising considering the short length of the series. The bias-adjusted LLE estimates are generally close to 0 which suggests that there is some predictability in the series. It appears that the rising part of each cycle is more predictable than the falling part. The asymmetry of the cycles is evidence of irreversibility, itself a well-known feature of nonlinear time series.

Figure 7 is based on similar calculations for the annual sunspot numbers, 1700–1990. In this case the bias correction has again been computed, but is much less important. The estimates of $\lambda_1^{(\ell)}$ are mostly positive, suggesting unpredictability, with greater level of unpredictability on the rising part than on the falling part of each roughly 11-year cycle. Note that this is the opposite pattern from the lynx example. The period around 1955–1961 is unusual and this has been partly compensated for by using a larger bandwidth ($h = 110$) for this portion than for the rest of the series ($h = 60$). The bandwidth for the bias correction estimate was $h_1 = 120$. The period around 1957 is much more unstable than the rest of the series and is omitted from plots c and d in view of the very large standard errors.

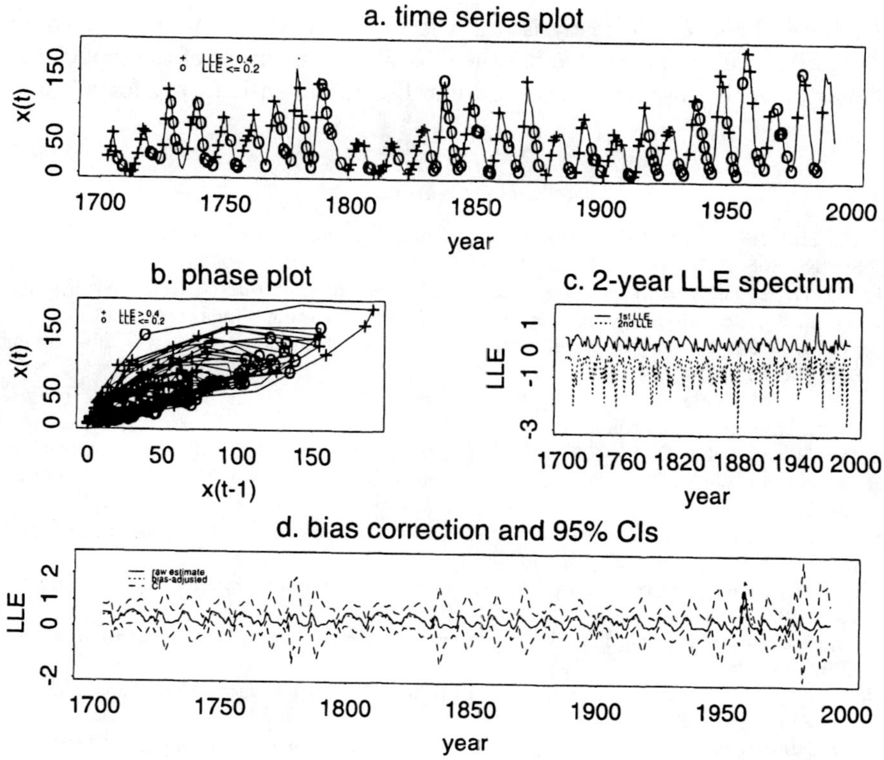

Figure 7 Annual sunspot numbers. o denotes a point for which the 2-year largest LLE is ≤ 0.2; $+$ denotes a point for which it is > 0.4. The estimation parameters are $p = 2$, $h = 80$, $h_1 = 120$.

Once again there is strong evidence of irreversibility of the time series and this is reinforced by the LLE estimates.

7 Summary and Conclusions

Local Lyapunov exponents are an important indicator of the dynamical behavior of a nonlinear system. For large ℓ they approximate the global Lyapunov exponents, and in particular the sign of the largest LLE is an indicator of chaos. For small ℓ, however, they can also be valuable, indicating more or less predictable regions of the phase space.

In this paper, we have presented one method for their estimation, via local polynomial regression, and given theoretical formulae for the bias and variance of the resulting estimators. These formulae help to determine theoretically optimal values for the bandwidth parameter h, and also the overall accuracy of the estimation procedure.

Practical experience with these methods is still very limited, but we have discussed one approach towards bandwidth selection. Of course there are a number of others which could be investigated. Preliminary analyses of two well-known time series, the lynx and sunspots data, have shown the ability of the LLE spectrum to distinguish different portions of the phase space which appear to correspond to different levels of predictability.

However there remain many issues where further work is needed. Two which may be of particular importance are the selection of the order of the model, p, and the behavior of the estimation procedure in higher dimensions, say for $p \geq 5$.

References

Cheng, B. and Tong, H. [1992] *Consistent nonparametric order determination and chaos*, J. R. Statist. Soc. B **54**, 427–449.

Cutler, C. D. [1993] *A review of the theory and estimation of fractal dimension*, Dimension Estimation and Models, edited by H. Tong, World Scientific, Singapore, pp. 1–107.

Cutler, C. D. [1994] *A theory of correlation dimension for stationary time series*, Phil. Trans. R. Soc. (Lond.) **A348**, 343–355.

Drazin, P. G. and King, G. P. [1992] *Interpretation of Time Series from Nonlinear Systems*, North-Holland, Amsterdam.

Eaton, M. L. and Tyler, D. E. [1994] *The asymptotic distribution of singular values with applications to canonical correlations and correspondence analysis*, J. Multivariate Analysis **50**, 238–264.

Eckmann, J.-P., Oliffson Kamphorst, S., Ruelle, D. and Ciliberto, S. [1986] *Lyapunov exponents from time series*, Phys. Rev. A **34**, 4971–4979.

Fan, J. [1993] *Local linear regression smoothers and their minimax efficiency*, Ann. Statist. **21**, 196–216.

Grassberger, P. and Procaccia, I. [1983] *Measuring the strangeness of strange attractors*, Physica D **9**, 189–208.

Grenfell, B. T., May, R. M. and Tong, H. (eds.) [1994] *Discussion meeting on chaos and forecasting*, Phil. Trans. R. Soc. (Lond.) **A348**, 325–530.

Isham, V. [1993] *Statistical aspects of chaos: a review*, Networks and Chaos — Statistical and Probabilistic Aspects (ed. O. Barndorff-Nielsen, J.L. Jensen and W.S. Kendall), Chapman and Hall, London, pp. 124–200.

Jensen, J. L. [1993] *Chaotic dynamical systems with a view towards statistics: a review*, Networks and Chaos — Statistical and Probabilistic Aspects (ed. O. Barndorff-Nielsen, J.L. Jensen and W.S. Kendall), Chapman and Hall, London, pp. 201–250.

Lorenz, E. N. [1963] *Deterministic nonperiodic flow*, J. Atmos. Sci. **20**, 130–141.

Lu, Z.-Q. [1994] *Statistical estimation of local Lyapunov exponents: Towards characterizing predictability in nonlinear systems*, Preprint, University of North Carolina at Chapel Hill.

Lu, Z.-Q. [1996] *Multivariate weighted polynomial fit and partial derivative estimation*, Journal of Multivariate Analysis (to appear).

McCaffrey, D., Ellner, S., Gallant, A. R. and Nychka, D. [1992] *Estimating the Lyapunov exponent of a chaotic system with nonparametric regression*, J. Amer. Statist. Assoc. **87**, 682–695.

Nychka, D., Ellner, S., Gallant, A. R. and McCaffrey, D. [1992] *Finding chaos in noisy systems*, J. R. Statist. Soc. B **54**, 399-426.

Ruppert, D. and Wand, M. [1994] *Multivariate locally weighted least squares regression*, Ann. Statist. **22**, 1346–1370.

Smith, L. A. [1988] *Intrinsic limits on dimension calculations*, Physics Letters A **133**, 283–288.

Smith, R. L. [1992] *Estimating dimension in noisy chaotic time series*, J. R. Statist. Soc. B **54**, 329–352.

Tong, H. [1990] *Non-linear Time Series: A Dynamical Systems Approach*, Oxford University Press, Oxford.

Tong, H. [1995] *A personal overview of non-linear time series analysis from a chaos perspective*, Scand. J. Statist. **22**, 399–421.

Tong, H. and Smith, R. L. (eds.) [1992] *Royal Statistical Society meeting on chaos*, J.R. Statist. Soc. B **54**, 301–474.

Wolf, A., Swift, J. B., Swinney, H. L. and Vastano, J. A. [1985] *Determining Lyapunov exponents from a time series*, Physica D **16**, 285–317.

Wolff, R. C. L. [1992] *Local Lyapunov exponents: Looking closely at chaos*, J. R. Statist. Soc. B **54**, 353–371.

Wolff, R. C. L. [1994] *Independence in time series: another look at the BDS test*, Phil. Trans. R. Soc. (Lond.) **A348**, 383–395.

Yao, Q. and Tong, H. [1994] *Quantifying the influence of initial values on nonlinear prediction*, J.R. Statist. Soc. B **56**, 701–725.

Fields Institute Communications
Volume 11, 1997

Defining and Measuring Long-Range Dependence

Peter Hall
Centre for Mathematics and its Applications
Australian National University
Canberra, ACT 0200, Australia

Abstract. The classical definition of long-range dependence of a time series is in terms of the sum of the autocovariances — if the sum diverges then the series is long-range dependent. This is particularly appropriate if one is examining the problem of estimating a mean, in which context the sample mean converges at a slower rate than $O(n^{-1/2})$, where n denotes sample size, if and only if the time series is long-range dependent. In other situations, however, the classical definition is not necessarily closely related to convergence properties, and is arguably inappropriate. An alternative prescription is to declare the time series to be long-range dependent if and only if the convergence rate of the statistic of interest is strictly slower than it would be in the context of independent data. This definition associates long-range dependence with a particular problem, as well as with a particular time series; by way of contrast, the classical definition is not problem-specific. In this note we discuss the alternative definition, and outline examples from nonparametric curve estimation where the new approach is particularly relevant. We also describe empirical methods, valid for both the classical and alternative definitions, for determining whether a time series is long- or short-range dependent.

1 Classical Definition of Long-Range Dependence

The classical definition of long-range dependence is founded on the problem of estimating the mean, μ, of a second-order stationary time series $\{X_1, X_2, \dots\}$. Assuming that the variance is finite, and writing $\gamma(i) = \text{cov}(X_j, X_{i+j})$ for the autocovariance function (representing the covariance of data values lagged i apart), the variance of the sample mean, $\bar{X} = n^{-1} \sum_{i \le n} X_i$, equals

$$E(\bar{X} - \mu)^2 = n^{-1}\gamma(0) + 2n^{-1} \sum_{i=1}^{n} \left(1 - n^{-1}i\right) \gamma(i) .$$

Therefore, if the autocovariance is ultimately of unchanging sign then $\bar{X}$ is root-n consistent for μ if and only if

$$\sum_{i=1}^{\infty} |\gamma(i)| < \infty , \tag{1.1}$$

1991 *Mathematics Subject Classification.* Primary 62M10, 58F13; Secondary 62G05.

and converges to μ at a strictly slower rate than $n^{-1/2}$ if and only if

$$\sum_{i=1}^{\infty} |\gamma(i)| = \infty . \tag{1.2}$$

Conditions (1.1) and (1.2) are the defining relations for *short- and long-range dependence,* respectively. See for example Barndorff-Nielsen and Cox [1989, p. 13]. In these terms, the process $\{X_i\}$ is short-range dependent if and only if

$$E(\bar{X} - \mu)^2 = O(n^{-1}) . \tag{1.3}$$

Arguably the simplest way to parametrize long-range dependence is through the so-called *self-similarity index,* or via the *Hurst constant,* when they exist. Suppose

$$\gamma(i) = i^{-\theta} L(i) , \tag{1.4}$$

where the function L is slowly varying at infinity, and $\theta > 0$ is termed the self-similarity index. Then the time series $\{X_i\}$ is long-range dependent if $\theta < 1$, and short-range dependent if $\theta > 1$. (The type of dependence when $\theta = 1$ is determined by properties of L.)

The Hurst constant is defined as $H = 1 - \frac{1}{2}\theta$. It takes its name from the hydrologist H.E. Hurst, who noted twenty-five years ago the importance of long-range dependence to modelling water discharge in rivers. (Sometimes a symbol other than H is used; J is employed by B.B. Mandelbrot, in recognition of the biblical account, describing Joseph's interpretation of Pharaoh's dreams, of cycles of plenty determined by the ebb and flow of the river Nile.)

Long-range dependence may alternatively be defined in terms of the spectral density,

$$f(t) = (2\pi)^{-1} \gamma(0) + \pi^{-1} \sum_{i=1}^{\infty} \gamma(i) \cos(it) .$$

In regular cases, if (1.4) holds then $f(t) = t^{\omega} M(t)$ as $t \to 0$, where $\omega = \theta - 1$ and M is slowly varying at the origin. Therefore, excepting the marginal case $\omega = 0$, which may denote either short- or long-range dependence, long-range dependence is characterized by the condition $\omega < 0$.

Section 2 will discuss nonparametric estimation of θ. A definition of long-range dependence alternative to that at (1.2), and based on the particular estimation problem as well as on the time series, will be proposed in Section 3. Sections 4 and 5 will discuss some of the implications of this view, through examples of nonparametric density estimation and nonparametric regression.

2 Estimating the Self-Similarity Index

The majority of work on estimation of θ (or, virtually equivalently, H or ω) is in the context of structural models for γ and distributional assumptions about $\{X_i\}$. For example, Fox and Taqqu [1986] and Dahlhaus [1989] discuss optimal estimation of θ under the assumption that $\{X_i\}$ is a Gaussian process. Yajima [1985, 1988] addresses a closely related problem, and also treats least-squares estimation for non-Gaussian processes. All these approaches, however, require at least parametric knowledge of L (defined at (1.4)), or of M (in the definition of ω). By way of contrast, a particularly important recent paper of Robinson [1994a] provides an estimator of θ in a nonparametric setting where L need not be known completely.

He derives consistency of his estimator, with a convergence rate of $O(n^{-\epsilon})$ for some $\epsilon > 0$. Results of Robinson [1994b] may be used to deduce more explicit rates of convergence.

It is possible, however, to estimate θ by first estimating γ and then regressing $\log|\widehat{\gamma}(i)|$ on $\log i$. An argument along these lines is outlined by Hall, Koul and Turlach [1995], and may be described as follows. Define

$$\widehat{\gamma}(i) = (n-i)^{-1} \sum_{j=1}^{n-i} (X_j - \bar{X})(X_{i+j} - \bar{X})$$

and $x_i = \log i$. We regress $\log|\widehat{\gamma}(i)|$ on x_i for i in the range $m_1 + 1 \leq i \leq m_1 + m_2$, where m_1 and m_2 may increase with n. Put

$$\bar{x}_n = m_2^{-1} \sum_{i=m_1+1}^{m_1+m_2} x_i \quad \text{and} \quad y_{ni} = x_i - \bar{x}_n \, .$$

In this notation an estimator of θ is

$$\hat{\theta} = -\left\{ \sum_{i=m_1+1}^{m_1+m_2} y_{ni} \log|\widehat{\gamma}(i)| \right\} \Big/ \left(\sum_{i=m_1+1}^{m_1+m_2} y_{ni}^2 \right).$$

Under appropriate regularity conditions on $\{X_i\}$, which are framed solely in terms of lower-order moments and are not structural, and with assumptions on L that are similar to those imposed by Robinson [1994a], exact convergence rates of $\hat{\theta}$ may be derived. In particular, if

$$L(i) = c_1\left(1 + c_2\, i^{-\alpha}\right) + o\left(i^{-\alpha}\right) \quad \text{as} \quad i \to \infty,$$

where $\alpha > 0$, $c_1 \neq 0$ and c_2 are constants, then for appropriate choice of m_1 and m_2 it may be proved that the following convergence rates of $\hat{\theta}$ to θ obtain:

$$\text{when } \theta > 1: \qquad \begin{cases} n^{-\alpha/(\alpha+2\theta)} & \text{if} \quad 0 < \alpha \leq 1 \\ n^{-\alpha/(2\alpha+2\theta-1)} & \text{if} \quad \alpha > 1, \end{cases}$$

$$\text{when } \tfrac{1}{2} < \theta < 1: \qquad \begin{cases} n^{-\alpha/(\alpha+2\theta)} & \text{if} \quad \alpha \leq 2\theta - 1 \\ n^{-\alpha/(2\alpha+1)} & \text{if} \quad \alpha > 2\theta - 1, \end{cases}$$

$$\text{when } \theta < \tfrac{1}{2}: \qquad n^{-\alpha\theta/(\alpha+\theta)}.$$

(The rates when $\theta = \frac{1}{2}$ or 1 are similar, except that they involve factors of $(\log n)^c$, where c is a constant, multiplied by the rates obtained on letting $\theta \to \frac{1}{2}$ or 1.) For example, if $\theta > 1$ and $0 < \alpha \leq 1$ then it is possible to select m_1 and m_2 such that $\hat{\theta} - \theta = O_p(n^{-\alpha/(\alpha+2\theta)})$. Thus, the optimal convergence rate depends on α, which plays a major role in determining the bias of $\hat{\theta}$, as well as on θ.

3 More Generally, How Should We Define Long-Range Dependence?

In the context of estimating μ, the formal definition (1.2) of long-range dependence may be reinterpreted as

> the time series $\{X_i\}$ is long-range dependent if and only if the mean square convergence rate of the estimator is slower than in the context of independence. (3.1)

If the time series is Gaussian then μ may be replaced by any other sample moment, without upsetting the equivalence of the two definitions. Indeed, using a Taylor-expansion argument we may show that the two definitions are equivalent if the parameter of interest may be expressed as a sufficiently regular function of a vector of moments of the marginal distribution of $\{X_i\}$, provided this time series is Gaussian.

For other time series, however, the mean square convergence rate of a sample moment higher than the first depends on more than the autocovariance, and so one must generally abandon the notion of equating the classical definition (1.2) of long-range dependence with that at (3.1). Furthermore, if the parameter of interest is not a moment or a smooth function of moments, then even under conditions of independence, convergence rates can have forms very different from those in the case of the mean. In such settings, we argue, the classical definition of long-range dependence becomes largely irrelevant to the problem at hand. We propose that the definition at (3.1) be used instead. Under this specification (as well as the classical one), a time series that is not long-range dependent would be called short-range dependent. The suggestion is motivated particularly by problems of nonparametric curve estimation, and will not be appropriate in all circumstances, for example those where convergence rates are governed by tail-weight properties.

If this course is agreed to then the problem of analysing long-range dependence becomes one of determining just what (3.1) entails in a particular problem, rather than showing (as is the case in a range of interesting problems involving time series data) that definition (1.2) means relatively little. A drawback to the new approach, however, is that the definition of long-range dependence is now phrased in terms of the problem being investigated as well as the time series generating the data. Under definition (3.1), the same time series — a Gaussian process, for example — might be long-range dependent in the context of estimating the mean, but short-range dependent in other settings, for example in the problem of nonparametrically estimating the marginal density. Formerly, under definition (1.2), a time series was intrinsically either long- or short-range dependent in *all* settings.

4 Long-Range Dependence in Problems of Nonparametric Density Estimation

In this example the definitions of long-range dependence at (1.2) and (3.1) are both applicable, but are contradictory. That at (3.1) is, we argue, more appropriate.

Assume that the marginal distribution of the time series $\{X_i\}$ is continuous with density f. A kernel estimator of f is given by

$$\hat{f}(x) = (nh)^{-1} \sum_{i=1}^{n} K\{(x - X_i)/h\},$$

where the kernel K is a compactly supported, symmetric probability density, and h is a bandwidth. Performance of $\hat{f}$ is typically described using mean integrated squared error (MISE), and we shall adopt that approach here. In the context of independence, MISE may be Taylor-expanded as

$$\begin{aligned}\mathrm{MISE}(h) &= \int E(\hat{f} - f)^2 \\ &= (nh)^{-1}\, R_1(K) + h^4\, \tfrac{1}{4}\, R_2(K)\, R_1(f'') + o\{(nh)^{-1} + h^4\}\,,\end{aligned} \tag{4.1}$$

where $R_1(K) = \int K^2$ and $R_2(K)^{1/2} = \int z^2\, K(z)\, dz$. See for example Silverman [1986, Section 3.3] or Wand and Jones [1995, Chapter 3]. Result (4.1) argues that minimum MISE is of order $n^{-4/5}$:

$$\inf_{0<h<\infty} \mathrm{MISE}(h) = O\big(n^{-4/5}\big)\,. \tag{4.2}$$

This formula may be viewed as the analogue, in the setting of density estimation, of (1.3) for point estimation of the mean. Therefore, if we were adhering to definition (3.1) of long-range dependence then we might say that *in the context of this particular estimation problem* the time series $\{X_i\}$ is long-range dependent if and only if (4.2) fails. Otherwise, it is short-range dependent.

Observe from (4.1) that the bandwidth h_0 which is optimal in the sense of minimizing MISE satisfies

$$h_0 \sim (C/n)^{1/5}\,, \tag{4.3}$$

where $C = R_1(K)/\{R_2(K)\, R_1(f'')\}$.

The problem of the rate of convergence of MISE for time series data has been investigated by, among many others, Hall, Lahiri and Truong [1995]. (There, the classical definition (1.2) of long-range dependence is employed when discussing results.) It is proved that if (a) the sequence $\{X_i\}$ is generated as a sufficiently regular function of a Gaussian process, and (b) the autocovariance satisfies $\gamma(i) \sim c\, i^{-\theta}$ for constants $c \neq 0$ and $\theta > 0$ [a slightly stronger form of (1.4)], then (4.2) fails if and only if $\theta < 4/5$. This inequality, then, is the definition of long-range dependence according to (3.1), in the context of this problem and this time series. On the other hand, according to (1.2) the time series $\{X_i\}$ is long-range dependent if and only if $\theta \leq 1$.

When $\{X_i\}$ is long-range dependent according to (3.1), we should replace (4.2) by

$$\inf_{0<h<\infty} \mathrm{MISE}(h) = O\big(n^{-\theta}\big)\,.$$

Like (4.2), this order-of-magnitude bound is best possible, in the sense that the ratio of the left- and right-hand sides is bounded away from zero as well as infinity.

Incidentally, the problem of optimal bandwidth choice assumes a form that is first-order identical to that for independent data, if and only if $\theta > 2/5$. That is, formula (4.3) for the first-order optimal bandwidth remains valid, with the same constant C, if and only if $\theta > 2/5$. When $\theta \leq 2/5$, optimal selection of h is a more complex matter involving second-order terms in an expansion of MISE.

In Hall, Lahiri and Truong [1995] these results are derived with respect to the Gaussian process $\{Y_i\}$ of which $\{X_i\}$ is a function. In particular, it is proved that (4.2) fails if and only if $\{Y_i\}$ is long-range dependent in the sense of (3.1). It may be shown, however, that under the conditions imposed by Hall, Lahiri and Truong [1995] this is generally equivalent to long-range dependence of $\{X_i\}$ in the sense of (3.1).

To further emphasize the problem-specific nature of definition (3.1) we should stress that the result that long-range dependence is equivalent to $\theta < 4/5$ depends heavily on the order of K. By assuming that K is a symmetric probability density we have effectively specified the kernel order to be 2. More generally, K might be an r'th order kernel, i.e. satisfy

$$\int x^i K(x)\,dx \quad \begin{cases} = 1 & \text{if} \quad i = 0 \\ = 0 & \text{if} \quad 1 \le i \le r-1 \\ \ne 0 & \text{if} \quad i = r, \end{cases}$$

where $r \ge 1$. In this case the time series is long-range dependent in the sense of (3.1) if and only if $\theta < 2r/(2r+1)$.

Recent, closely related work in the area includes that of Csörgő and Mielniczuk [1992], who discuss asymptotic distribution theory for kernel density estimators computed from a time series defined as a function of a Gaussian process. Castellana and Leadbetter [1986] and Castellana [1989] also consider kernel estimators under conditions that encompass long-range dependence. Hart and Vieu [1990] discuss empirical bandwidth choice in the context of short-range dependent time series.

5 Long-Range Dependence in Problems of Nonparametric Regression

In this example, definition (1.2) of long-range dependence is not really applicable, since the covariance function of the time series depends on n. Nevertheless, definition (3.1) is valid.

The time series in this example is a stationary process of errors with zero mean. Out of respect for conventional notation for errors in regression problems we denote it by $\{\epsilon_i\}$ instead of $\{X_i\}$. We suppose that the observed data $\{Y_i, 1 \le i \le n\}$ are generated by the model $Y_i = g(x_i) + \epsilon_i$, where $x_i = (i+c)/n$ for a constant c, and g is a smooth function. There is a variety of estimators of g, of which the one that we consider here is defined by local linear smoothing:

$$\hat{g}(x) = \sum_{i=1}^{n} w_i(x)\, Y_i \,,$$

where the weights w_i are given by

$$w_i(x) = v_i(x) \left\{ \sum_{j=1}^{n} v_j(x) + n^{-2} \right\}^{-1} ,$$

$v_i(x) = K\{(x - x_i)/h\}\,\{s_2 - (x - x_i)s_1\}$ and $s_k = \sum_{j \le n} K\{(x - x_j)/h\}\,(x - x_j)^k$ for $k = 1, 2$, and h is a bandwidth. (The term n^{-2} in the denominator in the definition of $w_i(x)$ is a ridge parameter, and may be selected within a very wide range without affecting our analysis.) The mean integrated squared error of $\hat{g}$ is

$$\text{MISE}(h) = \int_{\mathcal{I}} E(\hat{g} - g)^2 \,,$$

where $\mathcal{I}$ denotes the interval $(0, 1)$.

Local linear smoothing for independent errors has been discussed extensively in the literature; see Fan [1993] and Hastie and Loader [1993] for important recent contributions. Results similar to those that we shall describe may be obtained for a wide variety of alternative estimators, for example based on Nadaraya-Watson smoothing.

On the present occasion we model the time series $\{\epsilon_i\}$ by supposing that $\epsilon_i = Z(\lambda x_i)$ for $1 \le i \le n$, where Z is a second-order stationary stochastic process in the continuum. We assume that Z has zero mean and autocovariance β, and take $\lambda = \lambda(n)$ to be a sequence of positive numbers that would typically increase with n. Under this model, $E(\epsilon_i \epsilon_j) = \beta\{\lambda(x_i - x_j)\}$, and λ may be interpreted as a measure of the strength of dependence of the process $\{\epsilon_i\}$, with larger values of λ indicating weaker dependence. In particular, $\lambda = \infty$ corresponds to independence, and $\lambda/n \to \infty$ to asymptotic independence, in the sense that first-order asymptotic properties of $\hat{g}$ are identical to those under independence.

We assume that $\lambda \to \infty$ as $n \to \infty$. The classical description of dependence among errors in nonparametric regression arises when $\lambda \equiv n$, and results in so-called "time-series errors" with $E(\epsilon_i \epsilon_j) = \beta(i - j)$. This context has been extensively studied; see for example Altman [1990], Chu and Marron [1991] and Hart [1991].

It is shown by Hall, Lahiri and Polzehl [1995] that, under appropriate conditions on the second-order kernel K, the autocovariance β and the function g, which include the assumption that $\int |\beta| < \infty$ and $\int \beta \neq 0$, we have

$$\begin{aligned} \mathrm{MISE}(h) = {} & (nh)^{-1} R_1(K)\,\beta(0) + (\lambda h)^{-1} I \\ & + h^4\, \tfrac{1}{4}\, R_2(K)\, R_1(g'') + o\{(\Lambda h)^{-1} + h^4\}\,, \end{aligned} \tag{5.1}$$

where

$$I = \begin{cases} \int \beta & \text{if} \quad \lambda/n \to 0 \text{ or } \infty \\ \sum_{i \neq 0} \beta(Ci) & \text{if} \quad \lambda/n \to C \in (0, \infty), \end{cases}$$

and $\Lambda = \min(n, \lambda)$. Result (5.1) represents the analogue of (4.1) in the present setting, although unlike (4.1), (5.1) is valid for dependent data.

The mean square convergence rate in the case of independent data may be deduced by taking $\lambda = \infty$ in (5.1), and has exactly the form at (4.2). Simple calculations based on (5.1) show that this rate is achieved more generally if and only if λ/n is bounded away from zero. Thus, if we define long-range dependence by (3.1) then in the context of the problem of estimating g by $\hat{g}$, the time series $\{\epsilon_i\}$ is long-range dependent if and only if $\lambda/n \to 0$. Except in the case where λ equals a constant multiple of n, definition (1.2) of long-range dependence is not really applicable, since the autocovariance $\gamma(i) = \beta\{\lambda(n)i/n\}$ depends on n.

References

Altman, N. S. [1990] *Kernel smoothing of data with correlated errors*, J. Amer. Statist. Assoc. **85**, 749–759.

Barndorff-Nielsen, O. E. and Cox, D. R. [1989] Asymptotic Techniques for Use in Statistics, Chapman and Hall, London.

Castellana, J. V. [1989] *Integrated consistency of smoothed probability density estimators for stationary sequences*, Stoch. Proc. Appl. **33**, 335–346.

Castellana, J. V. and Leadbetter, M. R. [1986] *On smoothed probability density estimation for stationary processes*, Stoch. Proc. Appl. **21**, 179–193.

Chu, C. K. and Marron, J. S. [1991] *Comparison of two bandwidth selectors with dependent errors*, Ann. Statist. **19**, 1906–1918.

Csörgő, S. and Mielniczuk, J. [1992] *Density estimation under long-range dependence*, Manuscript.

Dahlhaus, R. [1989] *Efficient parameter estimation for self-similar processes*, Ann. Statist. **17**, 1749–1766.

Fan, J. [1993] *Local linear regression smoothers and their minimax efficiencies*, Ann. Statist. **21**, 196–216.

Fox, R. and Taqqu, M. S. [1986] *Large-sample properties of parameter estimates for strongly dependent stationary Gaussian time series*, Ann. Statist. **14**, 517–532.

Hall, P., Koul, H. A. and Turlach, B. A. [1995] *On estimating the self-similarity index for a long-range dependent process*, Manuscript.

Hall, P., Lahiri, S. N. and Polzehl, J. [1995] *On bandwidth choice in nonparametric regression with both short- and long-range dependent errors*, Ann. Statist. **23**, 1921–1936.

Hall, P., Lahiri, S. N. and Truong, Y. K. [1995] *On bandwidth choice for density estimation with dependent data*, Ann. Statist. **23**, 2241–2263.

Hart, J. D. [1991] *Kernel regression estimation with time series errors*, J. Roy. Statist. Soc. Ser B **53**, 173–187.

Hart, J. D. and Vieu, P. [1990] *Data-driven bandwidth choice for density estimation based on dependent data*, Ann. Statist. **18**, 873–890.

Hastie, T. and Loader, C. [1993] *Local regression: automatic kernel carpentry*, Statist. Science **8**, 120–143.

Robinson, P. M. [1994a] *Semiparametric analysis of long-memory time series*, Ann. Statist. **22**, 515–539.

Robinson, P. M. [1994b] *Rates of convergence and optimal spectral bandwidth for long range dependence*, Probab. Theory Relat. Fields **99**, 443–473.

Silverman, B. W. [1986] Density Estimation for Statistics and Data Analysis, Chapman and Hall, London.

Wand, M. P. and Jones, M. C. [1995] Kernel Smoothing, Chapman and Hall, London.

Yajima, Y. [1985] *On estimation of long-memory time series models*, Austral. J. Statist. **27**, 303–320.

Yajima, Y. [1988] *On estimation of a regression model with long-memory stationary errors*, Ann. Statist. **16**, 791–807.

Fields Institute Communications
Volume **11**, 1997

Modelling Nonlinearity and Long Memory in Time Series

Peter M. Robinson
and
Paolo Zaffaroni
Department of Economics
London School of Economics
London WC2 2AE, England

Abstract. We discuss models that impart a form of long memory in raw time series x_t or instantaneous functions thereof, in particular x_t^2, on the basis of a linear or nonlinear model. The capacity of linear models for x_t to imply long memory in nonlinear functions of x_t is discussed. Empirical observation motivates investigation of models which lead to a short memory (or even white noise) x_t, but a long memory x_t^2. One such model which we describe is based on the long memory generalized ARCH model introduced by Robinson [1991b]. The other is an extension of the nonlinear moving average model of Robinson [1977].

1 Introduction

Nonlinear modelling is increasingly finding use in the description and forecasting of time series arising in finance and economics, and in the physical and biological sciences. Linear time series models are appropriate in the case of Gaussian series. They also model the second-moment structure of non-Gaussian series, in the sense that Gaussian inference procedures can be asymptotically justified in many non-Gaussian environments (see e.g., Hannan [1973]). It is also possible to efficiently estimate linear models with non-Gaussian innovations (stochastic perturbations) when the latter have a parametric or even a nonparametric distribution. However, linear models can afford only a description of second moment properties or of one of the higher moment properties that is heavily dependent on the linear dynamics.

Despite the prevalence of linear modelling, methods of time series analysis motivated by possible nonlinearity have a long history. Two rival approaches to nonparametric modelling, using smoothed estimation, focus on higher-order spectra (e.g., Brillinger and Rosenblatt [1966]) on one hand, and probability densities and conditional expectations, on the other (Robinson [1983]). The properties and application

1991 *Mathematics Subject Classification.* Primary 62M10, 58F13; Secondary 62G05.
Research funded by ESRC Grant no.R000235892 and Banca d'Italia .

of a number of specific members of the limitless range of nonlinear stochastic processes have been explored, such as nonlinear functions of Gaussian processes, nonlinear autoregressive (Tong [1991]) or moving average processes (Robinson [1977]), and bilinear processes (Subba Rao and Gabr [1984]), especially in a parametric context. A popular recent theme in finance has been the interpretation of series which appear serially uncorrelated but not independent, in terms of chaos (Brock and Sayers [1988]) or autoregressive conditional heteroskedasticity (Engle [1982]).

Before employing a specific nonlinear approach, *memory* properties of the time series have to be borne in mind (as with linear analysis). For (stationary) Gaussian processes the autocorrelation function (equivalently the spectral density) unambiguously describes the memory properties. Autoregressive moving average processes can be said to have *short memory* in the sense that autocorrelations decay exponentially and the spectral density is analytic. It is convenient mathematically to adopt a much broader definition of short memory, such as that the autocovariances are absolutely summable or that the spectral density is bounded. *Long memory* processes, which violate these requirements, are increasingly of interest. Long memory typically implies that autocorrelations decay hyperbolically with exponent exceeding -1, or that the spectral density increases as a power law in the neighbourhood of a singularity, usually at the zero frequency. For a recent survey, see Robinson [1994].

For non-Gaussian processes there are no such "global" concepts of short and long memory. For linear transformations of non-Gaussian white noise it seems natural to define short memory as absolute summability of the Wold representation weights (implying summability of the autocovariances). For nonlinear processes we may be unable to analytically determine the rate of decay of autocorrelations, in which case we cannot even say whether there is short or long memory in a second order sense. On the other hand many nonlinear processes that have been studied appear to have short memory in the sense that they seem likely to satisfy some more or less strong form of mixing condition. For example many (finite order) nonlinear autoregressive schemes are of this type, in view of the Markov property. On the other hand, there has been little effort to combine nonlinearity and long memory in modelling, with such exceptions as Rosenblatt [1987], Taqqu [1987], and Robinson [1991b]. It is in any case possible that both short and long memory features can simultaneously be established, in the sense for example that, say, the raw series, or *levels*, can have short (long) memory but some instantaneous nonlinear function can have long (short) memory. In view of the vast array of possible nonlinear mechanisms, it is important to be guided by empirical observation, and financial data suggest that it is worth pursuing a version of the above theme. Not only can the levels appear uncorrelated with their squares exhibiting autocorrelation, as in the ARCH scheme, but there is a good deal of evidence that the squares have long memory autocorrelations (e.g., Ding, Granger, and Engle [1993]). In fact a generalized ARCH model which describes such behaviour had earlier been proposed by Robinson [1991b].

In the present paper we shall emphasize models with this dual character. We shall first show the way in which the coefficients of linear models determine the memory properties of the squared process, so that long memory in the levels is a precondition for long memory in the squares but does not guarantee it. We then study models with the opposite character, having white noise levels but long memory squares. One such are the generalized ARCH models first considered by Robinson [1991b]. These models can be used to test for non-ARCH against long memory ARCH and estimated by maximum likelihood (ML), but confident application of large sample inference rules is currently impeded by lack of a rigorous asymptotic distribution theory for the estimates. In any case, this is only one of many ways of describing white noise in levels and long memory in squares. Another class of models which we pursue are nonlinear moving averages. Robinson [1977] proposed a short memory nonlinear moving average model, and showed how to estimate it by ML and the method of moments (MM), as well as giving limit distribution theory for the MM estimates and some basic limit theory which is more generally relevant. We consider a long memory version, discussing its memory properties and its estimation. While limit distribution theory for ML estimates seems mathematically intractable, it seems possible to derive such a theory for Gaussian estimates based on the false assumption that the squares are Gaussian (but without of course assuming this in the theory).

2 Linear Models with Long Memory

Consider a fourth-order stationary process:

$$x_t = \sum_{j=-\infty}^{\infty} \alpha_j \epsilon_{t-j} , \tag{2.1}$$

where:

$$\sum_{j=-\infty}^{\infty} \alpha_j^2 < \infty , \tag{2.2}$$

and the ϵ_t satisfy:

$$E(\epsilon_t) = 0 , t = 0, \pm 1, \ldots \tag{2.3}$$

$$E(\epsilon_t \epsilon_s) = \begin{cases} 1, & s = t \\ 0, & s \neq t \end{cases} \tag{2.4}$$

$$E(\epsilon_s \epsilon_t \epsilon_u) = 0, \ \forall s, t, u \tag{2.5}$$

$$E(\epsilon_s \epsilon_t \epsilon_u \epsilon_v) = \begin{cases} 3 + \kappa & s = t = v = u \\ 1 & \begin{cases} s = t \neq v = u \\ s = u \neq t = v \\ s = v \neq t = u \end{cases} \\ 0 & \textit{otherwise.} \end{cases} \tag{2.6}$$

The ϵ_t behave like an independent identically distributed sequence up to fourth moments, κ denoting the fourth cumulant, which is zero in the Gaussian case. On the other hand (2.1)–(2.4) are implied if x_t is a covariance stationary process with absolutely continuous spectral distribution function, and (2.1)–(2.4) with $\alpha_j = 0$, $j < 0$, are implied if x_t is purely nondeterministic.

The lag-j autocovariance of x_t is given by

$$\gamma_j = \sum_{i=-\infty}^{\infty} \alpha_i \alpha_{i+j} , j = 0, \pm 1, \ldots , \tag{2.7}$$

and the spectral density f of x_t (if it exists) satisfies

$$\gamma_j = \int_{-\pi}^{\pi} f(\lambda) \cos(j\lambda) \, d\lambda , \ j = 0, \pm 1, \ldots . \tag{2.8}$$

The x_t process is uncorrelated if and only if $\alpha_j = 0,\ \forall j \neq 0$. On the other hand the mild ergodicity condition

$$\gamma_j \to 0\ ,\ \text{as}\ j \to \infty \tag{2.9}$$

is implied by integrability of $f(\lambda)$, in view of the Riemann-Lebesgue theorem. In particular, it is possible to contemplate γ_j that decay to zero at a variety of rates, and to choose the α_j accordingly.

One choice of particular interest is

$$\gamma_j \sim Cj^{2d-1},\ \text{as}\ j \to \infty, \tag{2.10}$$

for

$$0 < C < \infty,\ \ 0 < d < 1/2. \tag{2.11}$$

x_t is then a long memory process in the sense that while (2.9) holds,

$$\sum_{j=0}^{\infty} |\gamma_j| = \infty. \tag{2.12}$$

In the frequency domain we can consider

$$f(\lambda) \sim c\lambda^{-2d},\ \text{as}\ \lambda \to 0^+ \tag{2.13}$$

for

$$0 < c < \infty,\ \ 0 < d < 1/2, \tag{2.14}$$

so that x_t has long memory in the sense that $f(\lambda)$ is unbounded.

The properties (2.10) and (2.13) do not necessarily co-exist, but do so under suitable additional conditions (Yong [1974]), and in particular when x_t is a stationary and invertible fractionally integrated autoregressive moving average sequence (ARFIMA), given by

$$(1-L)^d a(L) x_t = b(L)\epsilon_t\ ,\ t = 0, \pm 1, \dots\ , \tag{2.15}$$

where L is the lag operator and $a(\cdot)$ and $b(\cdot)$ are finite order polynomials, all of whose zeros are outside the unit circle in the complex plane. In case of (2.15) we have

$$\alpha_j \sim K\, j^{d-1} \tag{2.16}$$

for some $K > 0$, and indeed more generally (2.16) implies (2.10).

Now consider the squared process

$$y_t = x_t^2\ , \tag{2.17}$$

and define

$$\delta_j = cov(x_t^2, x_{t+j}^2). \tag{2.18}$$

Under (2.1)–(2.6) it is easily shown that

$$\delta_j = 2\Big(\sum_{i=-\infty}^{\infty} \alpha_i \alpha_{i+j}\Big)^2 + \kappa \sum_{i=-\infty}^{\infty} \alpha_i^2 \alpha_{i+j}^2\ , \tag{2.19}$$

so that the autocovariance properties of y_t reflect κ as well as the α_j. When $\kappa = 0$, as in the Gaussian case, we have

$$\delta_j = 2\gamma_j^2 \tag{2.20}$$

so that under (2.10)

$$\delta_j \sim 2C^2 j^{4d-2}. \tag{2.21}$$

Then, within the range $0 < d < 1/2$, we have long memory in y_t if and only if there is sufficiently strong long memory in x_t, that is if $1/4 < d < 1/2$. For $0 < d < 1/4$, y_t has short memory, and the case $d = 1/4$ is indeterminate.

Similar considerations hold more generally. If $\kappa > 0$ the second term in (2.19) is non-negative so that under (2.16)

$$\delta_j \geq 2C^2 j^{4d-2} \tag{2.22}$$

for all sufficiently large j, providing a lower bound to the memory of y_t. On the other hand in general

$$\sum_{i=-\infty}^{\infty} \alpha_i^2 \alpha_{i+j}^2 = O(\max_{|i| \geq j/2} \alpha_i^2) \,, \tag{2.23}$$

applying (2.2). Under (2.16)

$$\sum_{i=-\infty}^{\infty} \alpha_i^2 \alpha_{i+j}^2 = O(j^{2d-2}) \tag{2.24}$$

so that the second term in (2.19) is dominated by the first, and the Gaussian long-run behaviour is manifested for all κ.

The above result can be compared with one for nonlinear functions of Gaussian processes (Taqqu [1975]). Consider a function $g(x_t)$ where x_t is stationary Gaussian and $Eg(x_t)^2 < \infty$. Then we can consider the Hermite expansion

$$g(x) = \beta_0 + \sum_{j=1}^{\infty} \frac{\beta_j}{j!} H_j(x) \tag{2.25}$$

where $H_j(x)$ is the jth Hermite polynomial. Taqqu [1975] defined the *Hermite rank* m of $g(\cdot)$ by

$$m = max\{i : \beta_j = 0, j < i\}. \tag{2.26}$$

Further, he showed that under (2.10) and (2.26), $g(x_t)$ has lag-j autocorrelation decaying like $j^{m(2d-1)}$. Thus $g(x_t)$ has long memory if and only if $d > 1/2 - (1/2m)$. Clearly $m = 2$ is case (2.17) so we have generalized the result for (2.17) in the Gaussian case of (2.1).

Taqqu [1975] showed that the limiting distribution of suitably normalized partial sums $\sum_{t=1}^{n} g(x_t)$ is governed by the Hermite rank m. In particular, if $m = 1$ the limiting distribution is Gaussian, whereas if $m > 1$ it is a nonstandard functional of fractional Brownian motion, varying with m. These results of Taqqu have been extended to *smoothed probability density estimation,* an important nonparametric technique for analysing possibly non-Gaussian time series. Let $p(y)$ be the probability density function of $y_t = g(x_t)$ and estimate it by

$$\hat{p}(z) = \frac{1}{nh} \sum_{t=1}^{n} k(\frac{y_t - z}{h}) \tag{2.27}$$

where the kernel $k(\cdot)$ integrates to 1, and $h = h_n$ is a bandwidth sequence tending to zero more slowly than $n \to \infty$. In case of short memory y_t, such as strongly mixing ones with a suitably decaying mixing number, and under other regularity conditions, for fixed distinct points z_i , $i = 1, \ldots, r$ the $\hat{p}(z_i)$ are known to be asymptotically normally and independently distributed (see Robinson [1983]). The limiting distributional behaviour of the $\hat{p}(z_i)$ in the long memory case was first considered by Robinson [1991a], observing that $k(\frac{y_t - z}{h})$ is an instantaneous nonlinear function of x_t if y_t is, and employing the Taqqu [1975] result for partial sums. The Robinson [1991a] result was extended by Cheng and Robinson [1994]. If $m = 1$ then the $\hat{p}(z_i)$ can be asymptotically normal, but with a singular limiting covariance matrix (of rank 1 for all r), in contrast to the short memory case, while for $m > 1$ the limiting distribution is nonstandard.

While probability density estimates contain information and can be extended to describe aspects of the joint distribution of finitely many y_t, they cannot satisfactorily describe the intrinsically infinite-dimensional long memory nonlinear dynamics, and so we turn to the estimation of explicitly nonlinear long memory models.

3 Long Memory Generalized Arch Models

We consider the modelling of processes x_t having zero autocorrelations but long memory in x_t^2, which, as we saw, linear models are incapable of describing. Denote by $\mathcal{F}_t$ the σ-field of events generated by $\{x_s : s \leq t\}$. We assume that

$$E(x_t|\mathcal{F}_{t-1}) = 0 \ a.s. \tag{3.1}$$

and that $\tau_t^2 = V(x_t|\mathcal{F}_{t-1})$ is not necessarily constant (as would be implied by (2.1)–(2.4)) but

$$\tau_t^2 = \tau^2 + \sum_{j=1}^{\infty} \psi_j (x_{t-j}^2 - \tau^2) \ a.s., \tag{3.2}$$

for some $\tau^2 > 0$. When $\psi_j = 0$ for all $j > p$, this becomes the ARCH(p) model of Engle [1982]. A special case with exponentially decaying ψ_j is the GARCH(p,q) model of Bollerslev [1986]. Both these models imply autocorrelation but short memory in x_t^2.

Robinson [1991b] proposed ψ_j in (3.2) which imply long memory in x_t^2. Write $\psi(L) = 1 - \sum_{j=1}^{\infty} \psi_j L^j$, so that

$$\tau_t^2 = \phi + (1 - \psi(L))x_t^2 \tag{3.3}$$

where

$$\phi = \psi(1)\tau^2. \tag{3.4}$$

We can rewrite this as

$$\psi(L)x_t^2 = \phi + \nu_t \tag{3.5}$$

where the $\nu_t = x_t^2 - \tau_t^2$ are martingale differences, $E(\nu_t|\mathcal{F}_{t-1}) = 0 \ a.s.$ and thus uncorrelated (assuming x_t has finite fourth moments). It follows that the x_t^2 have spectral density

$$f_2(\lambda) \propto |\psi(e^{i\lambda})|^{-2}, \ -\pi < \lambda \leq \pi \, . \tag{3.6}$$

Thus if

$$|\psi(e^{i\lambda})| \sim c_1 \lambda^d \ , \ as \ \lambda \to 0^+ \tag{3.7}$$

for $c_1 \neq 0$ and d satisfying (2.14), it follows that x_t^2 has long memory in the sense indicated in (2.13).

As a special case take

$$\psi(L) = (1-L)^d \frac{a(L)}{b(L)} \tag{3.8}$$

with $a(\cdot)$ and $b(\cdot)$ as defined in (2.15); then x_t^2 has the autocorrelation structure of an ARFIMA. Note that this implies $\phi = 0$. Robinson [1991b] employed the general model (3.1)–(3.2), with arbitrary finite parameterization, as a class of alternatives in deriving score tests of the no-ARCH hypothesis $\psi_j = 0, \ \forall j \geq 1$, stressing the special case of (3.8) with $a(L) = b(L) \equiv 1$ and the case when the ψ_j are AR coefficients of a "fractional noise" process with lag-j autocovariance $\frac{1}{2}\{|j-1|^{2d+1} - 2|j|^{2d+1} + |j+1|^{2d+1}\}$. Subsequently, Baillie, Bollerslev, and Mikkelsen [1994] considered the special case (3.8) of (3.2) but with ϕ a free parameter and not given by (3.4). When such a ϕ is non-zero, it follows that x_t^2 is not covariance stationary for any $d > 0$, due to the analogy with adding a drift to (2.15),

$$(1-L)^d a(L) x_t = \mu + b(L)\epsilon_t \tag{3.9}$$

where x_t is not covariance stationary for any $d > 0$ when $\mu \neq 0$. Nelson [1990] showed that $\tau_t^2 \to 0 \ a.s.$ when $\phi = 0$ in the short memory case $d = 0$ in (3.8).

We can describe a data generating process with properties (3.1) and (3.3) by

$$x_t = \epsilon_t \tau_t \tag{3.10}$$

where the $\epsilon_t | \mathcal{F}_{t-1}$ are independently distributed as

$$\epsilon_t | \mathcal{F}_{t-1} \sim N(0,1). \tag{3.11}$$

With a finite parameterization of the ψ_j, $\psi_j = \psi_j(\theta)$ for a p-dimensional vector θ, we can then estimate θ by maximum likelihood, as has been done with ARCH and GARCH models. However, irrespective of whether ϕ is given by (3.4) or as a free non-zero parameter, no rigorous asymptotic distribution theory for the maximum likelihood estimates seems to exist; even in the non-fractional cases $d = 0$ and $d = 1$ in (3.2)–(3.8), the asymptotics are known only for very special $a(\cdot)$ and $b(\cdot)$, and the techniques used to establish this do not seem to extend readily to the fractional case.

4 Long Memory Moving Average Models

A very wide class of $\mathcal{F}_{t-1}$-measurable functions τ_t^2 in (3.10) can give rise under (3.11) to x_t that are uncorrelated but not independent, and a number of these can imply long memory in x_t^2. The model (3.1) can be thought of as an (infinite-order) nonlinear AR,

$$x_t / \tau_t = \epsilon_t \ . \tag{4.1}$$

Alternatively we can think in terms of a nonlinear MA model

$$x_t = \epsilon_t h_t \ , \tag{4.2}$$

where $h_t = h(\epsilon_{t-1}, \epsilon_{t-2}, \dots)$. We can think of (4.2) as inverse to (4.1) if there is a well-defined solution of (4.1) for x_t of the form $x_t = g(\epsilon_t, \epsilon_{t-1} \dots)$, so that

$$\tau_t = \tau(x_{t-1}, x_{t-2}, \dots) = \tau(g(\epsilon_{t-1}, \epsilon_{t-2}, \dots), g(\epsilon_{t-2}, \epsilon_{t-3}, \dots), \dots) = h_t. \tag{4.3}$$

Commencing from the form (4.2) rather than (4.1) we have to choose a functional form for the h_t. In fact nonlinear MA models were considered by Robinson [1977], in particular he extended the linear MA(1) to

$$x_t = \epsilon_t + b\epsilon_{t-1} + c\epsilon_t\epsilon_{t-1}. \tag{4.4}$$

Robinson [1977] proposed estimates of b and c and established consistency and asymptotic normality, giving also limit theorems applicable to more general nonlinear MA models than (4.4). When $b = 0$ the x_t in (4.4) are uncorrelated but, when $c \neq 0$, not independent (in particular x_t^2 is autocorrelated) and we have a special case of (4.2) with

$$h_t = 1 + c\epsilon_{t-1}. \tag{4.5}$$

However, no instantaneous functions of such x_t have long memory, x_t being m-dependent.

We can introduce long memory, in particular in x_t^2, by allowing for infinitely many lagged ϵ_t. The algebraically simplest way of doing so is to extend the linear form of (4.5) to

$$h_t = \beta + \sum_{j=1}^{\infty} \alpha_j \epsilon_{t-j} \tag{4.6}$$

where (2.2) holds.

The question of whether or not β is zero significantly affects the autocorrelation structure of x_t^2. Suppose that $\beta = 0$. Under conditions (2.2)–(2.6)

$$\delta_j = cov(x_t^2, x_{t+j}^2) = 2\left[\alpha_j^2 \sum_{i=1}^{\infty} \alpha_i^2 + (\sum_{i=1}^{\infty} \alpha_i\alpha_{i+j})^2\right] \tag{4.7}$$

$$+ \kappa\left[\alpha_j^2 \sum_{i=1}^{\infty} \alpha_i^2 + \sum_{i=1}^{\infty} \alpha_i^2\alpha_{i+j}^2\right]. \tag{4.8}$$

So far as long memory is concerned, it is the behaviour of this as $j \to \infty$ that is of interest. Let us first dispose of (4.8). Suppose, (cf. (2.16)) that

$$\alpha_j = O(j^{d-1}), \text{ as } j \to \infty, \; 0 < d < \frac{1}{2}. \tag{4.9}$$

Then it immediately follows that (4.8) is $O(j^{2d-2})$ so (4.8) is summable. Thus, (4.8) cannot be a source of long memory, so from this point of view there is no distinction between the case of normal ($\kappa = 0$) or non-normal ϵ_t. Moreover, the first component of (4.7) is also $O(j^{2d-2})$ under (4.9). The other component of (4.7) can be written as $2\gamma_j^2$ where $\gamma_j = \sum_{j=1}^{\infty} \alpha_i\alpha_{i+j}$ is the autocovariance function of h_t in (4.6). Assuming that h_t is a long memory process with γ_j given as in (2.21) it follows that under (2.10)

$$\delta_j \sim 2C^2 j^{4d-2}, \quad \text{as } j \to \infty. \tag{4.10}$$

Of course a sufficient condition that both (2.10) and (4.9) hold is (2.16). The exponent in (4.10) can take any value in $(-1, 0)$ by choosing $1/4 < d < 1/2$, so that a sufficiently strong degree of long memory in h_t bestows long memory in x_t^2. Note the similarity with our discussion of the squares of a linear process. However, in that case x_t was itself autocorrelated whereas, under (4.2), x_t is not.

It is possible to estimate d in (4.10) by applying one of several "semiparametric" methods to the x_t^2 sequence, these being based on either sample autocovariances at long lags or periodograms at low frequencies. Some of these estimates have been shown to be consistent and asymptotically normal, so that approximate statistical inference can be carried out (see e.g. Robinson [1995a,b]). Such methods have the advantage of not requiring a finite parameterization of the α_j. However given n observations these estimates are less-than-$n^{1/2}$-consistent, whereas if the α_j are finitely parametrized, $\alpha_j = \alpha_j(\theta)$ for a p-dimensional θ that includes d, we can $n^{1/2}$-consistently estimate θ. Obvious parameterizations of the α_j are obtained by taking them to be the MA coefficients (apart from the zeroth) of the ARFIMA (2.15), or algebraically simpler ones such as $\alpha_j = j^{d-1}$ (cf. (2.16)).

Assuming also that the ϵ_t are normal, as in (3.11), it is possible to consider ML estimates of θ. However, as indicated by Robinson [1977] in case (4.4), the likelihood is rather intractable, in the sense that it can only be computed recursively and, more importantly, it seems very hard to handle mathematically in order to develop asymptotic statistical properties. It seems that these may depend on invertibility of the model (4.2)–(4.6) (cf. our discussion of (4.1) and (4.2)). Thus Robinson [1977] proposed MM estimation of (4.4), based on second and third order lagged moments of x_t, whose asymptotic distributional properties he was able to establish. In the present infinite-order nonlinear MA, however, there is considerable ambiguity as to which moments to use and it seems appropriate to use information on long lags. Accordingly, we propose pseudo-ML estimates (or approximations thereof) under the pretense that $(x_1^2, \dots x_n^2)$ is Gaussian. As $n \to \infty$ such estimates will use information in δ_j for all j. Moreover, while the conditions of Fox and Taqqu [1986] and Giraitis and Surgailis [1990] for asymptotic theory of Gaussian estimates of long memory models do not hold for the x_t^2, it seems a relatively straightforward, if very lengthy, task to establish $n^{1/2}$-consistency and asymptotic normality under (4.2) and (4.6), given this and other earlier literature on Gaussian estimation.

References

Baillie, R. T., Bollerslev, T. and Mikkelsen, H. O. A. [1994] *Fractionally integrated generalized autoregressive conditional heteroskedasticity*, preprint.

Bollerslev, T. [1986] *Generalized autoregressive conditional heteroskedasticity*, Journal of Econometrics, **31**, 302–327.

Brillinger, D. R. and Rosenblatt, M. [1966] *Asymptotic theory of higher-order spectra*, in Spectral Analysis of Time Series, (B. Harris, ed.), Wiley, New York, 189–232.

Brock, W. A. and Sayers, C. L. [1988] *Is the business cycle characterized by deterministic chaos?*, Journal of Monetary Economics, **22**, 71–90.

Cheng, B. and Robinson, P. M. [1994] *Semiparametric estimation from time series with long-range dependence*, Journal of Econometrics, **64**, 335–353.

Ding, Z., Granger, C. W. J. and Engle, R. F. [1993] *A long memory property of stock market and a new model*, Journal of Empirical Finance, **1**, 83–106.

Engle, R. F. [1982] *Autoregressive conditional heteroskedasticity with estimates of the variance of the United Kingdom*, Econometrica, **50**, 987–1007.

Fox, R. and Taqqu, M. [1986] *Large sample properties of parameter estimates for strongly dependent staionary Gaussian time series*, Annals of Statistics, **14**, 517–532.

Giraitis, L. and Surgailis, D. [1990] *A central limit theorem for quadratic forms in strongly dependent linear variablse and its application to asymptotically normality of Whittle's estimate*, Probability Theory & Rel.Fds., **86**, 105–129.

Hannan, E. J. [1973] *The asymptotic theory of linear time series models*, Journal of Applied Probabability, **10**, 130–145.

Nelson, D. B. [1990] *Stationarity and persistence in the GARCH(1,1) model*, Econometric Theory, **6**, 318–334.

Robinson, P. M. [1977] *The estimation of a nonlinear moving average model*, Stochastic Processes and their Applications, **5**, 81–90.

Robinson, P. M. [1983] *Nonparametric estimators for time series*, Journal of Time Series Analysis, **4**, 186–207.

Robinson, P. M. [1991a] *Nonparametric function estimation for long-memory time series*, in Semiparametric and Nonparametric Methods in Statistics and Econometrics, (W. Barnett, et al. ed.), Cambridge University Press, 189–232.

Robinson, P. M. [1991b] *Testing for strong serial correlation and dynamic conditional heteroskedasticity in multiple regression*, Journal of Econometrics, **47**, 67–84.

Robinson, P. M. [1994] *Time series with strong dependence*, in Advances in Econometrics Sixth World Congress vol.1, (C. Sims, ed.) Cambridge University Press, Cambridge, 47–96.

Robinson, P. M. [1995a] *Gaussian semiparametric estimation of long range dependence*, Forthcoming Annals of Statistics.

Robinson, P. M. [1995b] *Log-periodogram regression of time series with long-range dependence*, Annals of Statistics, **23**, 1048–1072.

Rosenblatt, M. [1987] *Scale renormalization and random solutions of the Burgers equation*, Journal of Applied Probability, **24**, 328–338.

Subba Rao, T. and Gabr, M. [1984], *Introduction to Bispectral Analysis and Bilinear Time Series Models*, vol. 24 of Lecture Notes in Statistics, Springer Verlag, New York.

Taqqu, M. [1975] *Weak convergence to fractional Brownian motion and the Rosenblatt process*, Z.Wahrscheinlichkeitstheor. Verw. Geb., **31**, 287–302.

Taqqu, M. [1987] *Random Processes with long-range dependence and high variability*, J. Geophysical Res., **92**, 9683–9986.

Tong, H. [1991] *Non-Linear Time Series: a Dynamical System Approach*, Oxford University Press.

Yong, C. H. [1974] *Asymptotic Behaviour of Trigonometric Series*, Chinese University of Hong Kong.

Fields Institute Communications
Volume 11, 1997

Ergodic Distributions of Random Dynamical Systems

L. Mark Berliner
National Center for Atmospheric Research
NCAR Geophysical Statistics Project
P. O. Box 3000
Boulder, Colorado, USA 80307-3000
and
Ohio State University
Columbus, Ohio, USA 43210-1247

Steven N. MacEachern
Department of Statistics
Ohio State University
1958 Neil Avenue
Columbus, Ohio, USA 43210-1247

Catherine Scipione Forbes
Department of Econometrics
Monash University
Clayton, VIC 3168, Australia

Abstract. Typical analyses of data obtained from randomly perturbed dynamical systems rely on the assumption of a stationary ergodic distribution. Though rich theoretical results concerning existence are available, simulation-based approaches are usually used to determine whether a particular system is ergodic. In this note; we investigate a family of systems and analytically characterize which members of the family are ergodic. We also propose and implement a non-simulation based computational strategy for the investigation of ergodicity of more general systems.

1 Introduction

Among the crucial impacts of chaos theory on scientific modeling is the realization that purely deterministic, physically motivated, nonlinear dynamical systems can produce behavior as complex as that associated with traditional randomness. Hence, statistical properties of even deterministic models are of interest. However, an increasingly popular model is a stochastic difference equation for successive values of the modeled process, denoted x_t:

$$x_{t+1} = f(x_t) + h(x_t, e_{t+1}). \tag{1.1}$$

1991 *Mathematics Subject Classification.* Primary 58F13, 60J20; Secondary 62M10.
The research of the first author was supported in part by NSF Grant DMS9312686.

In this representation the function f is viewed as a "primary" driving function of the dynamics of the x-process. To account for a variety of sources of uncertainty (e.g. external "noise"), the dynamics of the system are corrupted by a possibly state dependent error modeled by $h(\cdot,\cdot)$ where $\{e_t\}$ is a stochastic noise process. Such models are also of interest in accounting for numerical errors arising in computer models of deterministic systems; for examples, see Diamond, Kloeden, and Pokrovskii [1994] and Hammel, Yorke, and Grebogi [1987]. Typically, the e_t's are assumed to be independent, identically distributed (*i.i.d.*) random variables. (This assumption is questionable in the numerical error analysis setting.) A common special case of (1.1) is

$$x_{t+1} = f(x_t) + e_{t+1}. \tag{1.2}$$

The problem of establishing conditions under which the x-process is stationary and ergodic has been fundamental in the theory of dynamical systems. In particular, it is of special interest to study the "noise tends to zero" limits of these ergodic distributions. See Ruelle [1981, 1989] and Kifer [1988]. In addition to the "small noise" limiting arguments, other reasons for consideration of stationarity arise. Much of the statistical literature involves data analysis based on an observed time series of x's assumed to arise from (1.2). For recent examples and further discussion, particularly in the context of chaos, see Berliner [1992], Casdagli, des Jardins, Eubank, Farmer, Gibson, Theiler, and Hunter [1992], Chan and Tong [1994], and Nychka, Ellner, Gallant, and McCaffrey [1992]. A goal in some analyses is to learn about the dynamics of the process without making strong assumptions about the driving function f. To proceed, however, many analyses involve the assumption that the x-process is a stationary, ergodic stochastic process. Under this assumption, techniques are developed to estimate a variety of meaningful characteristics of the hypothesized stationary distribution such as geometric invariants (dimension) and Liapunov exponents. Also, to efficiently perform such estimation and enhance understanding of the function f, statistical procedures are often employed to estimate f from the data. Some of these procedures involve "nonparametric" function estimation, based on the slightest assumptions on the true f (see Nychka et al. [1992]), while others assume f lies in a large, flexible class of functions (see Tong [1990]). The presence or absence of stationarity of such estimated models determines the sort of inference or predictions the analyst is willing to assert. Given the strong role of the assumption of ergodicity as a foundation, we agree with the appraisal of Chan and Tong [1994], who suggest that "... the fundamental problem of identifying an appropriate stochastic perturbation term in [our (1.1)] so that the model is ergodic" warrants particular attention.

In the context of data analysis, one might also model "measurement errors." Specifically, suppose that the observed data is of the form $y_t = x_t + m_t$ at each observation time t. The m's represent physical measurement errors, modeled as random variables. Note that unlike the model error process, the m's do not actually affect the dynamics of the underlying process. As is typically assumed, if the m-process is (1) stationary and (2) independent of the x-process, then stationarity of the x-process implies stationarity of the y-process. Hence, we ignore the measurement error problem in this article.

The general area described has a long history in probability theory. The theory proceeds by recognizing the basic model (1.1) as generating a Markov chain. The structure of stationary, ergodic Markov processes is rich in applicable results; see Nummelin [1984]. Adaptations of this theory directed specifically to models such as

(1.1) comprise the major discipline of randomly perturbed dynamical systems; Kifer [1988] is particularly relevant in our context. There is a large literature involving continuous time models, stochastic differential equations, and stochastic stability; see Freidlin and Wentzell [1984] and Lasota and Mackey [1994]. We consider only discrete time, stochastic difference equations.

We address the issue of how the analyst can decide whether or not a particular specification of f and associated error model is indeed ergodic. The two most natural approaches are (1) to attempt a fully analytic analysis of the particular model, and (2) employ Monte Carlo simulation of the process. In this article we present a third approach involving a general computational strategy which can be used to actually *prove* ergodicity. This strategy employs a technique, *interval analysis*, typically associated with computer science. Also, the methods employed in Sections 3 and 4 can be used to develop bounds on the support of the x-process even if stationarity cannot be asserted.

The organization of the paper is as follows. In Section 2, we present probabilistic results which suggest when ergodic distributions can exist. The crucial assumptions are that (1.2) holds with *i.i.d.* errors with compact support equal to the interval $[-\theta, \theta]$. While proofs of such results are already available (see Chan and Tong [1994], Kifer [1988], and Lasota and Mackey [1994]), we offer a proof relying on less technical methods than is typical, yet directly relating mathematical properties of dynamical systems to ergodicity of Markov chains. Section 3 describes a general interval calculation analysis for implementing the theory. In Section 4, we present analytical results which describe conditions under which ergodicity can be claimed in the case of corrupted logistic maps. Our analyses offer extensions to the previous results of Mayer-Kress and Haken [1981]. The goal of Section 5 is to elucidate and compare analytical, numerical, and simulation based methods for deciding on the ergodicity of a system. A class of bimodal maps, displaying interesting asymptotic behaviors, are introduced to serve as test problems for the interval calculation. Extensions and comments are included in Section 6.

2 Probabilistic Results

We assume the model in (1.2), where f is continuous and has a compact, transitive attractor, say $\Lambda \subseteq G$, where G is a trapping region of f. Specifically,

$$\Lambda = \bigcap_{t \geq 0} f^t(G).$$

Transitivity implies that Λ cannot be decomposed into noncommunicating subsets which have nonempty interiors. For discussion of these definitions, see Devaney [1989, pp. 202-204]. (We use the Euclidean distance on $\mathcal{R}$ for closure and openness of sets.)

Next, it will be assumed that the errors in (1.2) are *i.i.d.* and have a continuous (with respect to Lebesgue measure) distribution whose density function, say $g(\cdot)$, has support equal to the interval $[-\theta, \theta]$, for a fixed, nonnegative constant θ. Hence, we may view x_t as a Markov chain with stationary transition probability density $p(x_{t+1}|x_t)$. Note that $p(\cdot|x_t)$ is simply g recentered at $f(x_t)$. (We make no notational distinction between densities and corresponding measures. Also, "measurable sets" are the usual Borel sets in $\mathcal{R}$.)

The key arguments used in this article involve the following definition: A P-trapping region, G_θ, of the chain x_t is a measurable, compact subset of $\mathcal{R}$ such that if $P(x_0 \in G_\theta) = 1$, then $P(x_t \in G_\theta) = 1$, $\forall t > 0$. We claim that if G, Λ, and G_θ all exist as described above, the condition $G_\theta \subseteq G$ is sufficient to imply that the chain is "uniquely" ergodic (Theorem 2.3).

The following preliminary results are used.

Lemma 2.1 *An invariant distribution π, that is, one satisfying*

$$\pi(A) = \int p(A|x)\pi(dx)$$

for all measurable sets A, must be continuous.

Proof The result is clear after noting that π is representable as the convolution of a distribution (the marginal distribution of $f(x_t)$) with the *continuous* error distribution. □

Lemma 2.2 *An invariant distribution exists.*

Proof Proposition 1.4 of Kifer [1988] says that the result is implied by the compactness of G together with the property that the transition measures, in our case based on $p(\cdot|x)$, depend on x continuously in the topology of weak convergence. To verify this last requirement is satisfied, recall that the Lévy distance metrizes the weak topology. Further, this distance, $d_L(F, H)$, between two distribution functions, F and H, has the property that "$\sqrt{2}\, d_L(F, H)$=the maximum distance between the graphs of F and H, measured along a 45°-direction." ([Huber, 1981, p. 25].) Suppose $|x - y| < \delta$. Continuity of f implies that $|f(x) - f(y)| < \varepsilon$, with ε (possibly depending on δ and y) going to 0 as δ goes to 0. A simple geometric argument then implies that $d_L(F_x, F_y) < \varepsilon$, where F_x and F_y are the distribution functions corresponding to $p(\cdot|x)$ and $p(\cdot|y)$, respectively. □

An alternative argument can be used employing the first conclusion of Proposition 12.7.1 of Lasota and Mackey [1994].

Theorem 2.3 *The invariant distribution, π, described in Lemmas 2.1 and 2.2 is unique and the x-process started by generating x_0 from π is ergodic.*

Proof The result stems from application of Theorem 7.16 of Breiman [1968]. To apply the result, we need two of his definitions. First, a measurable subset A of G_θ is *closed* if $p(A|x) = 1, \forall x \in A$. We refer to closure of this sort as *B-closure*. Next, a chain is *indecomposable* if there are no two disjoint B-closed measurable subsets of G_θ. Breiman's result requires that the chain be indecomposable. To verify this condition, we complete an indirect proof.

Assume that A_1 and A_2 are two non-empty, B-closed, disjoint subsets of G_θ. Define $B_i = \{x \mid |x - f(y)| < \theta \text{ for some } y \in A_i\}$, for $i = 1, 2$. It can be shown that these B_i are B-closed, and that since A_1 and A_2 are disjoint, so are B_1 and B_2.

Since Λ is an attractor of f, and G a trapping region, $\Lambda = \cap_{t=1}^{\infty}\{\cup_{x\in G} f^t(x)\}$. Since B_1 and B_2 are subsets of G with the property that $\{\cup_{x\in B_i} f(x)\} \subset B_i$, we have $C_i \doteq \cap_{t=1}^{\infty}\{\cup_{x\in B_i} f^t(x)\} \subset \Lambda$. Also, $C_i \neq \phi$.

Since Λ is transitive, for any U, V open with respect to Λ, we have that there exists some $k > 0$ for which $\{\cup_{x\in U} f^k(x)\} \cap V \neq \phi$. B_1 and B_2 are open with respect to $\mathcal{R}$, and so $B_1 \cap \Lambda$ and $B_2 \cap \Lambda$ are open with respect to Λ. Then there exists $k > 0$ for which $\{\cup_{x\in B_1} f^k(x)\} \cap B_2 \neq \phi$. But $\{\cup_{x\in B_1} f^k(x)\} \subset B_1$, and so $B_1 \cap B_2 \neq \phi$, implying a contradiction. Hence, no two B-closed sets can be disjoint. □

3 Interval Calculations

In this section, we describe a computational strategy for determining whether an ergodic distribution does exist. The computational strategy involves a search for a set, I, with the following invariance property: If the support of x_t is I, then the support of x_{t+1} is also I. One can imagine building up such a set, beginning with a single x_0 that is in I, and examining the support of $x_t|x_0$. In aperiodic settings, under mild regularity conditions, the support of x_t will converge to I as t grows. The issue becomes one of tracking these supports through time. We borrow the technique of interval calculation from computer science to track these supports.

A large literature on interval calculations exists (see, in a general context, Moore [1979], and for one use of these calculations in the context of dynamical systems, Hammel, Yorke, and Grebogi [1987]). This literature has often been motivated by the need to assess the accuracy of computer calculations. Conceptually, when each calculation is performed on the machine, the result is recorded with some round-off error. The interval calculation accounts for this error in a worst-case fashion, obtaining bounds within which the true answer must lie. Our setting is analogous when we view an iterate of the uncorrupted system as the true function, and view the error due to noise in the system as the round-off error.

When the algorithm, described in the balance of this section, is implemented on a machine with finite accuracy, the results are subject to an additional component of numerical round-off error. Impacts and remediations of this difficulty are discussed in Section 5.4.

The mechanics of the interval calculation are to define a set-valued function, h, that maps subsets of $\mathcal{R}$ into subsets of $\mathcal{R}$: Let I_t denote the current set. Applying the map yields $I_{t+1} = h(I_t) = \cup_{x\in I_t, e\in\Theta}\{f(x)+e\}$, where $f(x)$ denotes the function and Θ is the support of the error distribution. A function h of this sort is essentially what is referred to as a *united extension* of f in the interval calculation literature. In the case of a continuous map and interval I_t, the value of $h(I_t)$ is the interval

$$I_{t+1} = (\inf_{x\in I_t} f(x) - \theta, \sup_{x\in I_t} f(x) + \theta). \tag{3.1}$$

The next result lays the basis for the computational strategy. Before we present it, we recall the definition of a periodic Markov chain, where we assume that the stationary distribution has support equal to the state space, A.

Definition 3.1 A Markov chain is *periodic* if there exists a partition of the state space A into sets $A_1, \dots, A_p$, with $p > 1$, such that for all $i = 1, \dots, p$, if $x_0 \in A_i$, then $x_{tp} \in A_i$ with probability one. The *prime period* of the Markov chain is the smallest period p for which the chain is periodic.

Proposition 3.2 *Assume that $\{x_t\}$ is an irreducible Markov chain with a stationary distribution, π, that is absolutely continuous with respect to Lebesgue measure and has support equal to the state space A. Also, assume that the transition probability distributions of the chain are absolutely continuous with respect to π. Let x_0 be any element of A. If the chain is aperiodic, the support of x_t converges to A. If the chain is periodic with prime period p, then as $t \to \infty$, the support of $x_{tp+i}, i = 1, \dots, p$, converges to some set A_i, where the A_i's form a partition of A.*

Proof When the chain is aperiodic, Theorem 7.18 of Breiman [1968] implies that for all $x_0 \in A$, $\lim_{t\to\infty} P(x_t \in B|x_0) = \pi(B)$, for all π-measurable subsets B of A. It directly follows that the supports of $x_t|x_0$ converge to A.

When the chain is periodic with prime period p, π is representable as a mixture $\pi = p^{-1}\sum_{i=1}^{p}\pi_i$, where π_i has support A_i. Consider the subsequences $\{x_{tp+i}\}_{t=0}^{\infty}$. Each of these subsequences forms an irreducible, aperiodic Markov chain with state space A_i and stationary distribution π_i. Applying the previous argument in each case completes the proof. □

Our computational strategy is based on the behavior of sequences of intervals $I_1, I_2, \ldots$, obtained by applying the interval calculation described above. The strategy consists of three steps:

Initialization step. Find a value, x_0, that is within θ of the attractor of the uncorrupted system. Set $I_0 = [x_0, x_0]$.

Interval step. Iterate the interval calculation, from I_t to I_{t+1}.

Termination step. Terminate the interval step when (i) $I_{t+1} = I_t$, (ii) $I_{t+p} = I_t$, or (iii) the interval explodes.

The key justification of the algorithm is Proposition 3.2. Specifically, the output of the interval calculation is precisely the support sets alluded to in the proposition. To argue that the termination step indeed leads to containment of the desired limiting support, we employ the following result.

Proposition 3.3 *For the interval calculation as described above, if I_t contains I_{t-1}, then the sequence of intervals $I_t, I_{t+1}, \ldots$ is nested and non-decreasing. If I_t is contained in I_{t-1}, then the sequence of intervals is nested and non-increasing.*

Proof Consider the first statement. Assume that I_{t-1} is contained in I_t. Then

$$I_t = \cup_{x\in I_{t-1}, e\in\Theta}\{f(x)+e\}$$

and

$$\begin{aligned} I_{t+1} &= \cup_{x\in I_t, e\in\Theta}\{f(x)+e\} \\ &= \cup_{x\in I_{t-1}, e\in\Theta}\{f(x)+e\} \cup [\cup_{x\in I_t\setminus I_{t-1}, e\in\Theta}\{f(x)+e\}] \\ &= I_t \cup [\cup_{x\in I_t\setminus I_{t-1}, e\in\Theta}\{f(x)+e\}], \end{aligned}$$

and so we have that I_{t+1} contains I_t. Extending the argument shows that the sequence is non-decreasing. A similar argument shows that the second sequence mentioned is non-increasing. □

The strategy may be tailored to the case of an individual map. For example, in the case of a unimodal map with maximum value attained at x_m, if x_m is included in the support of I_t, it is clear that the upper endpoint of I_{t+1} will be the upper endpoint of the support (should the system have a stationary distribution). Recognizing this, one could directly investigate the lower endpoint of the support by iterating the map, $f(x) - \theta$, until it explodes, or until a fixed point is found.

Illustrations of the interval calculation technique are given in Section 5.

4 An Analytic Example: Logistic Map

4.1 Basic Results. In this section, we consider (1.2), where f is the logistic map, $f(x) = ax(1-x)$. Recall that for $a \in [0,4]$ and $x_0 \in [0,1]$, any orbit of the uncorrupted logistic system has all values in [0,1]. In general, for a given value of $a > 1$, a variety of asymptotic behaviors are possible, depending on x_0. The key, however, is that for each a only one of these possible behaviors is attracting in the sense that almost all x_0, in the sense of Lebesgue measure, lead to the same asymptotic behavior. Also note that iterates of the logistic map are typically not ergodic for a's outside [0,4], nor for x_0's outside the unit interval. For related review, see Berliner [1992]; in depth discussions may be found in Devaney [1989] and, with emphasis on ergodic theory, in Collet and Eckmann [1980].

Assumptions on the errors are as in Section 2. Our task is to find an upper bound, say θ^*, depending on a, that guarantees the existence of a P-trapping set G_{θ^*}. Our construction, which essentially traces "worst case scenarios," typically suggests that if $\theta > \theta^*$, then the corrupted process is almost surely transient. (Also, see Chan and Tong [1994].) In general, such an analysis can begin by considering the "domain of attraction" of the uncorrupted map. For the logistic map with $1 < a \leq 4$, the orbit corresponding to any $0 \leq x_0 \leq 1$ remains in the unit interval. However, initial conditions outside the unit interval result in orbits which rapidly decay toward $-\infty$; indeed the rate of this decay is sufficient to overpower additive errors. For $0 < a < 1$ the basic problem is the same, except that the curvature of f allows larger domains of attraction. Specifically, the orbit for any $1-1/a \leq x_0 \leq 1/a$ remains in this interval, suggesting that corruptions which take orbits outside, but not too far outside, the unit interval are permitted. A particularly interesting case occurs for $a = 1$; no corruptions of the type considered here can yield an ergodic process. (This fact is related to the notion of structural stability. Devaney [1989, p.57] notes that "One of the major ways a map can fail to be structurally stable occurs when there is a lack of hyperbolicity." This is precisely the case when $a = 1$.)

Consider fixed values of a and θ. Define the following functions:

$$f^+(x,\theta) = f(x) + \theta \text{ and } f^-(x,\theta) = f(x) - \theta.$$

As in the interval calculation method, we seek a compact interval of the form $I = [l_\theta, u_\theta]$ with the property that

$$\forall x \in I, \quad [f^-(x,\theta), f^+(x,\theta)] \subseteq I. \tag{4.1}$$

Once I is identified, we can then argue by Theorem 2.3 that the corrupted process is ergodic, with a distribution whose support is a subset of I.

The most obvious choice for u_θ is $a/4 + \theta$. Since f is globally maximized at .5 and $f(.5) = a/4$, the (easy) upper limit in (4.1) is clearly satisfied. (For $a > 1$ the curvature of f suggests the bound $1 - a/4 \geq \theta$. However, this bound turns out not to be sufficient.) Note that u_θ is a tight bound for all a which yield an attractor in the uncorrupted case that cannot be isolated from .5. For a large, this is typically the case. Even for moderate a in the so-called period doubling region, each region of period 2^k has an a value which contains .5 in its attractor.

Turning to the lower limit of I, the analysis focuses on the formula

$$f^-(x,\theta) = x. \tag{4.2}$$

Requiring $\theta \leq \theta_\circ \doteq \frac{(a-1)^2}{4a}$, we let $r_+(\theta)$ and $r_-(\theta)$ denote the real solutions

$$\frac{a - 1 \pm \sqrt{(a-1)^2 - 4a\theta}}{2a} \tag{4.3}$$

to (4.2). The key to establishing the lower limit of I involves the interplay between $r_+(\theta)$, $r_-(\theta)$, and $f^-(u_\theta, \theta)$.

Proposition 4.1 *Let* $I = [l_\theta, u_\theta]$, *where* $u_\theta = a/4 + \theta$ *and*

$$l_\theta = \min\{f^-(u_\theta,\theta), r_+(\theta)\}.$$

Define θ^* *by*

$$\theta^* = \frac{(a-1)^2}{4a}, \quad \forall a \in (0, 1 + .5\sqrt{6}),$$

and

$$\theta^* = .5(1 - a/2 + \sqrt{2 - 4/a}), \quad \forall a \in [1 + .5\sqrt{6}, 4).$$

If $x_0 \in I$ and $\theta \leq \theta^*$, then for all t, $x_t \in I$, with probability one.

Note that θ^* is tight in the following sense. If $\theta > \theta^*$ and realizations of the process can enter an arbitrarily small neighborhood of 0.5, then errors can, and therefore eventually will, occur which throw the process out of the unit interval. The arguments for transience as discussed at the beginning of this section then apply. Also, note that Proposition 4.1 is a strengthening of the result of Mayer-Kress and Haken [1981, p. 152].

4.2 Extensions.

4.2.1 Sharper analysis than that in Proposition 4.1 is possible for those a which yield attracting periodic behavior in the uncorrupted system. The effect of errors is quite interesting in such cases. Consider those a which form the now famous period doubling cascade for the logistic map; namely $3 < a < 3.57\ldots$. For almost all (in the sense of Lebesgue measure) $x_0 \in [0,1]$ and all $3 < a < 1 + \sqrt{6}$, uncorrupted x's are attracted to a period 2 cycle. As a increases, theory suggests that all periods of length $2^p, p > 1$ occur. Suppose a is such that the uncorrupted system has an (hyperbolic, as will defined below) attracting period of 2^p, $p > 1$. For the corrupted system, there exists a θ_{crit} such that for all $\theta < \theta_{crit}$, the corresponding ergodic distributions preserve the period 2^p behavior in the sense that the supports of these distributions are composed of 2^p disjoint intervals. Furthermore, as θ is increased beyond θ_{crit}, we observe a *mergence cascade* in which smaller periods, represented by $2^m, m < p$ disjoint intervals occur. This phenomenon, also discussed in Mayer-Kress and Haken [1981], is illustrated in Section 5.

The existence of θ_{crit} is established in Proposition 4.2. In settling for an existence result, we are able to show this fact for a general class of maps, including the logistic.

Proposition 4.2 *Assume the formulation of Section 2. Suppose f is continuously differentiable and there exists an attracting periodic orbit of minimal period p. Let $y_1, \ldots, y_p$ denote this orbit, where the labelling is such that $y_{i+1} = f(y_i)$, for $i = 1, \ldots, p-1$ and $y_1 = f(y_p)$. For $i = 1, \ldots, p$, define $d_i = |f'(y_i)|$. Assume that the attractor is hyperbolic; i.e., $\prod_{i=1}^{p} d_i < 1$. Then there exists $\theta_{crit} > 0$, such that for all $\theta \leq \theta_{crit}$, there exist p disjoint intervals, $I_1, \ldots, I_p$, each of which contains exactly one of the y_i's, such that if $x_0 \in \bigcup_{i=1}^{p} I_i$, then $\forall t$, $x_t \in \bigcup_{i=1}^{p} I_i$, with probability one.*

4.2.2 In his Section 4.2 Kifer [1988] considers the following random dynamical system based on the logistic map:

$$x_{t+1} = (a + \varphi_{t+1})x_t(1 - x_t), \tag{4.4}$$

where the φ's are *i.i.d.* continuous random variables with support $[-\epsilon, \epsilon]$. We can combine this form of perturbation of the map with the additive errors model, leading to

$$x_{t+1} = (a + \varphi_{t+1})x_t(1 - x_t) + e_{t+1}, \tag{4.5}$$

where the e_t's are *i.i.d.* continuous random variables with support $[-\theta, \theta]$.

To implement the basic program of this paper, the first step is to check that the results of Section 2 carry over to (4.5). Assuming that $0 < a < 4$, $a \neq 1$, this is not difficult. For example, the continuity of the transition measures in the topology of weak convergence, as discussed in Lemma 2.2, can readily be established after writing (4.5) as

$$x_{t+1} = f(x_t) + (\varphi_{t+1}/a)f(x_t) + e_{t+1}, \tag{4.6}$$

where f is the logistic map with parameter a. Obvious bounds on the φ_t's and the continuity of f imply the result. Once these arguments are completed, it remains to find bounds, depending on a, on θ and ϵ which imply the existence of a compact trapping set G, as in Section 2. To that end one could attempt analyses such as those in Section 4.1. Since the method is clear, we skip the analysis for brevity. We pause to note that a simple sufficient condition is that

$$\epsilon/4 + \theta \leq \theta^*, \tag{4.7}$$

where θ^* is as given in Proposition 4.1; any pair (θ, ϵ) satisfying (4.7) does indeed yield a compact trapping set. Consider the right-hand side of (4.6). Since $|(\varphi_{t+1}/a)f(x_t) + e_{t+1}| \leq \epsilon/4 + \theta$ with probability one, an interval calculation for the "singly" perturbed system ($\epsilon = 0$) traps the corresponding interval calculation for the "doubly" perturbed system under the assumption that (4.7) holds.

Pursuing the last argument, this system also offers a test for the interval calculation procedure of Section 3. The interval calculation takes I_t into

$$I_{t+1} = (\inf_{x \in I_t} f(x)(1 - \epsilon/a) - \theta, \sup_{x \in I_t} f(x)(1 + \epsilon/a) + \theta).$$

An example is reported in Section 5.

5 Implementations and Comparisons

5.1 Interval Calculations for Logistic Maps. We begin by illustrating the basic interval calculation strategy of Section 3 for the logistic map. First, set $a = 3.72$ and $\theta = 0.001$. To obtain an initial value near the attractor of the uncorrupted system, we chose a number at random (.7234123591) and iterated the uncorrupted system 1000 times. The resulting value was labeled x_0, leading to $I_0 = [x_0, x_0]$. Figure 1(a) presents a sequence of intervals, the result of the iteration step. The last interval remained unchanged under another iteration. Therefore, we have found a fixed point of the interval calculation. Note that at termination the endpoints of the interval are identical floating point numbers. Figure 1(b) presents results for the case of $a = 3.995$ and $\theta = 0.001$. In this instance the interval quickly overlaps the region where the uncorrupted system is explosive, and we conclude that the system has no stationary distribution. (Of course, this agrees with the suggestion of Proposition 4.1.) The computational strategy's behavior in these instances is typical of a wide range of values of a and θ, in that the convergence of the support interval to the fixed point (or periodic cycle) of the interval calculation is quite quick. Next, consider the mergence cascade discussed in Section 4.2.1. The phenomenon is illustrated in Figure 2. Finally, we illustrate the interval calculation applied to the doubly perturbed logistic map presented in Section 4.2.2. For example, an analysis, using the initialization procedure discussed above, for the case $a = 3.72$, $\theta = 0.001$, and $\epsilon = 0.002$ results in a single final interval (.231526, .932860). The algorithm converged after 10 iterates.

5.2 Interval Calculations for a Class of Bimodal Maps. As further tests of the interval calculation strategy, a class of bimodal maps are introduced and analyzed. Define a family of maps from the unit interval into the unit interval

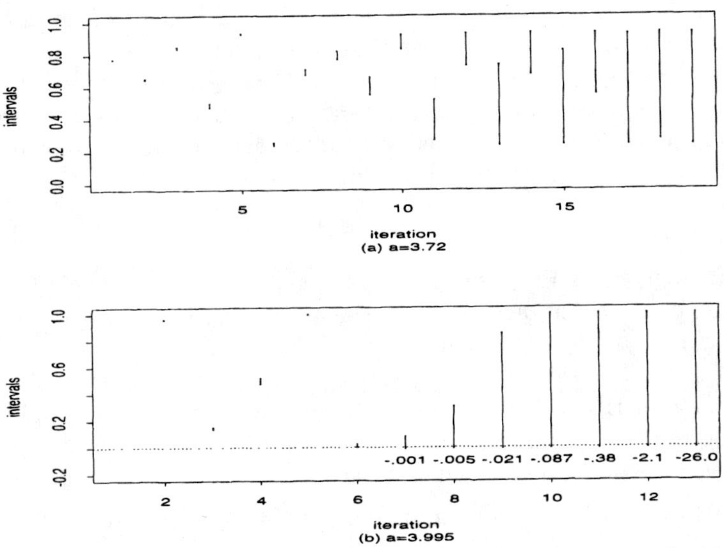

Figure 1 Results of the interval calculation for a logistic map. Intervals of support are graphed by iteration number. Panel (a) displays results for $a = 3.72$ and $\theta = 0.001$. The lower endpoints of intervals that extend below 0 appear beneath the intervals.

based on the functions

$$f(x) = \begin{cases} ax(1.0 - x/\tau) & \text{if } 0 < x \le t \\ \mu + \frac{b}{1.0-\mu}(x-\mu)(1.0-x) & \text{if } t < x \le 1.0, \end{cases} \tag{5.1}$$

where a, b are fixed constants between 0 and 4, and where $\mu, \tau,$ and t, are chosen to depend on the pair a, b as follows. The key is that the first component of the map is a rescaled logistic, while the second component is a rescaled and relocated logistic. The values $0 < \mu < 1.0$ and $0 < t < \tau < 1.0$ are chosen so that the resulting function is made continuous at the intersection of the two modified logistics.

Next, we assume that a stochastic map is generated, following (1.2), based on this map and continuous, compact errors, supported in the interval $[-\theta, \theta]$. For some choices of a and b, the remaining parameters of the map can be specified to permit the following claims: If $\theta \le \tau\theta_a^*$, there is a $P-$trapping set in the interval $(0, \tau)$ for the first component of the map. Also, if $\theta \le (1.0 - \mu)\theta_b^*$, there is a $P-$trapping set in the interval $(t, 1.0)$ for the second component of the map. (Here, the θ_a^* is the critical bound obtained in Proposition 4.1, with the additional subscript indicating the map parameter used.) Hence, a conservative result is that for any $\theta \le \min[\tau\theta_a^*, (1.0 - \mu)\theta_b^*]$, the corrupted bimodal map system admits a $P-$trapping set.

The following brief interval calculation experiment was performed. We set $a = 3.9, b = 3.8, \tau = 0.5, \mu = 0.51,$ and $t = .4399898$. Results for three values, $.001, .016,$ and, $.018,$ of θ are reported.

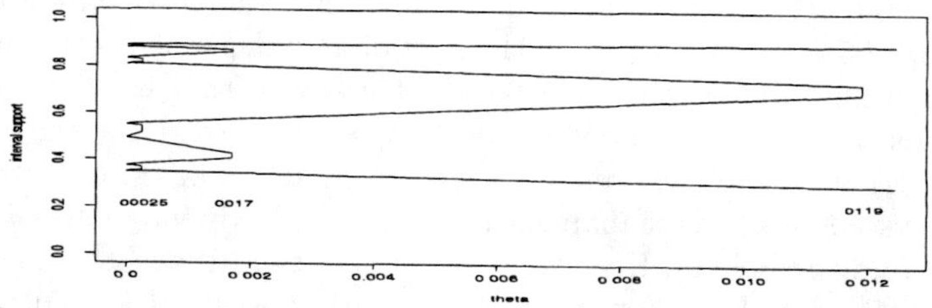

Figure 2 The mergence cascade based on interval calculations. The uncorrupted logistic map with $a = 3.56$ has an attractor of period 8, shown at the left of the figure. With corruptions, as the value of θ increases, the period 8 cycle collapses to a period 4 cycle near $\theta = 0.0017$. Indications of periodicity in the supports disappear for θ near 0.0119.

Table 5.1. Interval Calculation Results for a Bimodal Map.

θ	x_0	Final Interval
.001	.13	(.205.., .488..)
	.50	(.205.., .488..)
	.51	(.205.., .976..)
	.52	(.594.., .976..)
.016	.13	(.189.., .503..)
	.83	(.525.., .991..)
.018	.13	(.187.., .993..)
	.83	(.187.., .993..)

The interpretations of the results in Table 5.1 are quite interesting. For $\theta =$ .001, we see that small x_0 lead to trapping of the system in the lefthand peak of the map, while realizations beginning at large x_0 are trapped in the righthand peak. However, there is a range, containing the value .51, of initial conditions for which systems end up in either the left or right peak, depending on the chance shocks. Similar phenomena occur until θ is too large, for example, .018, to allow either peak to be stable, yet small enough for the full system to be trapped. For even larger θ, the stability breaks down completely.

We find the results here quite illuminating, primarily for the obvious cautions they suggest for analysts. Specifically, there may be severe dependence of the form of asymptotic behavior on initial conditions, or at least on the "basins of attraction in which they lie," as well as the magnitude of errors in even simple systems.

5.3 Simulation. Simulation based approaches to assessing stationarity dominate the literature, currently. Selected examples, along with the maps considered, include Chan and Tong [1994] ("polynomial autoregressions"), Lichtenberg and Lieberman [1983] (" 'Simplified' Ulam Mapping"), and McCaffrey, Ellner, Gallant, and Nychka [1992] ("Hénon Map"). We begin our considerations with another example taken from the literature. Yao and Tong [1994, p. 712] write "We begin with the simple one-dimensional model $Y_t = .246Y_{t-1}(16 - Y_{t-1}) + \varepsilon_t$, $t \geq 1$, where ε_t, $t \geq 1$, are independent random variables with the same distribution as the random variable 0.05η, and η is equal to the sum of 48 independent random variables each uniformly distributed on $[-0.5, 0.5]$." They note that each ε_t has support $[-1.2, 1.2]$, and continue "Bounded support of ε_t is necessary for the stationarity of the time series." They also remark that this model is based on a transformed version of the standard logistic map with $a = 3.936$. Multiplying both sides of their model by 16, and redefining the result in our notation, they have the corrupted logistic map with $\theta = .075$. For this value of a, Proposition 4.1 yields the critical value of $\theta^* \approx 0.012$. Hence, their model is in fact not stationary. (Of course, they never claimed that it was stationary.) They actually simulated 1200 observations from their model. We performed the following numerical experiment. We replicated 100 independent paths, each of length 1200, from this model. We found that 20 of the 100 realizations had not exploded (yet!).

We believe that this example points to some of the difficulties in using simulation based arguments. First, simulation requires specification of a particular form for the distribution of errors, as opposed to either the analytic or interval calculation approaches. Although this specification does not affect the stationarity of the system, it can have a dramatic effect on the apparent stationarity in a simulation experiment. Second, nonexploding realizations always leave a doubt that the simulation may be too short. This problem is exacerbated if either (1) the simulated errors have a sharp, centralized density, as in the example of Yao and Tong; or (2) if the value of θ is in fact near the critical value, θ^*. Finally, the example discussed in Section 5.2 also indicates potential problems with reliance on simulation. The example illustrates the possibility that highly dissimilar asymptotic, though apparently ergodic, behaviors are possible, depending on the early random shocks imparted to the system. Simulation methods may discover some, but not all asymptotic behaviors of the system.

5.4 Numerical Inaccuracy of Interval Analysis. Interval analyses designed to track the support of a Markov chain are subject to round-off error when implemented on a finite precision machine. Instead of obtaining $I_{t+1}|I_t$ as in (3.1), we obtain an approximation to I_{t+1} which, through round-off error, is forced to match the precision of the machine. In our applications the size of round-off error is typically much smaller than θ values of interest. Hence, a conservative calculation can be made by implementing the algorithm with a value of θ large enough to cover both the noise in the system and round-off error. Calculations for a slightly smaller value of θ could be used to confirm explosive behavior.

For examples terminating in a periodic regime, there is a subtle problem. Namely, round-off can in principle generate false periodic behavior. Consider the construction of a mergence analysis such as in Figure 2. A particular stochastic system may display a period 8 cycle, which merges directly to a period 2 cycle as θ grows past a critical value. It is possible that a finite precision interval analysis shows a small window of θ values displaying a period 4 cycle. For practical implementations, such a false period could only occur for a tiny (on the order of machine round-off) interval of θ. Hence, for example, the basic features of Figure 2 are correct, though the indicated critical θ values are approximate.

6 Comments and Extensions

Remark 6.1 An important comment involves the *i.i.d.* and continuity assumptions for the errors used to this point of the paper. Specifically, these requirements are not needed for either analytical methods or interval analyses if the goal of such analyses is the establishment of a P-trapping set for any realization of the system under study. These requirements can be replaced by the assumption that the errors all have common, compact support. We employed the *i.i.d.* assumption to permit convenient application of Section 2 in the inference of ergodicity.

Remark 6.2 The analysis that we have presented is aimed at settling the question of the existence of a stationary distribution for a proposed system. We note, however, that nonstationary systems are also of great interest. Many times, these systems will show behavior over long periods that appears to be stationary, even though the system has no invariant distribution; this phenomenon is often known as *persistence* in the literature. When matching experimental data to mathematical models, there is no guarantee that the experimental system is stationary.

Define an explosive region, E, so that E^c is compact, and a stopping time for the chain as the time when the chain first enters the explosive region. Two distinct chains are of interest. The first relies on the chain never entering the explosive region. Create a transition probability for the modified chain by replacing $p(.|x_t)$ with $p^*(.|x_t)$ where $p^*(x|x_t) = p(x|x_t)/P(X_{t+1} \in E^c|x_t)$ for $x \in E^c$ and $p^*(x|x_t) = 0$ for $x \in E$. This new transition probability can be created as long as $P(X_{t+1} \in E^c|x_t) > 0$ for all $x_t \in E^c$. This chain, with a modified error structure, when coupled with the conditions of Section 2, produces a parallel version of Theorem 2.3. The chain will have a unique invariant distribution when initialized in E^c. The aperiodicity (or period in the case of a periodic system) of the system may be determined through interval calculations.

The second chain is a direct representation of the nonstationary system, while retaining the basic structure of the original transition probability $p(.|x_t)$. The behavior of the system can be studied under a variety of distributions for x_0. One particularly appealing distribution is that obtained from using the invariant distribution under $p^*(.|x_t)$. This connects the two chains since for the t-step transition $P(X_t \in A|x_0; x_i \in E^c, i = 1, \ldots, t) = P^*(X_t \in A|x_0)$, provided $P(X_{t+1} \in E^c|x_t) > 0$ for all $x_t \in E^c$. Hence, for systems that operate for long time spans, the first chain may prove useful for analysis. We note that for these nonstationary systems, the stopping time contains information about the previous path of the system through the requirement that the previous path remain in E^c.

7 Appendix

Proof of Proposition 4.1 We begin with the arguments for $a > 1$ in detail. The following properties of the quantities defined are important. First, $\forall x \in (r_-(\theta), r_+(\theta))$, $f^-(x,\theta) > x$, and for $x < r_-(\theta)$, $f^-(x,\theta) < x$. Further, since u_θ is increasing in θ, it follows that $f^-(u_\theta, \theta)$ is decreasing in θ.

Note that if we establish that $r_-(\theta) \leq f^-(u_\theta, \theta)$, it then follows that

$$I = [\min\{f^-(u_\theta,\theta), r_+(\theta)\}, u_\theta] \tag{7.1}$$

satisfies (4.1).

The crucial step is to analyze the inequality $r_-(\theta) \le f^-(u_\theta, \theta)$. Note that $r_-(\theta)$ is increasing in θ (and $r_+(\theta)$ is decreasing), while $f^-(u_\theta, \theta)$ is decreasing. Simple calculation shows that $f^-(u_0, 0) > r_-(0)$. These remarks indicate that we need to determine if and when $f^-(u_\theta, \theta) = r_-(\theta)$. To that end, define the function

$$w(a) = f^-(u_{\theta_\circ}, \theta_\circ) - r_-(\theta_\circ). \tag{7.2}$$

Computational analysis of w demonstrates that $w(a)$ is positive if $1 < a < a_c$, where $a_c = 1 + .5\sqrt{6} \approx 2.22474$, $w(a_c) = 0$, and $w(a)$ is negative for $a_c < a < 4$. (Study of the sign of $w(a)$ can be reduced to analysis of a fourth degree polynomial in a. This polynomial has real roots equal to $1 \pm .5\sqrt{6}$ and has two imaginary roots.) Hence, for $1 < a \le a_c$, since $f^-(u_\theta, \theta) \ge r_-(\theta)$ for all $\theta \le \theta_\circ$, we have the desired result.

For $a > a_c$, the key to determining θ^* involves finding θ such that $r_-(\theta) = f^-(u_\theta, \theta)$. This equation can be solved by noting that it is equivalent to $f(r_-(\theta)) = f(u_\theta) = f(1 - u_\theta)$. Hence, we must solve $r_-(\theta) = 1 - u_\theta$ or

$$1 - \theta - a/4 = \frac{a - 1 - \sqrt{(a-1)^2 - 4a\theta}}{2a}. \tag{7.3}$$

Note that (7.3) directly implies the solution is less than $\theta_\circ$. Hence, the proof is complete if $a > a_c$, for all $\theta \le \theta^*$ where

$$\theta^* = .5(1 - a/2 + \sqrt{2 - 4/a}).$$

Turning to the case $0 < a < 1$, the analysis is easier than above. Specifically, $w(a)$ defined in (7.2) is positive on $0 < a < 1$, and so by the same arguments as used above, (7.1) does the job for all $\theta \le \theta^*$ where $\theta^* = \theta_\circ$. □

Proof of Proposition 4.2 The proof given is constructive. It essentially uses approximations to the interval calculation procedure of Section 3.

Recall the definition of the interval map h from Section 3. By Proposition 3.3, it is sufficient to show that disjoint I_i's exist with the properties that for each i, $y_i \in I_i$ and $h(I_i) \subseteq I_{i+1}$, where $I_{p+1} = I_1$.

For $i = 1, \ldots, p$ define intervals $B_i(\varepsilon) = (y_i - \varepsilon, y_i + \varepsilon)$. Let $d_i(\varepsilon) = \sup_{B_i(\varepsilon)} |f'(x)|$. Continuity and the hyperbolicity assumption insure the existence of some $\varepsilon > 0$ for which $D(\varepsilon) < 1$, where $D(\varepsilon) = \prod_{i=1}^{p} d_i(\varepsilon)$. Choose $\varepsilon < .5\times$the smallest absolute difference between any two of the y_i's, in order to insure that the $B_i(\varepsilon)$'s are mutually disjoint. Also, let $d(\varepsilon) = \max_{1,..,p} d_i(\varepsilon)$.

Next, note that for any $\varepsilon' < \varepsilon$, $\forall\theta > 0$, and any $x \in B_i(\varepsilon')$, the Mean Value Theorem implies that

$$|f(x) - f(y_i)| \le \varepsilon' d_i(\varepsilon').$$

This fact, combined with the result in (3.1), implies that

$$h[B_i(\varepsilon')] \subseteq [y_{i+1} - (\varepsilon' d_i(\varepsilon') + \theta), y_{i+1} + (\varepsilon' d_i(\varepsilon') + \theta)]. \tag{7.4}$$

Next, for ε' and θ sufficiently small, we iterate the process a total of p times, implying that

$$h^p(B_i(\varepsilon')) \subseteq [y_i - bound, y_i + bound], \tag{7.5}$$

where $bound = \varepsilon' D(\varepsilon) + (p-1)\theta \max(1, d(\varepsilon))^{p-1}$. To complete the proof, we must find bounds on ε' and θ that ensure that the above argument is valid *for every*

$i = 1, \ldots, p$. This is accomplished by first considering any $\varepsilon' < \varepsilon / \max(1, d(\varepsilon))^p$. For any such ε', it is not hard to check that the analysis is valid for any

$$\theta < [\min(\varepsilon - \varepsilon' \max(1, d(\varepsilon))^p, \varepsilon'(1 - D(\varepsilon)))]/[(p-1)\max(1, d(\varepsilon))^{p-1}].$$

□

References

Berliner, L. M. [1992] *Statistics, probability, and chaos,* Statist. Sci. **7**, 69-122 (with discussion).

Breiman, L. [1968] *Probability,* Addison-Wesley, Reading, MA.

Casdagli, M., des Jardins, D., Eubank, S., Farmer, J. D., Gibson, J., Theiler, J., and Hunter, N. [1992] *Nonlinear modeling of chaotic time series: Theory and applications,* In *Applied Chaos* (J. H. Kim and J. Stringer, eds.), Wiley, New York, NY, pp 335-380.

Chan, K. S. and Tong, H. [1994] *A note on noisy chaos,* J. Roy. Statist. Soc., Series B **56** 301-312.

Collet, P. and Eckmann, J.-P. [1980] *Iterated Maps of the Interval as Dynamical Systems,* Birkhäuser, Boston, MA.

Devaney, R. L. [1989] *An Introduction to Chaotic Dynamical Systems, 2/e,* Addison-Wesley, New York, NY.

Diamond, P., Kloeden, P., and Pokrovskii, A. [1994] *An invariant measure arising in computer simulation of chaotic dynamical system,* J. Nonlinear Sci. **4**, 59-68.

Freidlin, M. I. and Wentzell, A. D. [1984] *Random Perturbations of Dynamical Systems,* Springer-Verlag, New York, NY.

Hammel, S. M., Yorke, J. A., and Grebogi, C. [1987] *Do numerical orbits of chaotic dynamical systems represent true orbits?,* J. Complexity **3**, 136-145.

Huber, P. J. [1981] *Robust Statistics,* Wiley, New York, NY.

Kifer, Y. [1988] *Random Perturbations of Dynamical Systems,* Birkhäuser, Boston, MA.

Lasota, A. and Mackey, M. C. [1994] *Chaos, Fractals, and Noise 2/e,* Springer-Verlag, New York, NY.

Lichtenberg, A. J. and Liebermman, M. A. [1983] *Regular and Stochastic Motion,* Springer-Verlag, New York, NY.

Mayer-Kress, G. and Haken, H. [1981] *The influence of noise on the logistic map,* J. Statist. Phys. **26**, 149-171.

McCaffrey, D. F., Ellner, S., Gallant, A. R., and Nychka, D. W. [1992] *Estimating the Lyapunov exponent of a chaotic system with nonparametric regression,* J. Amer. Statist. Assoc. **87**, 682-695.

Moore, R. E. [1979] *Methods and Applications of Interval Analysis,* S.I.A.M. Studies in Applied Mathematics, Philadelphia, PA.

Nummelin, E. [1984] *General Irreducible Markov Chains and Non-negative Operators,* Cambridge University Press, Cambridge.

Nychka, D., Ellner, S., Gallant, A. R., and McCaffrey, D. F. [1992] *Finding chaos in noisy systems,* J. Roy. Statist. Soc., Series B **54**, 399-426.

Ruelle, D. [1981] *Small random perturbations of dynamical systems and the definition of attractors,* Commun. Math. Phys. **82**, 137-151.

Ruelle, D. [1989] *Chaotic Evolution and Strange Attractors,* Cambridge University Press, Cambridge.

Tong, H. [1990] *Non-linear Time Series: A Dynamical Systems Approach,* Oxford University Press, New York, NY.

Yao, Q. and Tong, H. [1994] *Quantifying the influence of initial values on non-linear prediction,* J. Roy. Statist. Soc., Series B **56**, 701-726.

Fields Institute Communications
Volume 11, 1997

Detecting Structure in Noise

Lisa Borland
Department of Physics
University of California at Berkeley
Berkeley, CA, USA 94720

Abstract. The problem of simultaneously modeling the underlying deterministic and stochastic nonlinear dynamics of an observed time series is discussed. We do this by way of a technique called Sequin (Stochastic EQUation INference), which we have developed extensively in earlier work (Borland and Haken [1992a,b,1993a,b], Borland [1996]). In this paper we present a short synopsis of the theory behind Sequin. Then we use Sequin to analyze a set of simulated time series differing only in the functional form of their underlying nonlinear stochastic dynamics. We see that Sequin is capable of separating deterministic structure from stochastic noise, and furthermore successful in detecting nonlinear structure within the noise

1 Introduction

Finding a model consistent with an observed time series is a problem of interest within many fields of science. Often, a priori knowledge about the type of system at hand – together with a substantial amount of intuition – help guide the scientist into coming up with a plausible mathematical model of the observed phenomenon. If, however, little is known about the underlying system, it is very useful to try and extract as much information as possible from the observed data itself.

There are a variety of existing techniques which have proven very useful for modeling and predicting time series. Whereas the goal of some methods, such as ARMA techniques (Box and Jenkins [1976], Tjøstheim [1992]) is to fit the best possible curve to the data, the goal of others is to extract the underlying dynamics of the process giving rise to the data.

In this paper, we review a technique belonging to the latter category. Most of the other methods within this category have been developed for nonlinear deterministic chaotic systems (see Packard et al. [1980], Casdagli and Weigand [1992]), although some attempts have been made to deal with small additive stochastic noise as well (Crutchfield and McNamara [1987]). However, a vast class of systems may consist of a nonlinear deterministic (perhaps chaotic) part *and* a nonlinear stochastic part. The questions that we pose are: how can we separate these from

1991 *Mathematics Subject Classification.* Primary: 62M05, 58F13; Secondary: 60H10.

each other, how can we detect deterministic "structure" through the noise, and how can we detect "structure" in the noise?

An attempt to answer these questions for the class of *stochastic Markov processes* has been the focus of our previous work (Borland and Haken [1992a,b,1993a, b], Borland [1996]). The main point of this paper will be to review that technique, which we call Sequin (Stochastic EQUation INference). In addition, we use Sequin to analyze simulations of processes differing only in the functional structure of their stochastic parts. These examples demonstrate that the method can indeed separate determinism from noise, and furthermore find appropriate nonlinear models for the underlying stochastic dynamics.

2 Stochastic Equations

For simplicity, we restrict our current discussion to that of one dimensional stochastic processes. The particular class of processes which we have in mind to characterize are those which can be described by a stochastic Ito-Langevin equation (see Gardiner [1983]) of the form

$$\dot{q} = K(q) + g(q)\eta(t) \tag{2.1}$$

In this equation, $K(q)$ represents a deterministic force, while $g(q)$ is a stochastic force modulated by the random variable $\eta(t)$ which has the properties

$$< \eta(t) >= 0 \tag{2.2}$$

and

$$< \eta(t)\eta(t') >= \delta(t-t') \tag{2.3}$$

where $< \cdot >$ denotes expectation, and $\delta(\cdot)$ denotes the usual delta function. $\eta(t)$ represents white noise and is a rapidly varying irregular function, uncorrelated in time. The averages in (2.2) and (2.3) are ensemble averages of values $\eta(t)$. However, for stationary ergodic systems they equate with averages taken over time.

Written in a differential form, it can be shown that (2.1) becomes

$$dq = K(q)dt + g(q)dw \tag{2.4}$$

where

$$< dw >= 0 \tag{2.5}$$

and

$$< dw^2 >= dt \tag{2.6}$$

which implies

$$dw \propto \sqrt{dt} \tag{2.7}$$

Both $K(q)$ and $g(q)$ may be nonlinear functions of q.

Assume that we have observations of one particular realization of a stochastic process of type (2.4). Due to the noise, a second realization will hardly give us the same observations. In trying to model the underlying dynamics of the process it is therefore useful to study properties of the evolution of the probability density function of realizations. Indeed, an equivalent representation of the process defined by (2.4) is given by the Fokker-Planck equation

$$\frac{d}{dt}P(q(t)) = -\frac{d}{dq}(K(q)P(q)) + \frac{1}{2}\frac{d^2}{dq^2}(G^{-1}(q)P(q)) \tag{2.8}$$

with

$$2g^2(q) = G^{-1}(q) \tag{2.9}$$

which describes the temporal evolution of the probability density function $P(q(t))$ of values of q.

3 Sequin – A Review

Starting from a stationary time series of a stochastic process denoted by $q(t)$, it is our goal to explicitly reconstruct the underlying dynamics in the form of an Ito-Langevin equation (2.4). This is a different problem to the one of fitting a curve through the time series; our approach (Borland and Haken [1992a,b,1993a,b], Borland [1996], see also Haken [1988]) attempts to capture the structural form of a generating equation for the time series.

The time series $q(t)$ is sampled at $T+1$ time points spaced at equal intervals τ from each other. We let q_i denotes the value of the ith observation, $i = 1, \dots, T+1$ and, in order to emphasize the dependence on the time increment τ, usually denote its successor q_{i+1} by $q_{i+\tau}$. The procedure Sequin starts out by estimating the joint probability density function $P(q_{i+\tau}, q_i)$ which, for small but finite τ, corresponds to a short-time solution of (2.8). We then show that it is possible to extract explicit formulas for the underlying dynamics of (2.4) from $P(q_{i+\tau}, q_i)$. The way this is done consists of 3 steps, which are presented below.

3.1 Step 1: The Probability Density Function. Our first task is to make an unbiased estimate of the joint probability density function $P(q_{i+\tau}, q_i)$, which can be written as the product of a conditional and a marginal density function:

$$P(q_{i+\tau}, q_i) = P(q_{i+\tau} \mid q_i)P(q_i) \tag{3.1}$$

We choose to estimate $P(q_{i+\tau}, q_i)$ by way of the Maximum Information Principle (MIP) of Jaynes [1957]. Here, the information caliber

$$I = -\int P(q_{i+\tau}, q_i) \ln P(q_{i+\tau}, q_i) dq_i dq_{i+\tau} \tag{3.2}$$

is maximized in conjunction with appropriate constraints. The type of constraints which we choose to use are certain short-time correlations of the type

$$< q_i q_{i+\tau} >_P \tag{3.3}$$

which should be seen as averages according to

$$< q_i q_{i+\tau} >_P = \int\int q_i q_{i+\tau} P(q_{i+\tau}, q_i) dq_i dq_{i+\tau} \tag{3.4}$$

Explicitly, the set of constraints used here are of the form

$$< q_i^{\nu} >_P \qquad \nu = 0, \dots, N, \tag{3.5}$$

$$< q_i^{\mu} q_{i+\tau} >_P \qquad \mu = 0, \dots, M, \tag{3.6}$$

$$< q_i^{\kappa} q_{i+\tau}^2 >_P \qquad \kappa = 0, \dots, L, \tag{3.7}$$

where each of these is an expression of type (3.4). We shall alternatively refer to the set of constraints (3.5) – (3.7) as

$$< \Omega_j(q_{i+\tau}, q_i) >_P \qquad j = 0, \dots, R \tag{3.8}$$

with $R = N + M + L + 3$, or simply as $< \Omega_j >_P$.

The form of the constraints defines the model class with which we wish to represent the underlying dynamics. The set used in (3.5) – (3.7) confines our final representation to the family of polynomials. The polynomial orders N,M and L correspond to the polynomial degree with which we choose to characterize the problem. As we have discussed in previous work, it is sufficient to choose $N = M + 1$ and $L = M - 1$. To capture the leading nonlinearities, we often use $N = 4$, $M = 3$, and $L = 2$. Note, however, that we need not characterize the variable $q_{\imath+\tau}$ by more than its second moment. We have shown (Borland and Haken [1992b,1993a]) that these constraints allow us to study processes with additive and weak multiplicative noise. They can however be extended to deal with other classes of processes, for example processes where strong multiplicative noise may occur (Borland and Haken [1992a]).

The formal MIP solution for $P(q_{\imath+\tau}, q_\imath)$ is found by maximizing I of (3.2) with respect to P under the constraints (3.5) – (3.7). The result obtained by using the method of Lagrange multipliers is given in the form of (3.1) (Borland and Haken [1992b]) with

$$P(q_\imath) = \exp\left(-\sum_{\nu=0}^{N} \lambda_\nu^{(1)} q_\imath^\nu\right) \tag{3.9}$$

and

$$P(q_{\imath+\tau} \mid q_\imath) = N(q_\imath) \exp\left(- \mid C(q_\imath) \mid \left(q_{\imath+\tau} - \frac{B(q_\imath)}{2C(q_\imath)}\right)^2\right). \tag{3.10}$$

Here,

$$B(q_\imath) = \sum_{\mu=0}^{M} \lambda_\mu^{(2)} q_\imath^\mu \tag{3.11}$$

$$C(q_\imath) = \sum_{\kappa=0}^{L} \lambda_\kappa^{(3)} q_\imath^\kappa. \tag{3.12}$$

$N(q_\imath)$ must be chosen so that $P(q_{\imath+\tau} \mid q_\imath)$ is normalized to 1. For simplicity, we assume in this paper that $C(q_\imath) = \mid C(q_\imath) \mid > 0$. The general case where this is not necessarily valid is discussed in detail in our previous work.

In (3.9) – (3.10), the parameters $\{\lambda_\jmath\}, \jmath = 0, \ldots, R$ are Lagrange multipliers

$$\lambda_\nu^{(1)}, \quad \nu = 0, \ldots, N, \tag{3.13}$$

$$\lambda_\mu^{(2)}, \quad \mu = 0, \ldots, M, \tag{3.14}$$

$$\lambda_\kappa^{(3)}, \quad \kappa = 0, \ldots, L \tag{3.15}$$

which correspond respectively to the constraints $< \Omega_\jmath >_P$ of (3.5) – (3.7).

3.2 Step 2: Estimating the Parameters. In the above section, we wrote down a formal solution for $P(q_{\imath+\tau}, q_\imath)$ of (3.1) by way of (3.9) and (3.10). This result is a function of the unknown parameters $\{\lambda\}$ which are yet to be determined.

To determine the $\{\lambda\}$, we argue in the following fashion. There is a "true" underlying probability density function $\tilde{P}(q_{\imath+\tau}, q_\imath)$ according to which the values of $q_\imath$ and $q_{\imath+\tau}$ are distributed. This is the probability density function which we are trying to estimate in the form of $P(q_{\imath+\tau}, q_\imath)$ of (3.1) with (3.9) and (3.10). If we knew the values of the constraints of (3.5) – (3.7) calculated as averages over $\tilde{P}$, then we could determine the $\{\lambda\}$ such that the numerical values of the integrals

$$<\Omega_j>_P = \int\int \Omega_j(q_{i+\tau}, q_i) P(q_{i+\tau}, q_i) dq_{i+\tau} dq_i \tag{3.16}$$

are equal to those known values $<\Omega_j>_{\tilde{P}}$. We do not however know these values $<\Omega_j>_{\tilde{P}}$. But if the time series is very long ($T \to \infty$) and stationary, then we can estimate them as temporal averages $<\cdot>_T$ which can be calculated from the time series according to

$$<\Omega_j(q_{i+\tau}, q_i)>_T = \frac{1}{T}\sum_{i=1}^{T} \Omega_j(q_{i+\tau}, q_i) \tag{3.17}$$

Equating $<\Omega_j>_{\tilde{P}} = <\Omega_j>_T$, the criterion for finding the $\{\lambda\}$ thus becomes

$$<\Omega_j>_P = <\Omega_j>_T, \tag{3.18}$$

with $<\Omega_j>_P$ given by (3.16) and $<\Omega_j>_T$ given by the values of (3.17). $P(q_{i+\tau}, q_i)$ is a function of the $\{\lambda\}$ as given by (3.9) and (3.10). Ideally, we would like to solve (3.18) for the $\{\lambda\}$. This is not always trivial so instead we find these by way of a gradient strategy technique resulting in

$$\dot{\lambda}_j = <\Omega_j>_P - <\Omega_j>_T \qquad j = 0, \ldots, R. \tag{3.19}$$

Derivatives here refer to the iteration time of the gradient dynamics in (3.19) and are not to be confused with the time associated with the dynamics of the time series. Initial values of the $\{\lambda\}$ are chosen at random or as described in Borland [1996]. The integrals in $<\Omega_j>_P$ are calculated using these $\{\lambda\}$. At each time step of the gradient dynamics (3.19), the $\{\lambda\}$ are then updated by an amount proportional to the r.h.s. of (3.19). This procedure is continued until the estimated correlations $<\Omega_j>_P$ reproduce the $<\Omega_j>_T$, thus yielding the values of $\{\lambda\}$ which correspond to the stationary solution of (3.19) which is equivalent to (3.18).

Note that the values $<\Omega_j>_T$ are in practice calculated for finite T and are only approximations of the "true" values $<\Omega_j>_{\tilde{P}}$. This introduces errors into the final estimates of the parameters $\{\lambda\}$. These finite data effects are discussed more extensively in Borland [1996].

3.3 Step 3: The Equations of Motion. Now that we have an estimate of the probability density function $P(q_{i+\tau}, q_i)$ we can extract the underlying equations of motion. To this end we argue in the following manner: If the conditional probability density of (3.10) behaves as a δ-function with respect to the variable $q_{i+\tau}$ for $\tau \to 0$, it can be shown that it must be a short-time propagator to the Fokker-Planck equation of (2.8) (see Haken [1988]). This is the case if $P(q_{i+\tau} \mid q_i)$ takes on the form

$$P(q_{i+\tau} \mid q_i) = N \exp\left(-\frac{G}{\tau}(q_{i+\tau} - q_i - \tau K(q_i))^2\right). \tag{3.20}$$

By expanding the parameters $\{\lambda\}$ occurring in the expression of (3.10) for $P(q_{i+\tau} \mid q_i)$ into power series of τ we can identify (3.10) with (3.20). By then taking the limit $\tau \to 0$, we are able to extract the following equations for the functions $K(q)$ and $G(q)$ (Borland and Haken [1992b]):

$$G(q_i) = \tau \sum_{\kappa=0}^{L} \lambda_\kappa^{(3)} q_i^\kappa \tag{3.21}$$

and

$$K(q_\imath) = \frac{1}{2\tau} \frac{\lambda_0^{(2)} + \sum_{\mu=1}^{M} (\lambda_\mu^{(2)} - 2\lambda_{\mu-1}^{(3)}) q_\imath^\mu}{\sum_{\kappa=0}^{L} \lambda_\kappa^{(3)} q_\imath^\kappa} . \tag{3.22}$$

Using the equality of (2.9), we further obtain

$$g(q_\imath) = \frac{1}{\sqrt{2\tau \sum_{\kappa=0}^{L} \lambda_\kappa^{(3)} q_\imath^\kappa}} . \tag{3.23}$$

We have thereby obtained expressions for the deterministic force $K(q)$ and the stochastic force $g(q)$ occurring in the Ito-Langevin equation of (2.4).

The polynomial representation of $K(q_\imath)$ and $g(q_\imath)$ is closely linked with our choice of constraints in (3.5) – (3.7). We remind the reader that this representation can be used to describe processes with additive and weak multiplicative noise. To include processes with strong multiplicative noise a larger set of constraints must be used (Borland and Haken [1992a]). Other model classes, such as that of trigonometric functions, can be obtained by choosing yet another set of constraints (Borland and Haken [1993a]). For most practical purposes, however, we believe that a polynomial description can provide a good approximate solution to the problem.

3.4 Discrete Time Processes. The above presentation of Sequin is valid for continuous time processes. However, it turns out that analogous results are obtained for discrete time processes of the form:

$$q_{\imath+1} = q_\imath + K(q_\imath) + F(q_\imath) \tag{3.24}$$

Here we assume that $F(q_\imath)$ can be expressed as

$$F(q_\imath) = g(q_\imath)\omega_\imath , \tag{3.25}$$

where $g(q_\imath)$ is a function which may depend on $q_\imath$ as in the continuous case, and $\omega_\imath$ is Gaussian white noise so that

$$< \omega_\imath > \;=\; 0 \tag{3.26}$$

$$< \omega_\imath \omega_\jmath > \;=\; \delta_{\imath\jmath} , \tag{3.27}$$

Step 1 and Step 2 of Sequin can immediately be applied for discrete times by identifying $i + \tau$ with $i + 1$. Step 3 of Sequin, however, requires a different set of formal arguments. In particular, we argue that the following equality must hold:

$$P(q_{\imath+1} \mid q_\imath) = P_F(q_{\imath+1} - q_\imath - K(q_\imath)) . \tag{3.28}$$

P_F corresponds to the probability density according to which the noise term $F(q_\imath)$ of (3.24) is distributed. In the case of Gaussian noise, (3.28) becomes

$$P(q_{\imath+1} \mid q_\imath) = N \exp\left(-\frac{1}{2g^2(q_\imath)} \left(q_{\imath+1} - q_\imath - K(q_\imath)\right)^2 \right) \tag{3.29}$$

Taking for $P(q_{\imath+1} \mid q_\imath)$ the MIP estimate of (3.10), we can solve for $K(q_\imath)$ and $g(q_\imath)$, resulting in

$$K(q_\imath) = \frac{\sum_{\nu=0}^{M} (\lambda_\nu^{(2)} - 2\lambda_{\nu-1}^{(3)}) q_\imath^\nu}{2 \sum_{\gamma=0}^{L} \lambda_\gamma^{(3)} q_\imath^\gamma} \tag{3.30}$$

and

$$g(q_\imath) = \frac{1}{\sqrt{2 \sum_{\gamma=0}^{L} \lambda_\gamma^{(3)} q_\imath^\gamma}} . \tag{3.31}$$

We see that formally (3.31) and (3.30) are identical to (3.23) and (3.22) respectively, with $\tau = 1$. And yet there is a fundamental difference; in the discrete case, the mapping defined by $K(q_i)$ (3.30) and $g(q_i)$ (3.31) in conjunction with (3.24) and (3.25) is only valid at the particular discrete time sampling step of the data (here, this time-step is set to 1). In the continuous case, however, we derived (3.22) and (3.23) so that they are valid for any small time τ. In other words, we can insert the value of τ into the r.h.s. of those equations and the resulting dynamics will theoretically, up to an accuracy of $O(\tau^2)$, be valid on any other small but finite time scales.

4 Numerical Examples

We shall demonstrate Sequin by way of several simulated examples. The emphasis here will be to show that the method is capable of modeling the underlying dynamics of time series differing only in the functional structure of their stochastic components.

Example 4.1 A time series of 10,000 points was simulated according to the evolution equation

$$dq_i = K(q_i)dt + wg(q_i)\sqrt{dt} \tag{4.1}$$

with

$$K(q_i) = \alpha q_i + \beta q_i^3 \tag{4.2}$$

and

$$g(q_i) = \sqrt{Q_0 + Q_1 q_i} \tag{4.3}$$

Here, $\alpha = 2$, $\beta = -4$, $Q_0 = 2$, and $Q_1 = .5$. w is a Gaussian distributed random variable with zero mean and variance 1. We refer to this time series as A. It is plotted in Figure 1.

This process can be visualized as the trajectory of a particle jiggling in a double-well potential. The noise, which gives the particle momentum to move about, is dependent on the actual value of q_i, or in other words, on the location of the particle in the potential well.

The first step of Sequin requires that the short-time correlations $< \Omega_j >_P$ of (3.5) – (3.7) be calculated as temporal averages $< \Omega_j >_T$ from the time series A, according to (3.17). To capture the first few nonlinearities, we chose $N = 4$, $M = 3$ and $L = 2$. Numerical values of these correlations are shown in Table 1 – Table 3, obtained for $\tau = .01$.

The second step of Sequin involves determining the Lagrange multipliers by way of the gradient strategy of (3.19). We started with randomly chosen initial values of $\{\lambda\}$ lying between 0 and 1 and obtained the final $\{\lambda\}$ values which are also listed in Table 1 – Table 3.

Finally, the numerical values of $\{\lambda\}$ are plugged into (3.22) and (3.23) resulting in the following expressions for the underlying dynamics of the process A:

$$K(q_i) = .04 + 2.7q_i + .6q_i^2 - 4.8q_i^3 \tag{4.4}$$

and

$$g(q_i) = \sqrt{1.95 + .47q_i + .09q_i^2} \tag{4.5}$$

These results are in good agreement with the actual forms of $K(q_i)$ and $g(q_i)$ used in the simulation of (4.1). Most importantly, the main features of the problem

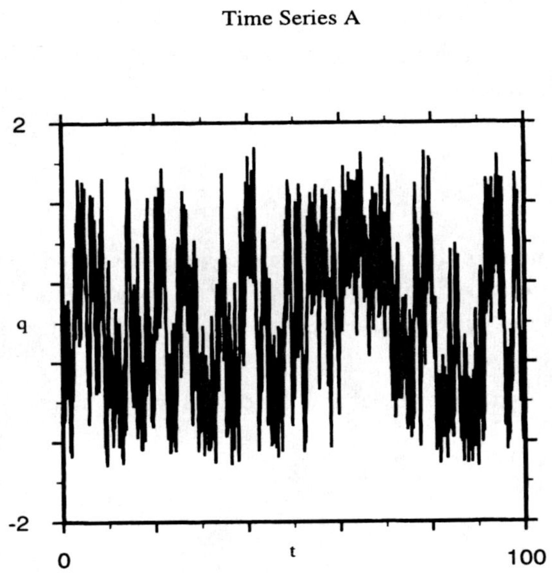

Figure 1 Time series generated by $dq = (2q - 4q^3)dt + \sqrt{2 + .5q}dw$.

are correctly reconstructed: A deterministic force of third order which corresponds to a double-well quartic potential, and a stochastic force which depends on q_i.

Errors in the parameters are mainly due to the finite number of data points in the simulation. Various stopping criteria can be incorporated into the learning dynamics to minimize this error and avoid overfitting. A more thorough error analysis has been performed theoretically in a previous paper (Borland [1996]) and will also be the focus of future work. In this paper we wish to emphasize the qualitative results rather than the exact numerical values of the results.

Example 4.2 A second time series B (see Figure 2) was created in exactly the same manner as the one of (4.1), with the same parameter values, except that the stochastic part of the dynamics was now

$$g(q_i) = \sqrt{Q_0 + Q_1 q_i^2} \tag{4.6}$$

Note that the only difference is that q_i is replaced by q_i^2. Again, the constraints of (3.5) – (3.7) with $N = 4$, $M = 3$ and $L = 2$ were calculated from the time series. These were used in conjunction with (3.19) to determine numerical values of the Lagrange multipliers $\{\lambda\}$. Random initial values of $\{\lambda\}$ lying between 0 and 1 were used. Results are listed in Table 1 – Table 3. These values were inserted into (3.22)

Time Series B

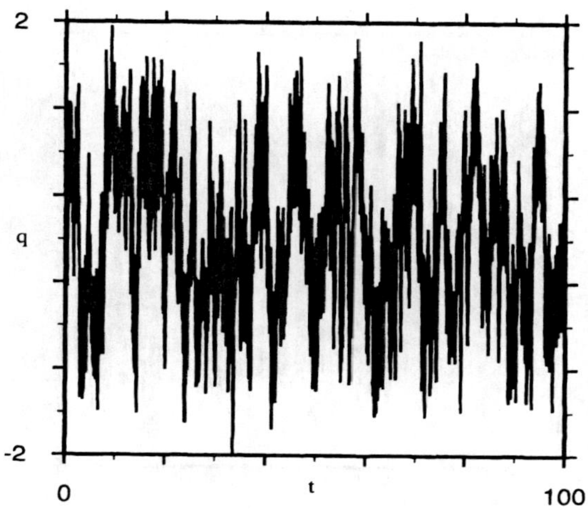

Figure 2 Time series generated by $dq = (21q - 4q^3)dt + \sqrt{2 + .5q^2}dw$.

and (3.23) to yield

$$K(q_\imath) = -.15 + 1.2q_\imath - .4q_\imath^2 - 2.7q_\imath^3 \tag{4.7}$$

and

$$g(q_\imath) = \sqrt{2.0 - .01q_\imath + .3q_\imath^2}. \tag{4.8}$$

In this example, Sequin does relatively well in reconstructing the underlying dynamics of the time series, although the absolute values of the coefficients in the deterministic force are a little low. We attribute this error to limited data. We do, however, wish to point out that Sequin succeeded in correctly detecting a $q_\imath^2$ dependence in the stochastic dynamics of the second time series B, as opposed to a $q_\imath$ dependence in the first one A.

Example 4.3 As a second set of examples we chose a slightly more complicated simulation, in the sense that the generating function does not belong to the model class of polynomials. Two time series were generated, again according to (4.1). We shall refer to them as C and D. The deterministic part of the equation was chosen as in (4.2) whereas the stochastic parts were chosen as

$$g(q_\imath) = \sqrt{Q_0 + Q_1 \sin(q_\imath)} \tag{4.9}$$

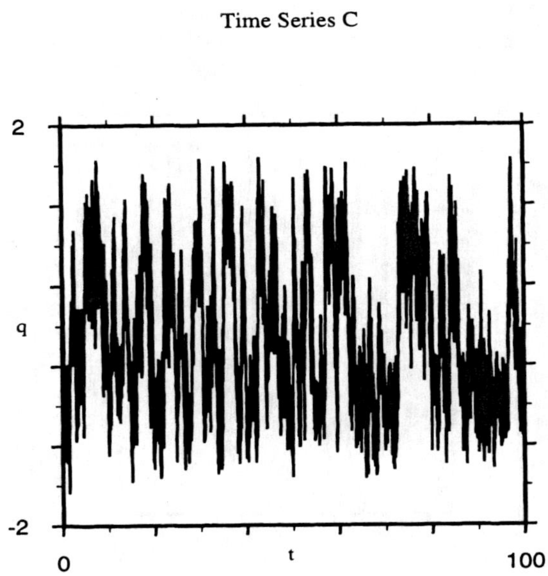

Figure 3 Time series generated by $dq = (2q - 4q^3)dt + \sqrt{2 + .5\sin(q)}dw$.

and

$$g(q_\imath) = \sqrt{Q_0 + Q_1 \cos(q_\imath)} \tag{4.10}$$

respectively. Parameter values $Q_0 = 2.$ and $Q_1 = .5$ were used. Again, the physical problem corresponds to one of a particle in a double well potential, with a random force functionally dependent on $q_\imath$. The two time series C and D are plotted in Figures 3 and 4. The constraints of (3.5) – (3.7) for $N = 4$, $M = 3$ and $L = 2$ were calculated as temporal averages according to (3.17) from each time series as shown in Table 1 – Table 3. Subsequently, (3.19) together with randomly chosen initial values of $\{\lambda\}$ lying between 0 and 1 was used to find values of the Lagrange multipliers. These were plugged into the equations (3.22) and (3.23). We obtain for time series C

$$K(q_\imath) = -.45 + 2.3q_\imath + .8q_\imath^2 - 4.2q_\imath^3 \tag{4.11}$$

and

$$g(q_\imath) = \sqrt{2.0 + .45q_\imath - .09q_\imath^2} \tag{4.12}$$

The reconstructed deterministic force is close to the original generating function of (4.2). A closer look at the stochastic force $g(q_\imath)$ shows that Sequin has correctly captured the form of (4.9), if we substitute $\sin(q_\imath)$ with its power series expansion

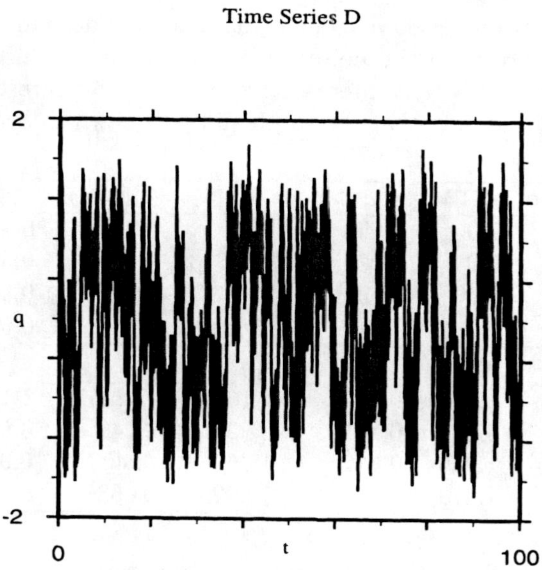

Figure 4 Time series generated by $dq = (2q - 4q^3)dt + \sqrt{2 + .5\cos(q)}dw$.

up to $O(q_\imath^2)$:

$$\sin(q_\imath) \approx q_\imath \tag{4.13}$$

Analogous results were obtained for time series D. Here, we obtained for the deterministic force

$$K(q_\imath) = 0.0 + 2.9q_\imath - .3q_\imath^2 - 4.9q_\imath^3 \tag{4.14}$$

and for the stochastic force

$$g(q_\imath) = \sqrt{2.5 - .05q_\imath - .22q_\imath^2}. \tag{4.15}$$

The structure of the deterministic force $K(q_\imath)$ is a good approximation to the generating function of (4.2). By substituting

$$\cos(q_\imath) \approx 1 - \frac{1}{2}q_\imath^2 \tag{4.16}$$

into the expression of (4.10) for the stochastic generating force, together with the values for Q_0 and Q_1, it becomes:

$$g(q_\imath) = \sqrt{2.5 - .25q_\imath^2} \tag{4.17}$$

This expression is in very good comparison with the polynomial representation of (4.15) as reconstructed by Sequin. We see that Sequin again manages to correctly capture the first few nonlinearities of the underlying stochastic force.

The above four examples serve to show that Sequin indeed manages to separate deterministic structure from the noise, as well as model the nonlinear structure inherent within the noise. It is interesting to visually compare the time series of Figure 1 – Figure 4 and note that the subtle differences which Sequin detected are not obvious to the eye.

Example	$<1>$	$<q_i>_T$	$<q_i^2>_T$	$<q_i^3>_T$	$<q_i^4>_T$
A	1.0	0.004	0.517	0.033	0.496
B	1.0	0.002	0.497	0.028	0.521
C	1.0	-0.128	0.543	-0.065	0.546
D	1.0	-0.039	0.580	-0.053	0.619
	$\lambda_0^{(1)}$	$\lambda_1^{(1)}$	$\lambda_2^{(1)}$	$\lambda_3^{(1)}$	$\lambda_4^{(1)}$
A	1.043	0.197	-1.163	-0.215	1.120
B	0.831	0.141	-0.222	0.140	0.591
C	1.085	0.652	-1.067	-0.409	0.990
D	1.113	0.001	-1.092	0.063	0.920

Table 1: Stationary Moments and Lagrange Multipliers

Example	$<q_{i+\tau}>_T$	$<q_i q_{i+\tau}>_T$	$<q_i^2 q_{i+\tau}>_T$	$<q_i^3 q_{i+\tau}>_T$
A	0.004	0.507	0.030	0.479
B	0.002	0.485	0.029	0.501
C	-0.128	0.533	-0.065	0.529
D	-0.039	0.568	-0.052	0.598
	$\lambda_0^{(2)}$	$\lambda_1^{(2)}$	$\lambda_2^{(2)}$	$\lambda_3^{(2)}$
A	0.023	52.629	-12.190	0.678
B	-0.077	49.967	-0.208	-9.011
C	-0.219	50.276	-10.208	2.057
D	-.001	41.239	0.802	1.699

Table 2: Short-time Correlations and Lagrange Multipliers

Example	$<q_{i+\tau}^2>_T$	$<q_i q_{i+\tau}^2>_T$	$<q_i^2 q_{i+\tau}^2>_T$
A	0.517	0.030	0.475
B	0.497	0.029	0.495
C	0.543	-0.065	0.524
D	0.580	-0.052	0.591
	$\lambda_0^{(3)}$	$\lambda_1^{(3)}$	$\lambda_2^{(3)}$
A	25.613	-6.258	1.615
B	24.702	-0.226	-3.837
C	24.532	-5.416	0.132
D	20.016	0.450	1.92

Table 3: Short-time Correlations and Lagrange Multipliers

5 Concluding Remarks

This paper gives a short review of the procedure Sequin for reconstructing the underlying dynamics of stochastic time series. The applicability of the technique was demonstrated by way of four simulated time series, differing only in the functional structure of their underlying stochastic dynamics. Sequin proved successful in modeling both the underlying deterministic and stochastic nonlinear dynamics, indicating its possible application to a variety of interesting real world problems.

References

Borland, L. [1996] *Simultaneous modeling of nonlinear deterministic and stochastic dynamics*, Physica D, to be published.

Borland, L. and Haken, H. [1992a] *Unbiased estimate of forces from measured correlation functions, including the case of strong multiplicative noise*, Ann. Physik **1**, 452.

Borland, L. and Haken, H. [1992b] *Unbiased determination of forces causing observed processes. The case of additive and weak multiplicative noise*, Z. Phys. B - Condensed Matter **81**, 95.

Borland, L. and Haken, H. [1993a] *On the constraints necessary for macroscopic prediction of stochastic time-dependent processes*, ROMP **33**, 35.

Borland, L. and Haken, H. [1993b] *Learning the dynamics of two-dimensional stochastic Markov processes*, Open Systems and Information Dynamics, **1** No. 3, 311.

Box, G. E. P. and Jenkins, G. M. [1976] *Time Series Analysis Forecasting and Control*, 2nd Ed., Holden-Day, San Francisco.

Casdagli, M. and Weigand, A. (eds.) [1992] *Nonlinear Modeling and Forecasting*, Santa Fe Institute Series XII, Addison-Wesley.

Crutchfield, J. P. and McNamara, B. S. [1987] *Equations of motion from a data series*, Complex System **1**, 417.

Gardiner, C. W. [1983] *Handbook of Stochastic Methods for Physics, Chemistry, and the Natural Sciences*, Springer-Verlag, Berlin.

Haken, H. [1988] *Information and Self-Organization, A Macroscopic Approach to Complex Systems*, Springer Ser. Synergetics, Vol. 40, Springer Verlag, Berlin, Heidelberg.

Jaynes, E. T. [1957] *Information theory and statistical mechanics*, I, II Phys. Rev , **106**, 620, **108**, 171.

Packard, N. H., Crutchfield, J. P., Farmer, J. D. and Shaw, R. S. [1980] *Geometry from a time series*, Phys. Rev. Lett. **45**, 712.

Tjøstheim, D. [1992] *Nonlinear time series: a selective review*, SFB 123, Universität Heidelberg.

Fields Institute Communications
Volume 11, 1997

Characterizing Nonlinearity in Weather and Epilepsy Data: A Personal View

Martin C. Casdagli
The Harrison Randall Laboratory of Physics
The University of Michigan
Ann Arbor, MI, USA 48109-1120
Current address: The Prediction Company
320 Aztec Street
Santa Fe, NM, USA 87501

Abstract. I present an informal summary of recent results on characterizing nonlinearities in time series data from weather and epilepsy. The weather data consists of daily barometric pressure recordings from 23 locations in Australia, allowing comparisons to be made between tropical and temperate regions. The epilepsy data consists of invasive EEG recordings from multiple depth and subdural electrodes in 7 patients with temporal lobe epilepsy, allowing comparisons to be made between normal and epileptogenic regions of the brain. Nonlinearities in these time series are characterized using correlation integrals, surrogate data, nonlinear prediction and residual delay maps. Practical and theoretical implications of the results are discussed, and directions for future research speculated upon.

The hardest thing in life to learn is which bridge to cross and which to burn.
David Russell

1 Introduction

There are a wide variety of techniques available for the analysis of aperiodic time series data. Until recently, it was thought that aperiodic behavior must be due to complicated underlying dynamics, and linear techniques motivated by the theory of stochastic processes dominated the field. Nonlinear generalizations of these linear techniques have been developed. For example, the power spectrum can be generalized to the bispectrum, and linear autoregressive models can be generalized to threshold autoregressive models (Tong [1990]). Also, it is now common knowledge that remarkably simple deterministic nonlinear systems can produce aperiodic *chaotic* time series. For example, a periodically driven pendulum can exhibit chaos. When considered in the underlying state space, chaotic time series from such systems correspond to motion on an attractor with a low fractal dimension. Low-dimensional attractors are also expected to occur near transitions from periodic to aperiodic behavior in more complicated "many degree of freedom" systems such as

1991 *Mathematics Subject Classification.* Primary 58F13, 62M20; Secondary 62M10.

fluids. These results have stimulated the development of techniques to test for and quantify chaos in time series data (Abarbanel et al. [1993], Eckmann and Ruelle [1985], Tsonis [1992]). Most of these techniques attempt to reconstruct the underlying attractor from the time series, and estimate invariants such as dimensions and Lyapunov exponents. Other techniques attempt to construct deterministic autoregressive models for the time series. Techniques have also been developed using a combination of ideas from the nonlinear deterministic and stochastic approaches. For example, various bootstrapping and surrogate data techniques are becoming increasingly popular (Brock et al. [1991], Grassberger [1986], Theiler et al. [1991]). Using the above techniques, it is now possible to obtain answers to the following questions about time series obtained from laboratory experiments or from observations of natural processes.

1. Is the time series linear or nonlinear? In particular, is it completely characterized by the power spectrum?

2. Is the time series chaotic? In particular, does it come from an approximately deterministic system with a low-dimensional attractor?

3. Can the time series be predicted more accurately with nonlinear autoregressive models than with linear autoregressive models?

4. If the time series is nonlinear, what is the functional form of an appropriate statistical model for characterizing the nonlinearity in the time series?

In this paper, I address the above questions for time series from weather and epilepsy that I have been studying with several collaborators. Special emphasis will be placed on the characterization of nonlinearities in these time series. The presentation will be informal, in the sense that an introductory overview of ideas and results will be given, with several personal comments on the motivation for the choice of techniques and interpretation of results. For further details, papers written with my collaborators may be consulted (Casdagli et al. [1996], Sugihara et al. [1995]). I will emphasize the interdisciplinary nature of this research, and discuss the practical and theoretical implications of obtaining answers to questions 1–4 above, within the context of weather and epilepsy. This research represents a few man-months of effort, excluding time spent in collecting data and debugging software. Consequently, the techniques used in this paper are of an exploratory nature, and were chosen to detect the "dominant" nonlinearities in these time series. Directions for future research are speculated upon below. It should also be noted that the approach taken in this paper reflects my scientific background, which is in deterministic dynamical systems theory. It is of course possible to take many other approaches to these problems.

2 Weather Recordings

I recently received a time series of daily barometric pressures recorded in Sydney, Australia during 1955-1989, from the good offices of George Sugihara. He had detected nonlinearity in the time series, and enquired whether I could confirm his result. This section first summarizes my investigations into the time series, and then summarizes the results of a collaboration with George Sugihara on pressure records from other locations in Australia.

The atmosphere is a nonlinear, approximately deterministic, dynamical system with many degrees of freedom, not experimentally controlled to lie near a transition to periodic behavior. Not surprisingly, mathematical models of the atmosphere exhibit high-dimensional chaos. Indeed, Lorenz has shown that the enormous variety of large-scale spatial patterns observed in the atmosphere is inconsistent with low-dimensional chaos (Lorenz [1969]). Reports of low-dimensional chaos in "weather" (short time scale) or "climate" (long time scale) recordings should be treated with suspicion, as cautioned by Grassberger [1986]. On the other hand, Palus recently reported strong statistical evidence for nonlinearity in a time series of daily barometric pressures recorded in Prague from 1789-1989 (Palus and Novotna [1994]). This is a non-trivial result, because although atmospheric dynamics is known to be highly nonlinear, the dynamics may be so complex that nonlinearity cannot be detected in a time series recorded from a single location. For example, Palus reports that temperature recordings from Prague failed to exhibit nonlinearity (Palus and Novotna [1994]).

Following Palus, I first tested the Sydney data for nonlinearity using a *surrogate data technique,* because this approach has the advantage of being sensitive to a wide range of nonlinearities. Surrogate data techniques compare a *discriminating statistic* of a time series to the discriminating statistic of randomized surrogate data sets with the same power spectrum as the time series. If significant differences are found, this is evidence that there is additional "nonlinear" structure in the time series beyond the linear structure captured in the power spectrum. A simple algorithm for generating surrogate data sets with a Gaussian distribution is summarized as follows (Theiler et al. [1991], Theiler and Rapp [1996]). First, the original time series is detrended to avoid the effect of edge discontinuities on the Fourier transform (FT). Second, the FT of the time series is computed. Third, the phases of the FT are uniformly randomized, and the amplitudes preserved. Fourth, the inverse FT is computed to obtain a surrogate data set. The randomization algorithm is then iterated to obtain several surrogate data sets. The resulting surrogate data sets will be referred to as *Gaussian surrogates* below.

Gaussian distributions are the unique distributions with maximum entropy for a fixed covariance (Shannon and Weaver [1964]). This suggests the use of entropy as the discriminating statistic in the surrogate data technique, so that it is sensitive to the widest possible range of nonlinearities, as follows. Let $x_1, x_2, .., x_N$ be the values of the original time series. If the joint distribution of the "embedding vectors" $\mathbf{v_t} = (x_t, x_{t-\tau}, .., x_{t-(m-1)\tau})$ is not Gaussian for some embedding dimension m and time delay τ, then the entropy of the joint distribution will be strictly less than for the Gaussian surrogates. Entropies of joint and conditional distributions have been used as discriminating statistics by Palus (Palus and Novotna [1994]). Correlation integrals are closely related to entropies, but can be estimated more efficiently from time series data (Prichard and Theiler [1995]). Following (Brock et al. [1991], Grassberger [1986], Theiler et al. [1991]), in this paper I will use correlation integrals as discriminating statistics in the surrogate data technique. The *correlation integral* $C(r, m, \tau, w)$ is the probability that two randomly chosen embedding vectors $\mathbf{v_t}$ and $\mathbf{v_s}$ with embedding dimension m and time delay τ lie within distance r of each other, and is estimated by computing a double sum (Grassberger [1990], Grassberger and Procaccia [1983]). An *exclusion window* w may be chosen to eliminate pairs of embedding vectors $\mathbf{v_t}$ and $\mathbf{v_s}$ with $|t - s| \leq w$, that are close due to temporal correlations in the time series (Theiler [1986]). Correlation integrals were originally developed to quantify low-dimensional chaos in a time series. This is because, if the time series is generated by a chaotic attractor of *correlation dimension* d, then the correlation integral satisfies the scaling property

$C(r, m, \tau, w) \sim r^d$ as r tends to zero, for embedding dimensions $m > d$ (Grassberger and Procaccia [1983]).

Figure 1(b) shows a scatter plot of embedding vectors $\mathbf{v_t}$ with embedding dimension $m = 2$ and time delay $\tau = 1$, for 45 years of Sydney data, one year of which is shown in Figure 1(a). The distribution of embedding vectors is clearly non-Gaussian, so that the surrogate data technique described above is practically guaranteed to detect nonlinearity in this data. It is possible to modify the surrogate data technique so that it is insensitive to any nonlinearities arising from the measurement function used to obtain the original time series, by rescaling the Gaussian surrogates. The resulting surrogate data sets will be referred to as *rescaled surrogates* below, and are obtained as follows. Let $g_1, g_2, .., g_N$ be a Gaussian surrogate of the original time series $x_1, x_2, .., x_N$. A rescaled surrogate $y_1, y_2, .., y_N$ is obtained by setting $y_i = f(x_i)$, $i = 1, 2, .., N$, where the nonlinear rescaling transformation f is constructed so that $y_1, y_2, .., y_N$ have the same marginal (one-dimensional) distribution as the original time series $x_1, x_2, .., x_N$. The rescaling transformation f is uniquely determined; for details see Theiler et al. [1991]. It follows that if the original time series was obtained from a nonlinear measurement function f of a Gaussian time series, then the rescaled surrogates will have approximately the same statistical properties as the original time series.

Figure 1(c) shows how the correlation integral $C(r, m, \tau, w)$ varies with r, for embedding dimension $m = 6$, time delay $\tau = 1$, and exclusion window $w = 12$, for the Sydney data and each of 10 Gaussian and 10 rescaled surrogates. The correlation integrals of the Sydney data are significantly larger than those of the surrogates, demonstrating that there is nonlinearity in the Sydney data which cannot be explained by a nonlinear measurement function. Figure 1(d) shows correlation dimensions $D(r)$ estimated from the slopes of the line segments in Figure 1(c). The correlation dimensions of the Sydney data are less than the correlation dimensions of the surrogate data, but only over a range of r where $D(r)$ is not constant (I obtained similar results using more sophisticated techniques for computing correlation dimensions from correlation integrals (Smith [1991])). This implies that the Sydney data has a correlation dimension of at least 6, as is to be expected from the introductory remarks of this section. The results shown in Figure 1(e) address the question of whether the Sydney data can be predicted more accurately with nonlinear autoregressive models than with linear autoregressive models, using a "deterministic versus stochastic" (DVS) technique (Casdagli [1992]). Since nonlinearity has already been detected in the Sydney data using the surrogate data technique, one might expect that nonlinear predictability is forced to exist in this data. However, the nonlinearity identified by the surrogate data technique only indicates that there is more "clustering" in a 6 dimensional version of the scatter plot shown in Figure 1(b) than for the surrogates. Clustering can be exploited in an information-theoretic sense, but does not necessarily imply that nonlinear predictability exists in the Sydney data. For example, most financial time series can be shown to be nonlinear using surrogate data techniques, but exhibit very little predictability in any conventional sense (Hseih [1991]). A solution to this apparent paradox is that financial time series exhibit "volatility" structure (large price moves tend to be followed by large price moves, but the direction of the moves are unpredictable), and so are predictable in an unconventional sense (Hseih [1991]).

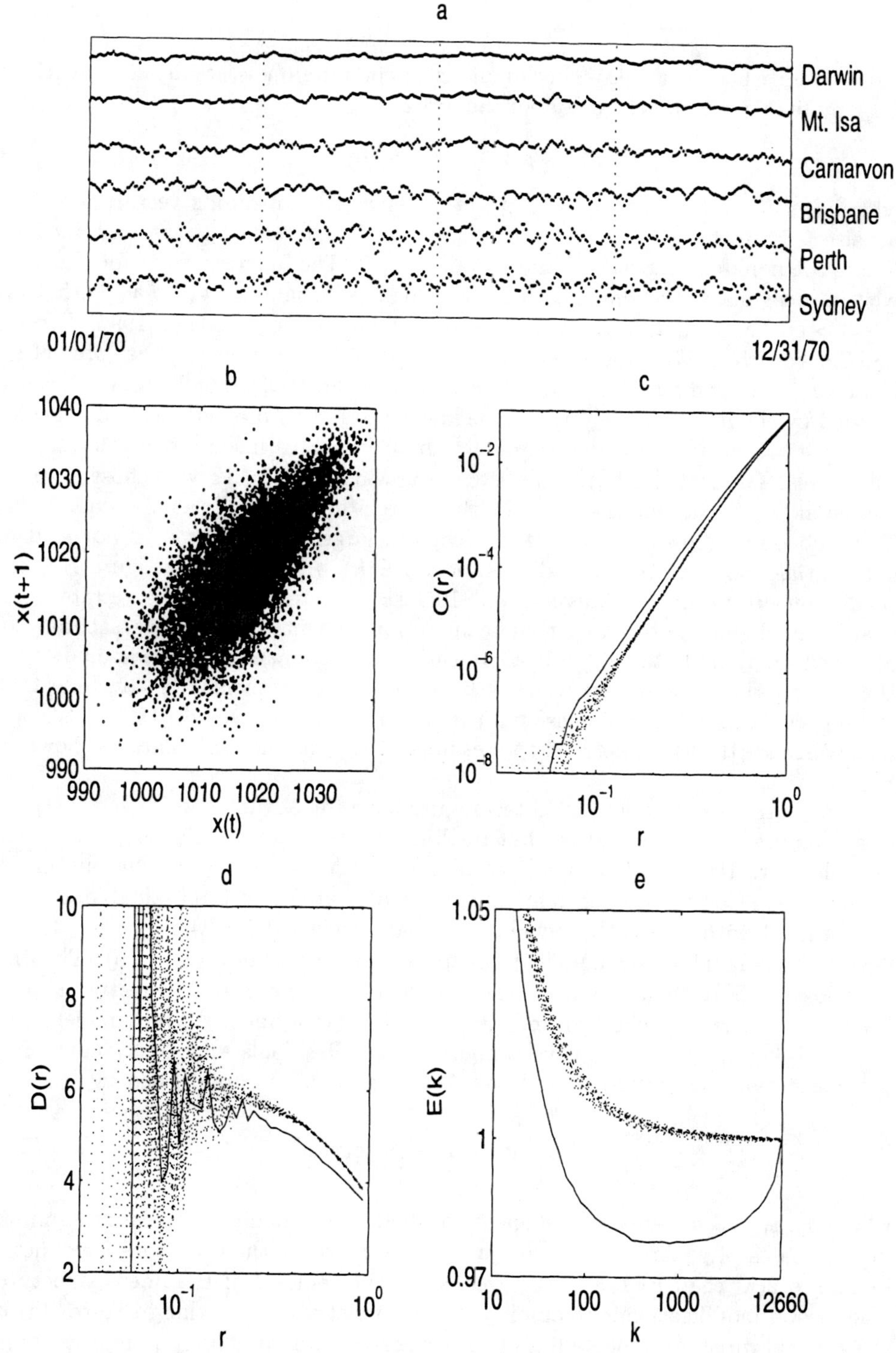

Figure 1 (a) Barometric pressures recorded daily in 6 locations in Australia during 1970. Latitudes of the locations are shown in Figure 2(c). (b) Scatter plot of pressures x_{t+1} against x_t for 45 years of data from Sydney. (c) Dependence of the correlation integral $C(r, m = 6, \tau = 1, w = 12)$ on r. Solid curve: Sydney data; Dots: 10 Gausssian and 10 rescaled surrogates. (d) Dependence of the correlation dimension $D(r)$ on r. (e) Dependence of the prediction error $E(k)$ on the number of neighbors k, with parameters $m = 4, \tau = 1, w = 12$.

The idea behind the DVS technique is to compute one-step ahead predictions $\hat{x}_{t+1}$ by fitting a nonlinear model of the form

$$x_{t+1} = f(\mathbf{v_t}) + n_t \tag{2.1}$$

where $\mathbf{v_t} = (x_t, x_{t-1}, .., x_{t-m+1})$ is an m-dimensional embedding vector, and n_t is a noise term to be minimized by least squares. Predictions $\hat{x}_{t+1}$ are obtained by fitting the nonlinear function f and setting $n_t = 0$. The function f is approximated with a local linear technique, using the k nearest neighbors $\mathbf{v_s}$ to $\mathbf{v_t}$ such that $|s-t| > w$. For similar local approximation techniques see (Casdagli [1992], Farmer and Sidorowich [1987], Sauer [1994], Sugihara [1994]). In Figure 1(e) the "prediction error" $E(k)$ is computed as the mean square error of the nonlinear predictions divided by the mean square error of the linear (largest k) predictions, and provides a comparison of nonlinear deterministic (small k) versus linear stochastic (largest k) models (Casdagli [1992]). The exclusion window $w = 12$ was chosen, to be consistent with the window used in the correlation integral calculations above. The embedding dimension $m = 4$ was chosen; larger values of m did not lead to substantial decreases in the prediction errors $E(k)$. Also shown in Figure 1(e) are prediction errors for 10 Gaussian and 10 rescaled surrogates. The results show that a small, but statistically significant, improvement in predictive accuracy of about 2% may be obtained with nonlinear models compared to linear models for the Sydney data. The improvement in accuracy is insufficient to conclude that there is evidence for low-dimensional chaos in this data (Casdagli [1992], Casdagli and Weigend [1993]), confirming the results of the dimension calculations shown in Figure 1(d).

The surrogate data and DVS techniques have the advantage of being sensitive to a wide range of nonlinearities, but the limitation that very little insight is gained into the characteristics of any nonlinearities identified (unless low-dimensional chaos is detected). Considerable insight into the nonlinearities in the Sydney data may be obtained with a simpler technique, called *residual delay maps* below, that I developed by making two modifications to the scatter plot in Figure 1(b). The first modification is to remove some linear structure. Figure 2(a) is a scatter plot of the vectors (x_t, r_{t+1}) where r_t are the residuals from a linear AR(m) model, with $m = 4$; other choices for m gave similar results. Residuals may be thought of as the one step ahead errors made by the AR(m) model, and are defined by

$$r_t = x_t - a_0 - \sum_{i=1}^{m} a_i x_{t-i} \tag{2.2}$$

where $a_0, a_1, \ldots, a_m$ are the coefficients of the AR(m) model obtained by minimizing the sums of squares of the residuals. Observe that there is more asymmetry in Figure 2(a) than Figure 1(b), revealed by the removal of the linear structure. The second modification is to amplify this asymmetry by smoothing Figure 2(a) in the vertical direction: the values of x_t are sorted into 25 equally populated bins, and the mean of the r_{t+1} is plotted against the mean of the x_t for each bin. The result is shown in Figure 2(b) as a solid curve, and is called a "residual delay map" (RDM). RDMs provide a simple estimate for the "lag-one nonlinearity" $f_1(x)$ in a time series model of the form

$$x_{t+1} = f_1(x_t) + a_0 + \sum_{i=0}^{m-1} a_i x_{t-i} + n_t \tag{2.3}$$

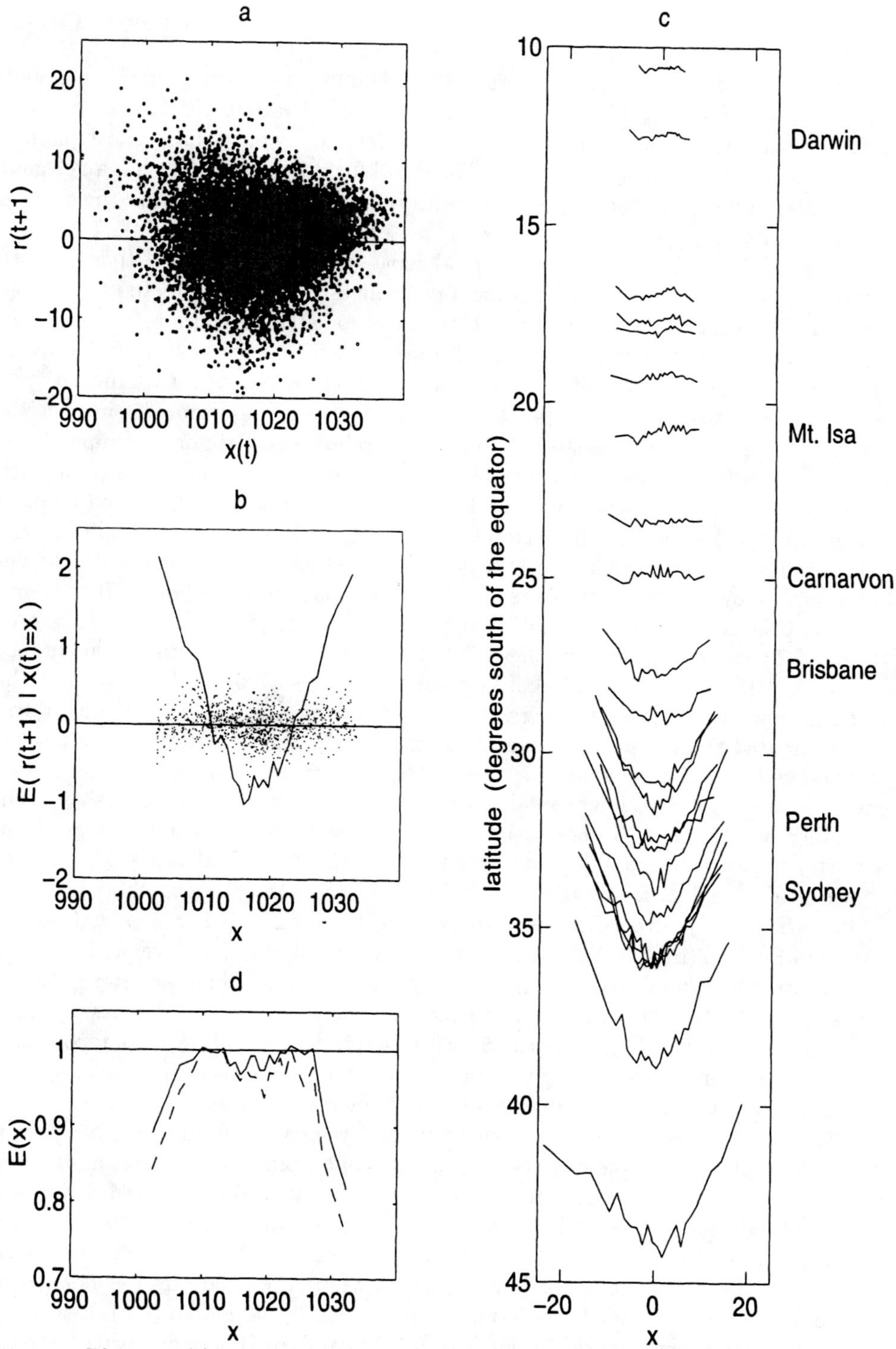

Figure 2 (a) Scatter plot of residuals r_{t+1} from an AR(4) model against pressures x_t for 45 years of data from Sydney. (b) Residual delay maps (RDMs). Solid curve: Sydney data; Dots: 10 Gaussian and 10 rescaled surrogates. (c) Dependence of RDMs on latitude for 23 locations in Australia. Pressures were normalized to have zero mean to compensate for low pressures at inland locations. (d) Dependence of the prediction error $E(x)$ on pressure x. Solid curve: predictions incorporating lag-one nonlinearity; Dashed curve: predictions for the fully nonlinear model (2.1) with $m = 4$ and $k = 500$.

where $a_0, a_1, .., a_m$ are the parameters in (2.2) defining the linear part of the model. Also shown in Figure 2(b) are RDMs for 10 Gaussian and 10 rescaled surrogates, demonstrating that the lag-one nonlinearity in the Sydney data cannot be explained by a nonlinear measurement function. The RDM for the Sydney data has an elegant "quadratic" functional form; inferior results were obtained by binning the vectors (x_t, x_{t+1}) or (r_t, r_{t+1}) instead of (x_t, r_{t+1}).

The results summarized above would have terminated my investigation into weather recordings had it not been for further investigations conducted by George Sugihara. He proceeded to compute RDMs for barometric pressure recordings from other locations in Australia. The results are shown in Figure 2(c), and provide remarkably clear evidence of a transition to lag-one nonlinearity as the latitude is varied from tropical to temperate extremes. Our collaborators from the Bureau of Meteorological Research Centre, Australia, who originally supplied the weather records, pointed out that the results in Figure 2(c) are consistent with growing theoretical evidence of a marked change in nonlinearity between tropical and extra-tropical regions. In particular, much of the large-scale variability in the extra-tropical atmosphere arises from baroclinic instability associated with the development of cyclones, anticyclones and fronts, and is highly nonlinear (Buizza and Palmer, Charney and Shukla [1981], Sardeshmukh and Hoskins [1995], Trenberth [1991]). The question of how these "spatio-temporal" nonlinearities can induce quadratic RDMs in pressure recordings remained unsolved until we utilized the fact that low pressure events, corresponding to storms, are typically of short duration compared to high pressure events. This causes predictions from linear models to underestimate the next day's pressure. The effect is apparent in Figure 2(a): the residuals r_{t+1} from linear models at low pressures x_t tend to be positive. On the other hand, it is well known that high pressure events persist for longer than average, causing linear models to again underestimate the next day's pressure, as shown in Figure 2(a) for large values of x_t. The structure in Figure 2(a) is amplified by the binning procedure used to produce Figure 2(b). The quadratic RDM shown in Figure 2(b) may thus be interpreted as quantifying the movement of storm fronts through Sydney, and the RDMs shown in Figure 2(c) may be interpreted as quantifying the lack of storms in regions of Australia north of 30 degrees latitude. This interpretation may be confirmed by a visual inspection of pressure recordings, some of which are shown in Figure 1(a). In particular, the recording from Sydney exhibits sharp lows and smooth highs throughout the 4 seasons.

The quadratic shape of the lag-one nonlinearity shown in Figure 2(b) has practical implications for the one day ahead prediction of barometric pressures at extreme low and high pressure events. It can be shown that inclusion of this feature in a well-regarded numerical weather forecasting model results in substantial improvements in prediction accuracy (Sugihara et al. [1995]). This is demonstrated for the case of linear prediction in Figure 2(d), in which the solid curve is defined to be the mean square error of predictions from the nonlinear model (2.3) divided by the mean square error of predictions from a linear AR(4) model, sorted into 25 equally populated bins by the current value of the pressure. The improvements in prediction accuracy obtained with the nonlinear model are concentrated at extreme events. The dashed curve in Figure 2(d) was obtained by fitting the fully nonlinear model (2.1) with $m = 4$, using $k = 500$ nearest neighbors, and gives even better predictions at extreme events. The opportunity for making substantial improvements to predictions at extreme events eluded the analysis of Figure 1(e), because a global measure of prediction error was used. Related techniques for detecting state-dependent predictability in nonlinear systems have been developed by Smith [1995].

The lag-one nonlinearity identified in Figure 2(b) deteriorates at time scales greater than 2 days. However, I have observed stronger annual periodicities in the τ dependence of correlation integrals for the Sydney data than for surrogates; similar results were reported by Palus for Prague (Palus and Novotna [1994]). I believe that this is due to seasonal effects, some of which are not adequately reflected by surrogates. For example, the variance of pressure recordings tends to be significantly larger in the winter than the summer, and this is not seen in surrogates. By contrast, there is very little seasonal variation in the quadratic structure of RDMs shown in Figure 2(c), except at locations near 30 degrees latitudes (Sugihara et al. [1995]). Such locations are known to have climates exhibiting temperate behavior in the winter and tropical behavior in the summer. More sophisticated time series models, incorporating seasonal effects, are required to fully describe the above phenomena.

Some questions raised by this research are as follows. From a practical point of view, it may be possible to obtain improved predictions by constructing more complicated time series models. For example, more sophisticated function approximation techniques such as neural nets (Weigend and Gershenfeld [1993]) could replace the local linear and binning techniques used above, and longer time scales could be incorporated. Nonlinear surrogate data, generated by iterating such models, could be checked for consistency with the Sydney data using a battery of statistics such as those presented above. Ultimately, one might expect further improvements in prediction accuracy from a spatio-temporal time series model describing the structure of pressure and velocity recordings from multiple locations. From a theoretical point of view, it would be interesting to know if simulated pressure data, produced by mathematical models of the atmosphere, exhibit RDMs with approximately the same functional form and latitudinal dependence shown in Figure 2(c). If not, some important physics may have been ignored in the construction of the mathematical model. The simple characterization of nonlinearities in real world pressure recordings by RDMs may provide a convenient basis for making such comparisons.

3 Epilepsy

The main project that I have been working on this year is the nonlinear time series analysis of a large database of invasive EEG recordings from patients with temporal lobe epilepsy (TLE). This section presents the results obtained to date. The medical background of TLE and some clinical objectives of quantitative EEG analysis are outlined in Section 3.1. Results obtained with surrogate data, prediction algorithms and residual delay maps are presented in Sections 3.2, 3.3 and 3.4 respectively. These sections may be skipped at a first reading. Finally, Section 3.5 summarizes the results and discusses the extent to which the medical objectives of Section 3.1 have been addressed.

3.1 Medical Background and Objectives. About 1% of the population suffers from some form of epilepsy (Engel [1989]). About half the cases are characterized by seizures which originate from a localized region of the brain, called the *epileptogenic focus,* and rapidly propagate to more normal regions of the brain. In many of these cases, called *temporal lobe epilepsy* (TLE), there is a unique epileptogenic focus located in one of the temporal lobes or hippocampi. Patients with

TLE usually respond well to drugs; the remainder are candidates for surgery. Noninvasive techniques such as MRI scans and scalp EEG often fail to unambiguously locate the epileptogenic focus. Invasive EEG recordings, which have much better spatial resolution and are relatively free of electromyogenic (muscle) artifact compared to scalp EEG recordings, are performed in a few hundred patients annually in the USA. The recordings may last for up to 10 days, during which a patient is hospitalized, the objective being to record several seizures. Electroencephalographers visually inspect the recordings, for the duration of a few minutes encompassing the seizures, in order to locate epileptogenic foci (King and Spencer [1995]). If the patient does indeed have TLE, a left or right temporal lobectomy is performed. Marked improvements in quality of life are achieved after about 70% of these operations, although side effects are difficult to quantify, and there is about a 2% risk of severe complications (Spencer and Pappas [1992]).

The use of computers in analyzing invasive EEG recordings is typically limited to the automated detection of seizures. Some medical objectives of a quantitative analysis of invasive EEG recordings are as follows.

1. *Localization.* Very little use is made of "interictal" (between seizures) recordings. If the epileptogenic focus could be reliably located from interictal recordings, this would drastically reduce the need for week-long recordings in hospitalized patients.

2. *Prediction.* It is notoriously difficult to predict the onset of seizures more than a few seconds in advance from a visual inspection of EEG recordings. If seizures could be predicted a few minutes in advance, this would raise the possibility of interrupting the onset of seizures, and the need for a temporal lobectomy might be avoided.

3. *Understanding.* The dynamical mechanisms responsible for generating seizures are not well understood. A quantitative analysis of invasive EEG recordings could potentially aid in the development and validation of mathematical models for TLE, and lead to a more scientific approach to the analysis of these recordings.

4. *Postoperative studies.* Approximately 30% of patients do not benefit from the temporal lobectomies they receive. If properties of invasive EEG recordings could be used to accurately discriminate between the 70% of patients with successful postoperative outcomes, and the 30% of patients with unsuccessful postoperative outcomes, then many unsuccessful temporal lobectomies could potentially be avoided in the future. To date, we have not attempted to address this objective, since it requires a larger database of recordings and postoperative results than is currently available to us.

It is natural to attempt to analyze invasive EEG recordings with nonlinear techniques, because the dynamics of neurons at the microscopic level is highly nonlinear. Moreover, in the past decade, convincing evidence for low-dimensional chaos has been found in various experiments on isolated neurons (Elbert et al. [1994]). Of particular relevance to TLE are results indicating that seizures induced in rats and rabbits exhibit low-dimensional chaos (Freeman [1987], Pijn et al. [1991]). The analysis of human EEG data is more problematic, due to the limited range of conditions under which recordings of the human brain can be made. Nevertheless, evidence for low-dimensional chaos has been reported in scalp EEG recordings of a petit-mal seizure (Babloyantz and Destexhe [1986]) , and in scalp EEG recordings of generalized seizures (Frank et al. [1990]). On the other hand, surrogate data techniques have been used to show that there is no evidence for low-dimensional chaos, and only weak nonlinearity, in scalp EEG recordings from healthy humans (Palus, Theiler and Rapp [1996]). Also, doubts have been raised about the evidence for low-dimensional chaos in generalized seizures (Theiler [1995]). Consequently, it is not easy to anticipate the outcome of a nonlinear analysis of invasive EEG recordings.

3.2 Surrogate Data and Spiking Activity. The surrogate data technique described in Section 2 is a natural technique to employ in an initial exploratory nonlinear time series analysis, because it is sensitive to a wide range of nonlinearities. Figures 3(a)-(d) show the results of applying the surrogate data technique to 85 minutes of an invasive EEG recording containing a seizure of 30 seconds duration. The recording, sampled at 200Hz, measures the potential differences between a scalp reference and 28 electrodes, 12 of which are "depth electrodes" placed within the hippocampal regions (referred to as electrodes LTD1,3,..,11 and RTD2,4,..,12 below), and 16 of which are "subdural electrodes" placed on the surface of the orbital frontal and subtemporal lobes (referred to as electrodes LOF1-4, ROF1-4, LST1-4 and RST1-4 below). For more details, see Casdagli et al. [1996]. A fundamental problem in the application of techniques of nonlinear time series analysis to EEG recordings is nonstationarity. The results shown in Figures 3(a)-(d) were obtained by dividing the recording into nonoverlapping segments of duration 10.24 seconds, in an attempt to make the signal approximately stationary within each segment, as well as to detect dynamical changes between segments. The recording was also split into 28 single electrode recordings to further simplify the analysis. This approach is similar to that taken by my collaborators in investigations of the spatio-temporal dependence of Lyapunov exponents in invasive EEG recordings (Iasemidis and Sackellares [1991], Iasemidis et al. [1990]). For a similar approach using correlation integrals, see Lerner [1996].

Figures 3(a) and (b) show how the correlation integral $C(r = 0.1, m = 4, \tau = 4, w = 12)$ varies over the duration of the recording from electrode LTD3 (other choices for the parameters m, r, τ, w did not lead to further insights). Also shown are 10 Gaussian surrogates. Significant nonlinearities are identified throughout the recording. The correlation integrals of the EEG segments and the Gaussian surrogates increase during the "postictal period" (immediately after the seizure) due to the increased autocorrelation in the recording during this period. The choices of 10.24 seconds for the segmentation time and $\tau = 4$ for the delay time are too short for the postictal period, which will be ignored in the remainder of this section. Figures 3(c) and (d) summarize the results of applying the analysis of Figures 3(a) and (b) to all 28 electrodes, using a measure of statistical significance defined as follows (Theiler et al. [1991]). Let C be the correlation integral of an EEG segment, and let μ and σ be the mean and standard deviation of the logarithm of the correlation integrals of the 10 Gaussian surrogates. The "significance" S is defined by $S = (\log_{10}(C) - \mu)/\sigma$. In Figures 3(c) and (d), any shaded location represents at least a "5 standard deviation effect" ($S > 5$), which is highly significant statistically. The most striking feature of Figures 3(c) and (d) is that specific electrodes exhibit significant nonlinearity consistently throughout the entire duration of the recording. The analysis of Figure 3 can be repeated with rescaled surrogates, different sampling rates and different embedding parameters, but the results remain qualitatively similar. These results can be interpreted by visual inspection of the recording, a few segments of which are shown in Figure 4(a). The nonlinearities

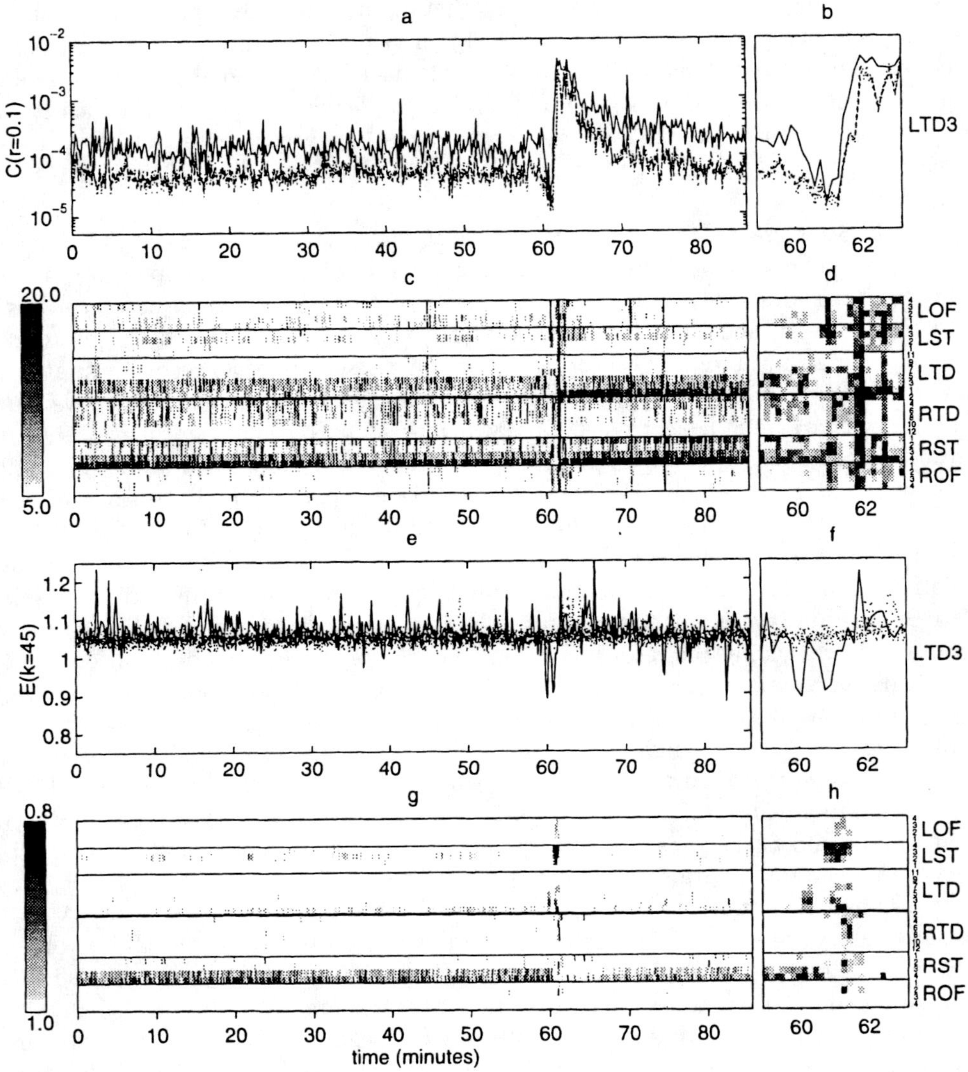

Figure 3 (a)(b) Dependence of the correlation integral $C(r = 0.1)$ on time. Solid curve: electrode LTD3. Dots: 10 Gaussian surrogates. (c)(d) Statistical significance S of nonlinearity as a function of time and electrode location. Magnitudes of S corresponding to the gray scale are illustrated by the bar to the left of the figure. (e)(f) Dependence of the prediction error $E(k = 45)$ on time. Solid curve: electrode LTD3; Dots: 10 Gaussian surrogates, (g)(h) Dependence of the prediction error $E(k = 45)$ on time for 28 electrodes. Magnitudes of the prediction error corresponding to the gray scale are illustrated by the bar to the left of the figure.

identified in specific electrodes by the correlation integral analysis in Figure 3(c) are almost entirely due to "spiking" activity, as illustrated in Figure 4(a). The segments from electrodes RST4 and LTD3, which were identified as nonlinear, exhibit spiking activity. By contrast, the segments from electrodes LST1 and ROF2, where no nonlinearity was identified before the seizure, do not exhibit spiking activity. This interpretation can be confirmed with a wavelet analysis (Casdagli et al. [1996]). The correlation integral analysis is sensitive to spiking activity because Gaussian surrogates appear "rougher" than an EEG segment with spikes in it, due to the high-frequency components of the EEG segment being spread uniformly across the Gaussian surrogates instead of being preserved within short time intervals.

It is well known that spiking in electrical activity, which is ubiquitous at the microscopic neuronal level, is not typically observed at the macroscopic (invasive or scalp) EEG level in humans, because the EEG measures electrical activity averaged over many thousands of neurons. The occurrence of spiking in EEG recordings signifies that there is pathological synchronization between neurons, and is often related to epileptogenesis (King and Spencer [1995]). Unfortunately, conventional spike detectors have not proven to be reliable indicators of the location of the epileptogenic focus itself (King and Spencer [1995]). Thus, the nonlinearities identified by the surrogate data technique in Figures 3(a)-(d) are only of minor medical interest, because they are predominantly due to spiking, a well known phenomenon that is of limited use in locating epileptogenic foci.

3.3 Prediction and Recurrent Activity. Unlike the case of pressure recordings in Section 2, the short term prediction of invasive EEG recordings is not of direct practical interest. However, results obtained from the nonlinear prediction algorithm (described in Section 2) for invasive EEG recordings provide a characterization of spatio-temporal activity in these data that is unobtainable from the surrogate data technique of Section 3.2, as follows. Figures 3(e),(f) show how the prediction error $E(k)$ varies over the duration of the recording from electrode LTD3, and for 10 Gaussian surrogates, and are obtained by dividing the recording into nonoverlapping segments of duration 10.24 seconds, as in Figures 3(a),(b). The recording was first resampled at 50Hz to save on computation time, and the parameters $k = 45$, $m = 4$, $\tau = 1$ and $w = 12$ were chosen. Figures 3(g),(h) show how the prediction error varies for all 28 electrodes (the results for the surrogates are not shown). By inspection of Figures 3(e),(f), significant nonlinear predictability is identified whenever the prediction error is less than one, the threshold used in the grayscale of Figures 3(g),(h). Note that the improvements in prediction over linear modeling shown in Figures 3(e)-(h) are at most about 20%, indicating that this data is not likely to have been generated by a low-dimensional chaotic system (Casdagli [1992]). More surprisingly, the results of the nonlinear prediction algorithm shown in Figures 3(e)-(h) are quite different to those of the surrogate data technique shown in Figures 3(a)-(d). First, several of the electrodes that consistently exhibit nonlinearity in Figures 3(a)-(d), do not exhibit nonlinear predictability in Figures 3(e)-(h). Second, a well-defined spatio-temporal pattern of nonlinear predictability occurs in Figure 3(h) that does not occur in Figure 3(d).

The spatio-temporal pattern of nonlinear predictability in Figure 3(h) may be interpreted as being due to "recurrent activity". By recurrent activity, I mean the tendency of an EEG segment to exhibit repeated patterns. Consider the segments from electrode LTD3 shown in Figure 4(b). The first 2 segments exhibit very little structure. The next 2 segments exhibit a repeated pattern of spikes and slow-waves, and were identified as nonlinear in Figure 3(h). Segments 6 and 7 exhibit high-frequency low-amplitude "β" activity, which is believed to be an important "precursor" to seizures (King and Spencer [1995]), but was not identified as nonlinear in Figure 3(h) due to its low amplitude and simple "linear" waveform. The last two segments exhibit repeated waveforms characteristic of seizure activity, and were identified as nonlinear in Figure 3(h). Also, Figure 4(c) illustrates that activity at electrode LST4 is more repetitive than that at electrode ROF2 during the seizure, as expected from Figure 3(h) using the above interpretation of recurrent activity. In principle, the surrogate data technique of Section 3.2 has the power to detect recurrent activity, but it is masked by the abundance of spiking activity in this patient. By contrast, nonlinear prediction algorithms exploit repeated patterns in a time series, and are less affected by infrequent spiking activity, as illustrated by Figures 3(e) and 4(a) for electrode LTD3 (if spiking is frequent enough, it can be exploited by a nonlinear prediction algorithm, as illustrated by Figures 3(g) and 4(a) for electrode RST4). Recurrent activity is more closely related to the location of the epileptogenic focus than spiking activity. Figure 5 extends the nonlinear prediction analysis of Figure 3(h) to 10 seizures in 7 patients. The patients are labeled A,B,..,G, and the numbers refer to the seizure number, including both clinical (with behavioral manifestations) and sub-clinical (without behavioral manifestations) seizures. All seizures displayed are clinical. The recording analyzed in Figure 3 contains seizure 15 from patient A. Visual inspection of the recordings from patients A,B,C,D (resp. E,F,G), for the duration of a few minutes encompassing the seizures in these patients, located epileptogenic foci in the regions LTD or LST (resp. RTD or RST). For example, electroencephalographers observed abnormally synchronized activity in recordings from electrodes LTD1,3,5 in patient A immediately prior to the onset of seizures. Left (resp. right) temporal lobectomies were subsequently performed on patients A,B,C,D (resp. E,F,G), removing most of the tissue from the left (resp. right) subtemporal and hippocampal regions. The results in Figure 5 can be interpreted as showing that, in the vicinity of the seizures, recurrent activity is more pronounced, and occurs earlier, in the hemisphere of the brain where the epileptogenic focus is believed to be located. A particularly clear example of recurrent activity is provided by electrode LTD3 in patient C. Recurrent activity was found to consistently occur in this electrode prior to seizure onset, and appears to be a precursor to seizures for this patient (Casdagli et al. [1996]). However, an absence of reliable precursors is found for the other patients in Figure 5.

Also shown in Figure 5 are the results of a linear analysis. The grayscale refers to the logarithm of the power in the "β band" (frequencies above 14Hz). Observe the strong correlation between recurrent activity and increased power in the β band. Thus, although the nonlinear prediction algorithm detects significant nonlinearities in well-defined spatio-temporal patterns, similar patterns can be found either by linear analysis or visual inspection. The results in Figure 5 may be thought of as quantifying important nonlinear phenomena used by electroencephalographers in identifying epileptogenic foci. However the results fail to uncover phenomena unknown to electroencephalographers, presumably due to their skill in detecting non-random patterns of activity in EEG recordings. This failure may also be due to the limited information contained in EEG segments of duration 10 seconds, constraining the complexity of the nonlinearities that may be identified with statistical significance.

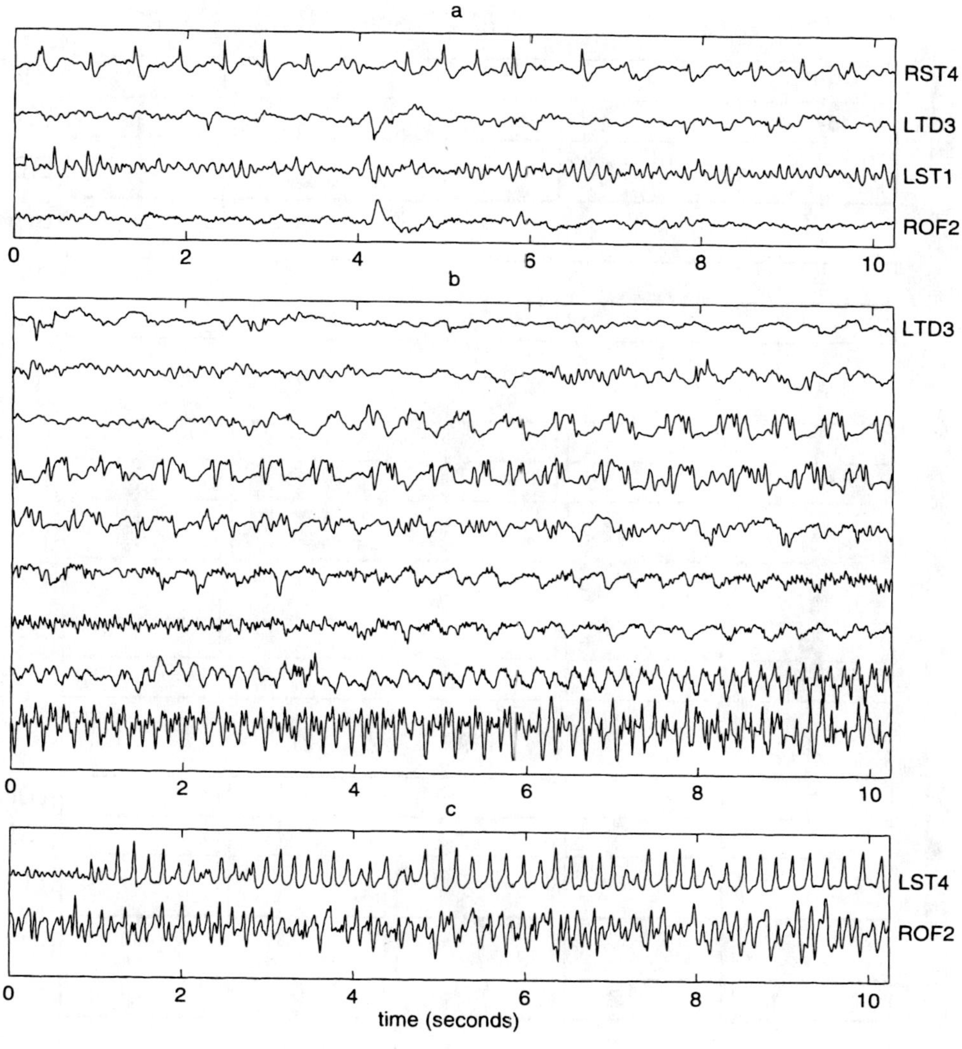

Figure 4 EEG segments of duration 10.24 seconds. (a) Electrodes RST4, LTD3, ROF2 and LST1 60 minutes before the seizure. (b) Electrode LTD3 leading up to the seizure. (c) Electrodes LST4 and ROF2 during the seizure.

3.4 Residual Delay Maps. As in the case of weather recordings, the most informative of the three techniques used to characterizing nonlinearities in invasive EEG recordings turned out to be the most simple, namely residual delay maps (RDMs). Figure 6 shows the results of the RDM technique described in Section 2, and were obtained by increasing the duration of the EEG segments from 10.24 seconds to 10 minutes. Figure 6(a) (resp. 6(b), 6(c)) shows 6 RDMs computed from electrode RST4 (resp. LST1, ROF2) during the first 60 minutes of the recording

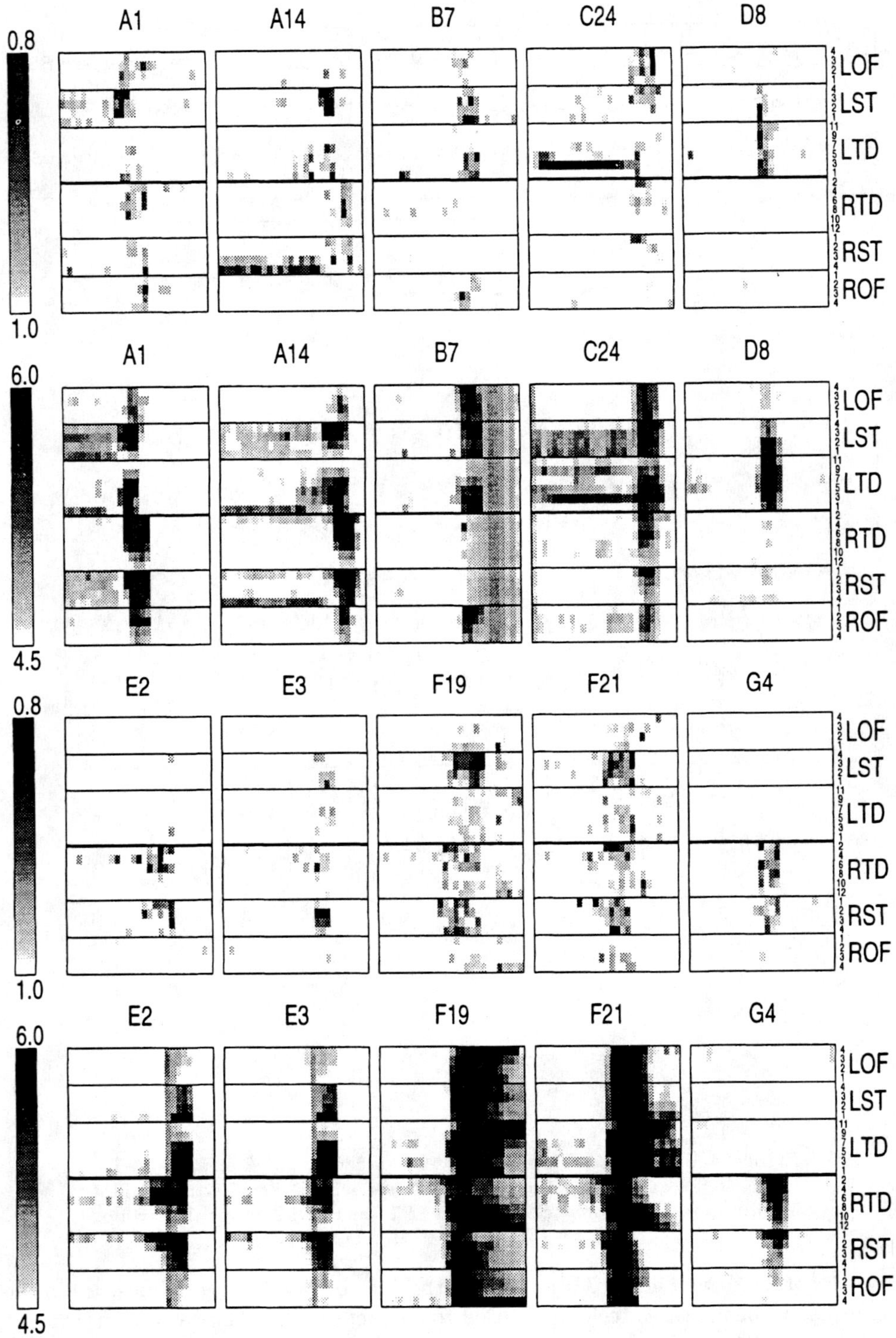

Figure 5 Rows 1 and 3: dependence of the prediction error $E(k = 45)$ on time and electrode location for 10 seizures in 7 patients. Rows 2 and 4: dependence of the power in the β band on time and electrode location.

analyzed in Figure 3. Also shown are RDMs for one Gaussian and one rescaled surrogate for each of the 6 EEG segments. The recording was sampled at 50Hz, and residuals were computed from an AR(m) model with $m = 4$; other choices for m gave similar results. The RDMs for a given electrode are remarkably stable, and show that the lag-one nonlinearities in the EEG recordings from these electrodes are stationary over long time scales before seizure onset. Lag-one nonlinearity is identified in the electrodes LST1 and ROF2, which were not found to be nonlinear in Figure 3. This is because the RDMs are computed from EEG segments of much longer duration than the segments used in Figure 3, and so are able to resolve much weaker nonlinear structure. The RDMs of the EEG segments are easily distinguishable from those of the surrogates, demonstrating that the lag-one nonlinearity cannot be explained by a nonlinear measurement function. In the case of electrode RST4, this may be confirmed by the asymmetry of the scatter plot in Figure 6(d) about the diagonal, indicating that the recording from this electrode is time-irreversible, and cannot be produced by rescaling the time-reversible Gaussian surrogates (Diks et al. [1995]).

The RDMs shown in Figures 6(a)-(c) are easy to interpret. A negative (resp. positive) slope in an RDM at extreme values of x implies that the EEG segment is less (resp. more) autocorrelated, i.e. sharper (resp. smoother) than a Gaussian random process at extreme values of x. Thus the "quadratic" RDMs in Figures 6(a),(b) indicate there is "up-spiking" in the recording from electrodes RST4 and LST1, although, in the case of electrode LST1, this is difficult to detect by visual inspection. Electrode LST1 is not considered to exhibit spiking in the conventional electroencephalographic sense; I will refer to such activity as "weak spiking" below. The "cubic" RDMs in Figure 6(c) indicate that the highs and lows of the recording from electrode ROF2 are smoother than for a Gaussian random process. This is consistent with a visual inspection of lengthy segments of the EEG recording from this electrode, which reveals the intermittent occurrence of high-amplitude low-frequency "slow-waves" on a background of low-amplitude high-frequency activity. A slow-wave occurs 4 seconds into the EEG segment from electrode ROF2 shown in Figure 4(a). The significance of slow-waves depends upon the clinical status of the brain. For example, slow-waves occur in scalp EEG recordings from sleep in normal patients. High-frequency waves reflect more complicated activity, such as solving mental problems. However, the normal frequency content for invasive EEG recordings is unknown, because invasive electrodes are not placed in normal human subjects.

Figure 6(e) shows how RDMs vary between electrodes for several EEG recordings from the 7 patients considered above. Recordings marked with a suffix "i" (resp. "p") were selected to terminate over two hours (resp. within a few minutes) of the occurrence of a seizure. With few exceptions, the RDMs for a given patient and electrode are remarkably stable. By contrast, there is considerable variation in the RDMs between different patients and electrodes. Also shown in Figure 6(e) are power spectra for some of this data. The RDMs appear to exhibit stronger, and more interesting, variations between different electrodes and patients than the power spectra. It is natural to enquire whether the variations in the RDMs are related to the location of the epileptogenic foci in these patients. Recall that visual inspection of EEG recordings, for the duration of a few minutes encompassing seizures in these patients, located epileptogenic foci in the regions LTD or LST (resp. RTD or RST) of patients A-D (resp. E-G). An inspection of Figure 6(e) reveals that, with the exception of patient A, quadratic RDMs, associated with

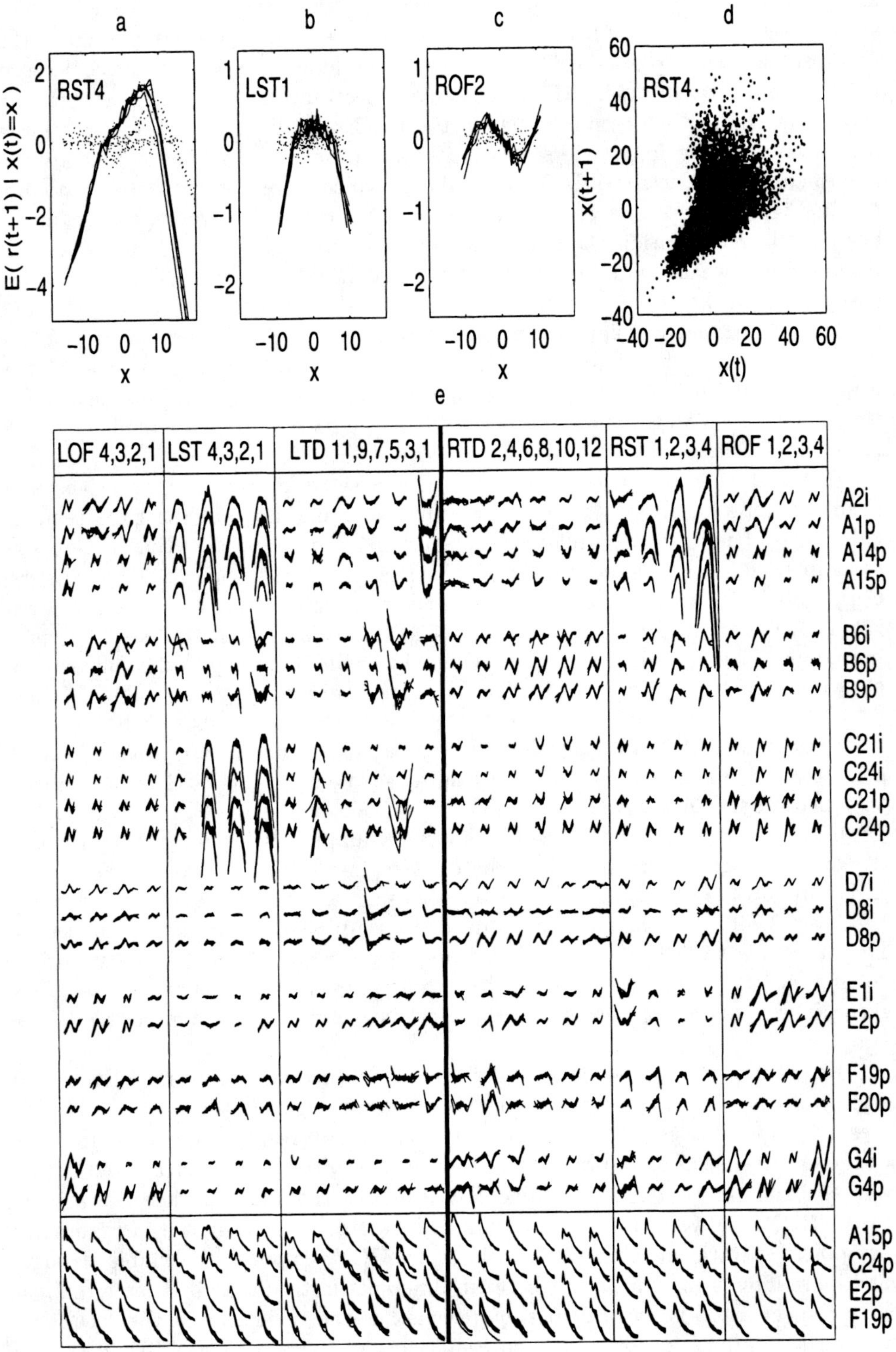

Figure 6 (a)-(c) Solid curves: 6 RDMs from electrodes RST4, LST1 and ROF2. Dots: Gaussian and rescaled surrogates. (d) Scatter plot of x_{t+1} against x_t for 10 minutes of the recording from electrode RST4, sampled at 50Hz. (e) Upper 20 rows: Dependence of RDMs on time and electrode location for several interictal recordings in 7 patients. Lower 4 rows: Power spectra for some of these recordings.

spiking (or "weak" spiking), occur predominantly in the hemisphere of the brain where the epileptogenic focus is located. Cubic RDMs, associated with slow-waves, occur in the orbital frontal regions (ROF and LOF), uniformly across patients. This is consistent with the clinical diagnoses in the patients, which reported a lack of epileptogenic activity in the frontal regions of the cortex.

A small number of instabilities in the RDMs occur in Figure 6(e), in particular at electrode LTD3 in C21p and C24p. Activity at this electrode was identified as a precursor to seizures in Figure 5 for patient C. Precursors correspond to non-stationary activity, and are expected to generate instabilities in RDMs. However, very few instabilities in the RDMs of other patients occur in Figure 12. This does not necessarily imply the non-existence of precursors to seizures for these patients, since there may be changes in linear properties of the recordings, or in the nonlinear properties at higher order lags, or in the conditional variance, or in the multivariate structure of the recordings, that do not affect the RDMs. Also, changes in dynamics that occur over short time frames are unlikely to be detected, since EEG segments of duration 10 minutes were used to compute the RDMs.

3.5 Summary of Results. The research presented in Sections 3.2-4 was intended to address the first three medical objectives stated in Section 3.1. The results, and potential directions for future research, may be summarized as follows.

1. *Localization of the epileptogenic focus.* RDMs were found to give a remarkably robust measure of nonlinearity for interictal (between seizures) recordings, and certain classes of RDMs were found to occur predominantly in the hemisphere of the brain where the epileptogenic focus is believed to be located. Moreover, RDMs were found to contain information that is difficult to obtain from power spectra or visual inspection. This raises the possibility that RDMs may be used to locate epileptogenic foci from interictal data alone, reducing the need for week-long recordings in hospitalized patients. The clinical implications of this need to be assessed by studying more patients.

2. *Prediction of seizures.* The results of the surrogate data and prediction algorithms presented in Sections 3.2 and 3.3 confirm the difficulty of this problem. Evidence for precursors (dynamical changes prior to seizure onset) was found in only one of the 7 patients studied. However, it is possible that new techniques will yield a better solution to this problem. The results from the RDM analysis suggest that the 10 second segmentation time used in Sections 3.2 and 3.3 may be much too short. The systematic study of nonstationarity and appropriate choices for the time scale of data segmentation are crucial issues that are only just starting to be addressed within a dynamical systems framework (Manuca and Savit). It is possible that monitoring changes in nonlinearities over appropriate time scales will allow one to gain insight into, and possibly predict, the onset of seizures, as suggested by Iasemidis et al. [1991], Iasemidis et al. [1990].

3. *Understanding the dynamical mechanisms responsible for generating seizures.* It is natural to hypothesize that the dynamical mechanism underlying transitions to the ordered activity observed in EEG recordings during seizures is a transition to low-dimensional chaos. However, low-dimensional chaos is difficult to detect in nonstationary systems, and the results of the prediction algorithm presented in Section 3.3 did not yield convincing evidence for low-dimensional chaos in invasive EEG recordings. An alternative approach to testing hypotheses about transitions to seizures is to simulate data from mathematical models of epilepsy (Freeman

[1987], Traub and Miles [1991]). The characterizations of nonlinearity presented above place constraints on the properties of such models for the case of temporal lobe epilepsy (TLE). Mathematical models for TLE should exhibit approximately stationary lag-one nonlinearities interictally, and should exhibit recurrent activity only in the immediate vicinity of the seizure. A model exhibiting a gradual transition to a chaotic attractor of dimension less than 4 is probably not a valid model for seizure onset in TLE.

4 Concluding Remarks

This paper presented the results of a variety of techniques of nonlinear time series analysis for data from two dynamical systems: weather and epilepsy. These systems have the common feature of being inhomogeneous spatially extended nonlinear systems that do not exhibit low-dimensional chaos. I conclude this paper with some remarks that may be of general relevance to the data analysis of other systems that belong to this important and complicated class of dynamical systems.

1. In spite of the lack of low-dimensional chaos observed in the two systems, the nonlinear techniques used were sufficiently general to extract usable and robust information not contained in the power spectrum.

2. Several of the results of the nonlinear analysis were straightforward to interpret in terms of phenomena of recognized importance in the two systems, for example storms and spiking. The results in some sense quantify such phenomena, and may prove useful in testing mathematical models for the systems.

3. The simplest technique, residual delay maps, gave the most informative results concerning the quantification of the spatial inhomogeneity in the two systems. A key result was the reduction of the voluminous quantities of complex data from the two systems into remarkably simple spatial patterns of lag-one nonlinearities.

4. Historically, the techniques used in this paper were largely motivated by simulations from low-dimensional systems, exhibiting dynamics at short time scales, for example the Hénon map and the Lorenz equations. It is possible that simulations from mathematical models with many degrees of freedom, such as those proposed in Freeman [1987], Traub and Miles [1991] for epilepsy, and for mathematical models of the weather, will motivate the development of new and improved techniques for time series analysis from such systems. For example, the mathematical models of epilepsy proposed in Freeman [1987], Traub and Miles [1991] exhibit dynamics at multiple time scales, a phenomenon which is closely related to nonstationarity.

Acknowledgements

I thank Edward Habjan, Dale Hess, Greg Holland, Richard Penner and George Sugihara for collaborating on the weather project, and Leo Iasemidis, Robin Gilmore, Steven Roper, Chris Sackellares and Bob Savit for collaborating on the epilepsy project. I also thank Edward Lorenz, Radu Manuca, Jonathan Raz, Thomas Schreiber and James Theiler for useful conversations. This work was supported by the National Institute for Health under grant R01NS31451, and by the Office of Naval Research under grant N00014-95-1-0034.

References

Abarbanel, H. D. I., Brown, R., Sidorowich, J. J. and Tsimring, L. S. [1993] *The analysis of observed chaotic data in physical systems*, Rev. Mod Phys., **65**, 1331–1392.

Babloyantz, A. and Destexhe, A. [1986] *Low dimensional chaos in an instance of epilepsy*, Proc. Nat. Acad. Sci. (U.S.A), **83**, 3513–3517.

Brock, W. A., Hseih, D. A. and LeBaron, B. [1991] *Nonlinear dynamics, chaos and instability: Statistical theory and economic evidence*, MIT Press.

Buizza, R. and Palmer, T N. [1996] *The singular-vector structure of the atmospheric general circulation*, J. Atmos. Sci., to appear.

Casdagli, M. C. [1992] *Chaos and deterministic versus stochastic nonlinear modeling*, J. Roy. Statist. Soc. B, **54**, 303–328.

Casdagli, M. C., Iasemidis, L. D., Gilmore, R. L., Roper, S. N., Sackellares, J. C. and Savit, R. S. [1996] *Characterizing nonlinearity in invasive EEG recordings from temporal lobe epilepsy*, Physica D, to appear.

Casdagli, M C. and Weigend, A. S. [1993] *Exploring the continuum between deterministic and stochastic modeling*, in Time Series Prediction (A. Weigend and N. Gershenfeld, eds.), Addison Wesley

Charney, J. G. and Shukla, J. [1981] in Monsoon Dynamics (J. Lighthill and R. Pearce, eds.), Cambridge University Press.

Diks, C., van Houwelingen, J. C., Takens, F. and DeGoede, J. [1995] *Reversibility as a criterion for discriminating time series*, Phys. Lett. A, **201**, 221–228.

Eckmann, J. P. and Ruelle, D. [1985] *Ergodic theory of chaos and strange attractors*, Rev. Mod. Phys., **57**, 617–656

Elbert, T., Ray, W. J., Kowalik, Z. J., Skinner, J. E., Graf, K. E. and Birbaumer, N. [1994] *Chaos and physiology: deterministic chaos in excitable cell assemblies*, Physiological Reviews, **74**, 1–47.

Engel, J. [1989] Seizures and Epilepsy, F. A. Davis Company.

Farmer, J. D. and Sidorowich, J. J. [1987] *Predicting chaotic time series*, Phys. Rev. Lett., **59(8)**, 845–848.

Frank, G. W., Lookman, T., Nerenberg, M. A. H., Essex, C., Lemieux, J. and Blume, W. [1990] *Chaotic time series analysis of epileptic seizures*, Physica D, **46**, 427–438.

Freeman, W. J. [1987] *Simulation of chaotic EEG patterns with a dynamical model of the olfactory system*, Biol. Cybern , **56**, 139–150

Grassberger, P. [1986] *Do climate attractors exist?*, Nature, **323**, 609–612.

Grassberger, P. [1990] *An optimal box-assisted algorithm for fractal dimensions*, Phys. Lett. A, **148**, 63–68.

Grassberger, P. and Procaccia, I. [1983] *Measuring the strangeness of strange attractors*, Physica D, **9**, 189–208.

Hseih, D. A. [1991] *Chaos and nonlinear dynamics: application to financial markets*, The Journal of Finance, **46**, 1839–1877.

Iasemidis, L. D. and Sackellares, J. C. [1991] *The evolution with time of the spatial distribution of the largest Lyapunov exponent on the human epileptic cortex*, in Measuring Chaos in the Human Brain (D. Duke and W. Pritchard, eds.), Singapore: World Scientific, 49–82.

Iasemidis, L. D., Sackellares, J. C., Zaveri, H. and Williams, W. [1990] *Phase space topography and the Lyapunov exponent of electrocorticograms in partial seizures.* Brain Topog., **2**, 187–201.

King, D. and Spencer, S. [1995] *Invasive electroencephalography in mesial temporal lobe epilepsy.* J. Clin. Neurophysiol., **12**, 32–45.

Lerner, D. E. [1996] *Monitoring changing dynamics with correlation integrals: Case study of an epileptic seizure.* Physica D, to appear.

Lorenz, E. N. [1969] *Atmospheric predictability as revealed by naturally occurring analogues.* J. Atmos. Sci., **26**, 636–646.

Manuca, R. and Savit, R. S. [1996] *Stationarity and nonstationarity in time series analysis.* Physica D, to appear.

Palus, M. [1994] *Nonlinearity in normal human EEG: cycles, temporal asymmetry, nonstationarity and randomness, not chaos*, Santa Fe Institute working paper 94-10-054.

Palus, M. and Novotna, D. [1994] *Testing for nonlinearity in weather records*, Phys. Lett. A, **193**, 67–74.

Pijn, J. P., Van Neerven, J., Noest, A. and Lopes da Silva, F. H. [1991] *Chaos or noise in EEG signals; dependence on state and brain site*, Electroencephalogr. Clin. Neurophysiol., **79**, 371–381.

Prichard, D. and Theiler, J. [1995] *Generalized redundancies for time series analysis*, Physica D, **84**, 476–493.

Sardeshmukh, P. and Hoskins, B. J. [1985] *Quart. Jour. Roy. Meteorol. Soc.*, **111**, 261–278.

Sauer, T. [1994] *Reconstruction of dynamical systems from interspike intervals*, Phys. Rev. Lett., **72**, 3811–3814.

Shannon, C. E. and Weaver, W. [1964] *The Mathematical Theory of Communication*, Urbana: University of Illinois Press.

Smith, L. [1995] *Locally optimized prediction of nonlinear systems: stochastic and deterministic*, in Chaos and Forecasting (H. Tong, ed.), World Scientific.

Smith, R. L. [1991] *Optimal estimation of fractal dimension*, in Nonlinear Modeling and Forecasting (M. Casdagli and S. Eubank, eds.), Addison Wesley.

Spencer, D. D. and Pappas, C. T. E. [1992] *Surgical decisions regarding medically intractable epilepsy*, Clinical Neurosurgery, **38**, 548–557.

Sugihara, G. [1994] *Nonlinear forecasting for the classification of natural time series*, Phil. Trans. R. Soc. Lond. A, **348**, 477–495.

Sugihara, G., Casdagli, M., Habjan, E., Hess, D., Holland, G. and Penner, R. [1995] *Nonlinearity in observed barometric pressure records: latitudinal gradients, dominant mechanisms, and improved forecasts.* preprint.

Sugihara, G. and May, R. M. [1990] *Nonlinear forecasting as a way of distinguishing chaos from measurement error in a data series*, Nature, **344**, 734–741.

Theiler, J. [1986] *Spurious dimension from correlation algorithms applied to limited time-series data.* Phys. Rev. A, **34**, 2427–2433.

Theiler, J. [1995] *On the evidence for low dimensional chaos in an epileptic electroencephalogram.* Phys. Lett. A., **196**, 335–341.

Theiler, J., Galdrikian, B., Eubank, S. and Farmer, J. D. [1991] *Using surrogate data to detect nonlinearity in time series*, in Nonlinear Modeling and Forecasting (M. Casdagli and S. Eubank, eds.), Addison Wesley.

Theiler, J. and Rapp, P. [1996] *Re-examination of the evidence for low-dimensional, nonlinear structure in the human electroencephalogram.* Electroenceph. Clin. Neurophysiol, to appear.

Tong, H. [1990] *Nonlinear Time Series Analysis: A Dynamical Systems Approach*, Oxford University Press.

Traub, R. D. and Miles, R. [1991] *Neuronal networks of the hippocampus*, Cambridge University Press.

Trenberth, K. E. [1991] Jour. Atmos. Sci., *48*, 2159–2178.

Tsonis, A. A. [1992] *Chaos: From Theory to Application*, Plenum.

Weigend, A. and Gershenfeld, N. [1993] *Time Series Prediction*, Addison-Wesley.

Fields Institute Communications
Volume 11, 1997

Assessment of Linear and Nonlinear Correlations Between Neural Firing Events

André Longtin and Daniel M. Racicot
Département de Physique
Université d'Ottawa
150 Louis Pasteur
Ottawa, Ontario, Canada K1N 6N5

Abstract. Point processes are often used to describe the sequence of times at which a neuron generates action potentials or "spikes". This paper contrasts the information revealed by certain linear and nonlinear analyses of correlations between interspike intervals. We illustrate our methodology by analyzing correlations between spikes recorded from a sensory neuron driven by a very accurate pacemaker rhythm. We find that linear correlations can account for some of the forecastability of interspike intervals, but that nonlinearity is also present. Our surrogate data analyses raise a key issue regarding the interpretation of null hypotheses for neural point processes. We argue that phase-randomized surrogate data can be used to assess the influence of linear correlations on the forecastability, but that in general one cannot associate a clear physical process with the null hypothesis. Finally, we analyze correlations in a phase-type "deviate" variable extracted from the neural point process. We also find evidence for nonlinearity in the dynamics of this variable.

1 Introduction

The main route by which neurons interconnect is via axons. The membrane potential of a neuron represents the spatially and temporally integrated actions of intrinsic and synaptic conductances across its membrane. In many neurons, this potential is converted into a sequence of action potentials or "spikes" which propagate down the axon to other neurons. The spikes are received by target neurons that decode information in the sequence of interspike intervals (ISI's). These sharp spikes are considered as "firing events" due to their very fast time scale (Moore et al. [1966]). Accordingly, sequences of spikes are typically considered as point processes (Cox and Isham [1980]). Generally, spikes occur when certain state variables governing the neural dynamics cross specific boundaries in phase space (these can be separatrices). In some experimental preparations, much is known about this dynamical picture. For many other preparations, and in particular for those where it is impossible to perform voltage measurements across the membrane, alternate non-dynamical descriptions have often been deemed more useful to quantify spike train properties. Stochastic point processes (Cox and Isham [1980], Sayers [1970]) and fractal noises (Teich [1992]) are examples of such descriptions.

To properly constrain theories of neural coding and information processing (see e.g., Abbott [1994], Konishi [1991]), it is important to characterize ISI patterning and correlations under relevant stimulus conditions. ISI patterning results from the often unknown dynamics of synaptic and intrinsic conductances and spike generation. A knowledge of these dynamics could indicate how correlations arise (if any), and whether actual spike times can in principal matter to the target neuron populations. Recent work (Sauer [1994], Racicot and Longtin [1995]) has further highlighted the fact that inputs to neurons can strongly determine correlations between ISI's and thus the predictability of ISI's.

We would like to use spike train analysis to discriminate between firing dynamics that produce different kinds of ISI correlations. Here are examples of questions we would like to answer. Is a stochastic point process description of the spike train appropriate, as it is when there is a significant amount of noise? Is there randomness of deterministic origin, i.e., chaos, as is expected when nonlinear systems such as neurons are periodically driven by currents? If the neuron is driven, does it fire in a complex phase-locked pattern with long period or long transient? Is there a mixture of deterministic and stochastic dynamics? What is the degree of predictability in the spike times? These questions are particularly difficult to answer when one can only measure firing times through extracellular recordings, as is often the case with small sensory receptor neurons (which can be encapsulated in other tissue) and with neurons that do not survive being impaled by a microelectrode.

The goal of this paper is to expose a methodology to assess the origin of ISI correlations, both linear and nonlinear. We illustrate our findings using data from a particular preparation of periodically stimulated neurons. Our work follows recent efforts to characterize nonlinearity in interval data sets using surrogate data techniques which allow rejection of specific null hypotheses about the dynamics governing the evolution of the intervals (Longtin [1993a], Schiff et al. [1994], Rapp et al. [1994a]). It also pays particular attention to the interpretation of surrogate data analyses for interval data, as well as to the dynamics of special variables, other than intervals, that are suggested by and can be extracted from a given point process.

The paper is organized as follows. In Section 2, we present the data set used to illustrate our methodology. This methodology relies on a battery of tests, some based on linear analysis techniques, others on nonlinear dynamics. The assessment of correlations using standard tests is the subject of Section 3. Nonlinear forecasting of intervals is described in Section 4. The use of surrogate data sets and their physical significance is discussed in Section 5. The correlations between firing phases is assessed in Section 6, and the paper concludes in Section 7 with a summary and outlook onto interesting problems in this area of dynamics and time series analysis.

2 Data From Electroreceptor Afferents

We focus our study on data from a primary afferent sensory neuron of the weakly electric fish *Apteronotus leptorhynchus*. This neuron is known as an electroreceptor. The fish produces a 660 Hz AC electric field known as the electric organ discharge (EOD) to electrolocate and electrocommunicate. This frequency varies from one fish to another. Electroreceptors transduce modulations and distorsions of this electric field into temporal patterns of action potentials. These action potentials can be recorded extracellularly from the afferent axon connected to this receptor. The recordings are done under the controlled condition where the receptors are driven only by the EOD of the fish, i.e., in the absence of modulations and distorsions. The EOD is generated by one of the most accurate biological pacemakers known. Thus we have a situation where the receptor is driven by an almost perfectly periodic electric field.

Under these conditions, this receptor is phase-locked to the EOD, i.e., it fires near a preferred phase of this stimulus. However, it skips an apparently random integer number of cycles of the stimulus between firings. This form of aperiodic phaselocking is known as "skipping". Figure 1 shows sequences of intervals and some of their first-order statistics. The raster plot in Figure 1(a) reveals the characteristic banded structure for skipping data: intervals are grouped around integer multiples of the firing period. An interspike interval histogram (ISIH) is shown in Figure 1(b). It can be seen as a measure of the density of points along the ordinate in Figure 1(a), after projecting all points onto this axis. An expanded view of the successive intervals in shown in Figure 1(c). A return map of successive intervals is shown in Figure 1(d).

We are ultimately interested in the biophysical mechanisms that underlie skipping. Validation of models for this firing behavior can benefit from the analyses described here and in previous work on other kinds of neurons that exhibit skipping (Longtin [1993a, 1993b, 1995]). In particular, we are interested in the correlations between the firing events. This is important for issues of neural coding.

3 Standard Tests to Determine the Presence of Correlations

Before determining the presence of correlations, it is important to verify that the data satisfy some criterion of stationarity. Nonstationarity can lead to erroneous results for linear or nonlinear correlation analyses. A suitable criterion of stationarity for the data studied here is the persistence of the characteristic banded pattern in the raster plot of the intervals. In this sense, our data appear stationary.

A standard test to determine the presence of correlations between intervals is to look at a return map of interval I_{n+1} vs interval I_n, as in Figure 1(d). In the neurophysiology literature, such plots are often referred to as "joint interval histograms" or "scatter plots". This plot displays correlations in the (normalized) density of events along vertical or horizontal lines. If this density is the same through all vertical lines, the probability of the next interval I_{n+1} is independent of the value of the present interval I_n. This can further be tested in the following way. Probability theory tells us that if intervals are successively independent and identically distributed (IID), the probability density of a variable equal to the sum of two successive intervals is the autoconvolution of the interspike interval density (Sayers [1970]). If we denote this interspike interval density by $P(I)$, and the sum of two successive intervals by $X_n = I_n + I_{n+1}$, the density $f_2(X)$ of the X_n's is given by

$$f_2(X) \equiv P(I) * P(I) = \int_{-\infty}^{\infty} P(X-I)\,P(I)\,dI\,.$$

$P(I)$ and $f_2(X)$ are known as, respectively, the K=1 and K=2 scaled interval histograms. They are shown in Figure 2(a) and 2(b). Figure 2(c) is the autoconvolution of $P(I)$. This latter histogram differs from the $K=2$ histogram, signifying that successive firing times, and thus successive intervals, are correlated. In contrast to the ISIH, which is a first-order statistic, independent of the ordering of

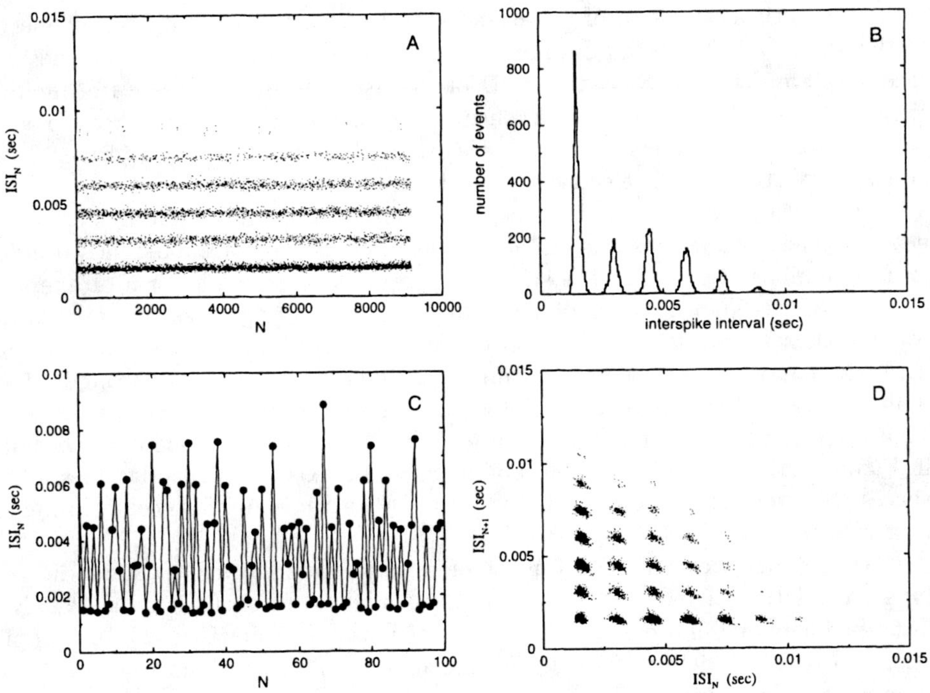

Figure 1 (a) Interspike interval versus interval number (raster plot) obtained from extracellular recordings of spikes from a primary afferent neuron of the weakly electric fish *Apteronotus leptorhynchus*. This receptor is stimulated by a 660 Hz electric field produced by the same fish. There are 9165 successive intervals. (Data provided courtesy of Joseph Bastian, U. Oklahoma). (b) Interspike interval histogram (ISIH) for the data shown in (a). This histogram plots the number of intervals between successive firings versus the interval duration, discretized in 300 bins between 0 and 15 msec. (c) Expanded view of a segment of the data shown in (a) (with points joined by lines). (d) Return map of successive intervals.

the ISI's, the autocorrelation functions shown in Figures 3 and 4 are second order statistics. Figure 3 shows the autocorrelation of the raw spike train comprised of zeroes and ones. The accurate phase-locked characteristic of the skipping pattern is responsible for the peaks at multiples of the EOD period seen up to high lag times. If the spike train is constructed from an ISI sequence obtained by randomly shuffling the raw ISI sequence, the phaselocking is still present, but it is not as accurate (Figure 3, lower panel). The peaks decay and overlap at higher lags.

This decay arises because phase-locked firing to a stimulus cannot fall too much out of step with the stimulus. Suppose, for example, that the firing probability is highest at the maximum of the EOD stimulus. Assume that spike $s1$ occurs at this maximum, and that, after one or more stimulus cycles, the next spike $s2$ occurs late, i.e., after the EOD reached its maximum. The corresponding interval I_1 between $s1$ and $s2$ is then greater than a multiple of the stimulus period, i.e., $I_1 > mT_o$, where m is an integer. The following spike $s3$ will again fire somewhere in the preferred phase region of the stimulus (again after zero, one or more EOD cycles are skipped). However, since $s2$ occurred late, the interval I_2 between $s2$ and $s3$ will tend to be shorter than a multiple of the period. An interval I_2 greater than a multiple of the period is unlikely to occur, as this would require that spike $s3$ occur

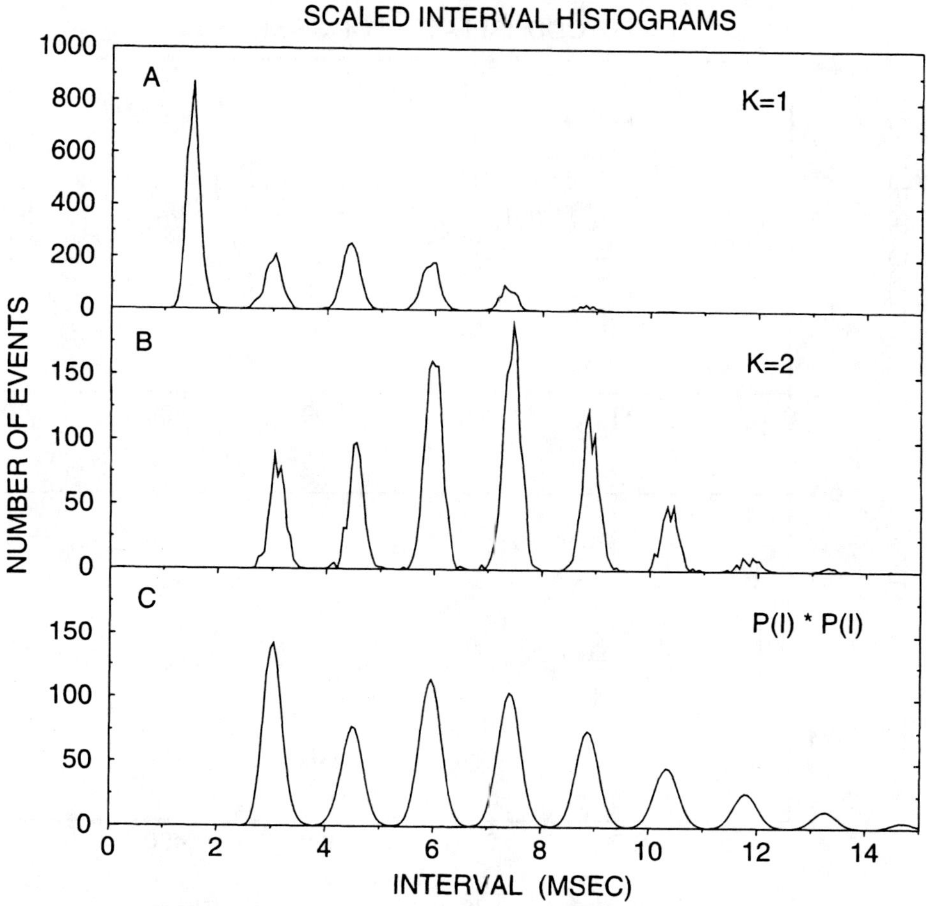

Figure 2 Determination of the existence of interval correlations using scaled interval histograms. (a) Same ISIH as in Figure 1(a), where intervals are measured between successive spikes (K=1). (b) Scaled interval histogram: intervals are measured between every second spike (K=2); in other words, an interval is formed by adding two successive intervals. (c) Autoconvolution $P(I) * P(I)$ of the (unnormalized) ISIH in Figure 2(a).

nearer to the trough of the stimulus, where the firing probability is very low for the noise levels encountered in these neurons.

Thus, the statistical fluctuations in the firing phases induce a negative serial correlation in a variable that measures the deviation of an ISI from a multiple of the period (see Figure 9 - this interval deviate is studied in more detail in Section 6). This is true even though the firing statistics, i.e., the firing probability versus stimulus phase, do not significantly change from one stimulus cycle to the next, as was found in a related study (Longtin and Racicot [1996]). From the point of view of the ISIH, these correlations imply that successive intervals have a tendency to alternate between left and right sides of the peaks, regardless of how many cycles are skipped (see Section 6). This correlation manifests itself also in the return map in Figure 1(d) as an elongation of the clusters in a direction of negative slope. This is a feature of phase-locked behavior (Longtin [1993a]). Random shuffling of the intervals removes the constraint that the firing remain in step with the EOD. This

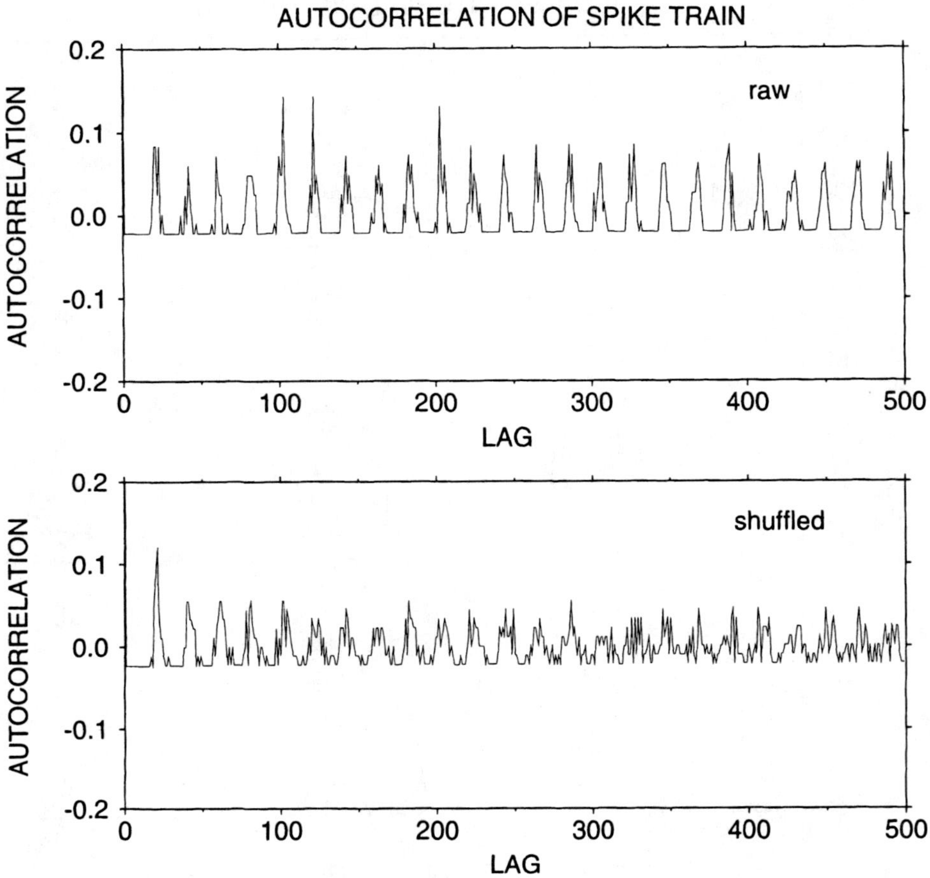

Figure 3 Autocorrelation of the spike train used in Figure 1. The autocorrelations are estimated using 500 lags over one spike train segment of $0.303 sec$ divided into 4096 samples of $7.4 \times 10^{-5} sec$. The lower panel is the autocorrelation of the spike train constructed from a randomly shuffled version of the interspike interval sequence of the raw spike train. Note the loss of phaselocking resulting from the random shuffling.

is seen in Figure 3, and in the return map in the right panel of Figure 5, where the clusters in the return map are now symmetric rather than elongated.

The ISI's also exhibit another type of correlation, in the form of an alternation between long and short skips. This is revealed by the autocorrelation function of intervals in Figure 4. The decorrelation of successive intervals is fast, as there is practically no correlation beyond one interval. Random shuffling of the intervals removes, as expected, all correlations between ISI's. Return maps for the raw and randomly shuffled ISI sequences are compared in Figure 5. The shuffling still produces a lattice of clusters, since the ISIH has not changed. However, it has produced more events along the diagonal away from the origin: the correlation between long-skip and short-skip events (and vice-versa) has disappeared. This type of correlation probably arises from the refractory period of the neuron, which is especially important at high frequencies such as that of the EOD (660 Hz). It implies that the neuron is more refractory to firing after a short ISI interval, thus lowering the probability of another spike until the neuron recovers. This lowered firing probability will favor a long skip, and thus a long interval.

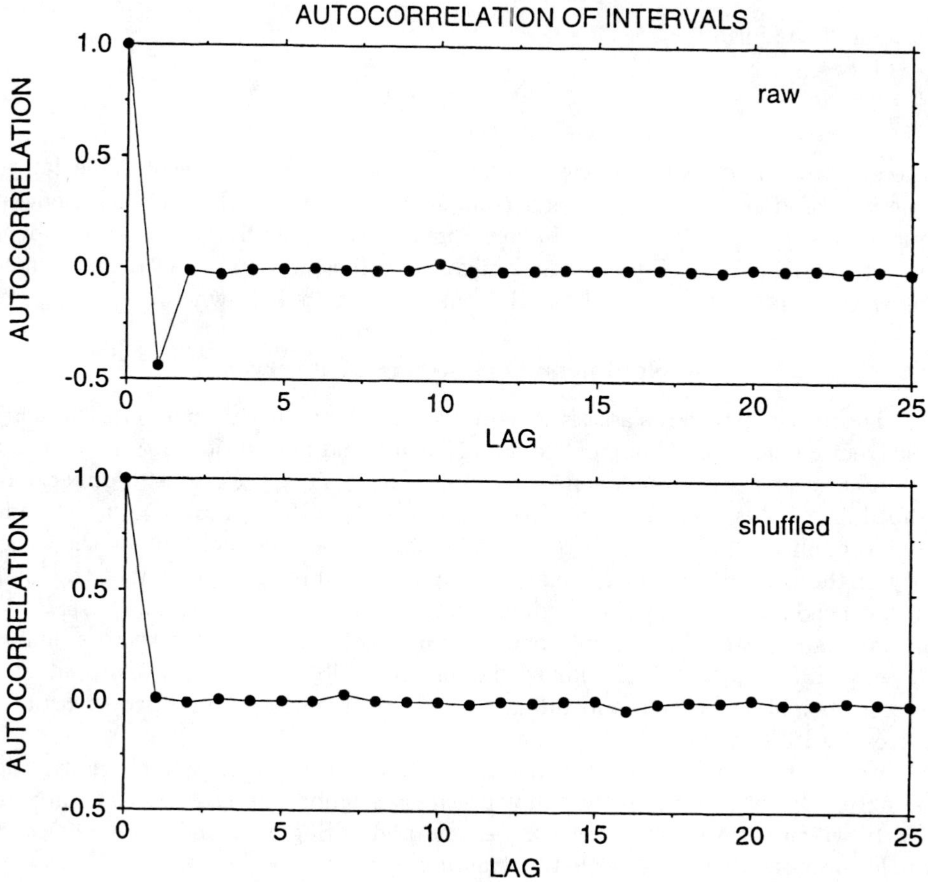

Figure 4 Autocorrelation of successive intervals, estimated from 50 lags over 9165 intervals. The shuffling has eliminated the negative correlation between successive intervals.

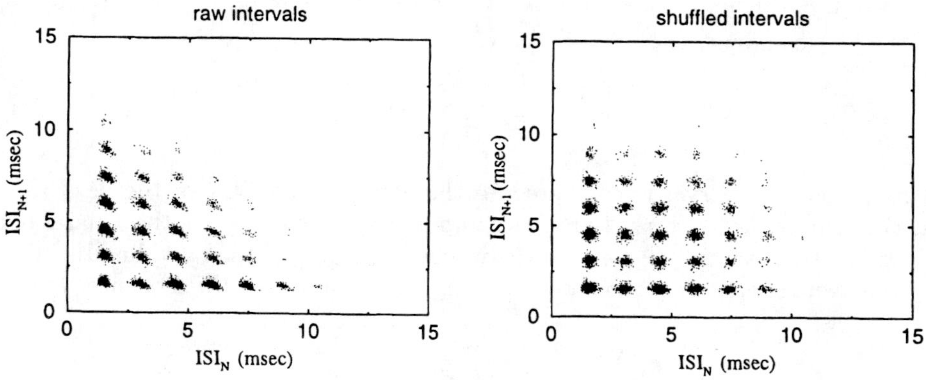

Figure 5 Return maps for the raw and shuffled ISI sequence.

If the ISI sequence has strong linear correlations, as can be concluded from our analysis so far, one can describe the dynamics of spike generation by a stochastic relation of the form

$$I_n = \sum_{k=1}^{q} a_k I_{n-k} + \sigma \xi_n \tag{3.1}$$

where ξ is an i.i.d. noise process. This noise would thus be responsible for the observed randomness, a conclusion compatible with recent biophysical modeling studies (Longtin [1993b, 1995]). In fact, Figure 4 suggests that a first-order moving average model (k=1 in Equation 3.1) might be sufficient to capture the main linear correlations in these interval data (this will be presented elsewhere).

4 Nonlinear Forecasting of Intervals

The first step towards assessing nonlinear forecastability in time series data is to construct a delay embedding of the data (Farmer and Sidorowich [1987]). The data in our case consist of the sequence of N ISI's $(I_1, I_2, I_3, ..., I_N)$. The m-dimensional embedding yields vectors of the form $\vec{x}_n = (I_n, I_{n-1}, I_{n-2}, ..., I_{n-m+1})$. The idea behind nonlinear forecasting is to find a function F, at each point along the trajectory in the training set, which maps a neighborhood of that point into the image of that neighborhood in the embedding space: $x_{n+1} = F(I_n, I_{n-1}, ..., I_{n-m+1})$. We are interested in predicting only one coordinate of the embedded vectors, namely the next ISI, I_{n+1}, from the knowledge of the m ISI's immediately preceding it. Nearest neighbors in the embedding space correspond to similar sequences of m successive ISI's in the data.

Rather than estimate local linear functions for F, as in our previous study (Longtin [1993a]), we adopt the simpler nearest-neighbor method described in Sauer [1994], which is a variant of the simplex method of Sugihara and May [1990]. The whole data set serves as both the training and the test set. For each $\vec{x}_n$, one identifies βN $(\beta < 1)$ nearest neighbors $\vec{x}_n^j \equiv (I_n^j, I_{n+1}^j, ..., I_{n-m+1}^j)$, $j = 1, ..., \beta N$. For all calculations in this paper, $\beta = 0.01$. As in Sauer [1994], we omit nearest neighbors which are close to $\vec{x}_n$ in time, to avoid biasing the prediction by insample interpolation. The prediction is based on the range of the $\vec{x}_n^j$ under one iteration, i.e., the $\vec{x}_{n+1}^j$. The predicted interval, which we denote by f_n, is simply the average of the first components of the $\vec{x}_{n+1}^j$, i.e.,

$$f_n = \frac{1}{\beta N} \sum_{j=1}^{\beta N} I_{n+1}^j . \tag{4.1}$$

This predicted interval is compared to the actual value I_{n+1} of the next ISI. A normalized prediction error (NPE) is computed by first averaging the squares of all the prediction errors $(f - I)^2$ over the whole data set, and then dividing this result by the average error incurred by predicting the mean of the ISI sequence, $\bar{I}$:

$$NPE = \left[\frac{\langle (f_n - I_{n+1})^2 \rangle}{\langle (\bar{I} - I_{n+1})^2 \rangle} \right]^{1/2} . \tag{4.2}$$

This NPE can be calculated as a function of the embedding dimension, i.e., of the number of intervals used to identify neighbors and carry out the predictions. This NPE is sensitive to nonlinear dynamical structure in the data (Casdagli [1992], Farmer and Sidorowich [1987], Sauer [1994]). An NPE value close to zero signifies that there is predictability in the ISI sequence; a value near one signifies there is little predictability, and one can do just as well by forecasting the mean $\bar{I}$. Also,

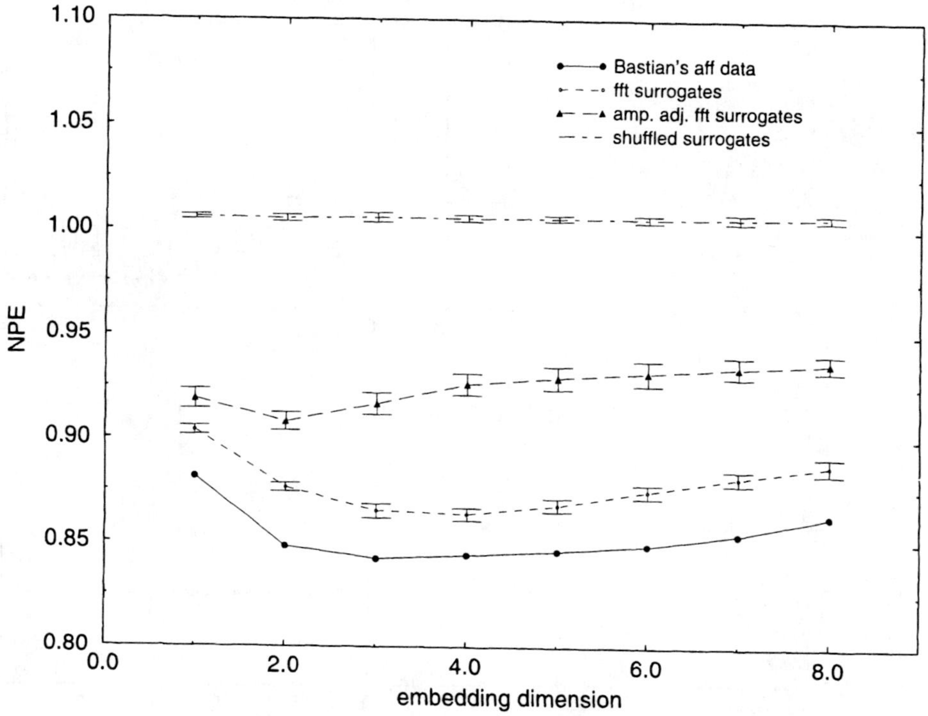

Figure 6 Normalized forecast error (NPE) versus embedding dimension for the 9165 successive intervals shown in Figure 1(a). The NPE is shown for the raw interval sequence, and for 10 surrogates of each of the following surrogate data types: shuffled, FFT phase-randomized, and amplitude-adjusted FFT phase-randomized. The error bars correspond to 2 standard deviations of the NPE's for the surrogates.

linear correlations, present e.g. in linearly correlated noise, can fool nonlinear forecasting algorithms by giving NPE values less than one (Tsonis and Elsner [1992]). It is important therefore to test the predictions against appropriate surrogate data sets (see Theiler et al. [1991] and references therein), as described below.

Figure 6 plots the NPE for the data in Figure 1 as a function of embedding dimension. We find that the NPE is around 0.85 for $m = 1$ to 8. This suggests that there is some structure in these data that leads to this small yet non-negligible forecastability. The linear correlations found in the previous section may be responsible for producing this value of the NPE. This is assessed in the next section.

5 Surrogate Data for Point Processes

The simplest form of surrogate data against which we have tested our forecasts is the shuffled surrogate, as was the case for the linear analyses. As expected, Figure 6 shows that shuffling produces a NPE just above one (the value converges to one as larger data sets are used). Clearly then, we can reject the null hypothesis that there are no correlations between the ISI's, as expected already from the linear analyses.

The next step is to verify whether linear correlations between ISI's are responsible for producing the forecastability picked out by our nonlinear predictor. We proceed by straightforward application of the FFT-phase-randomized surrogate

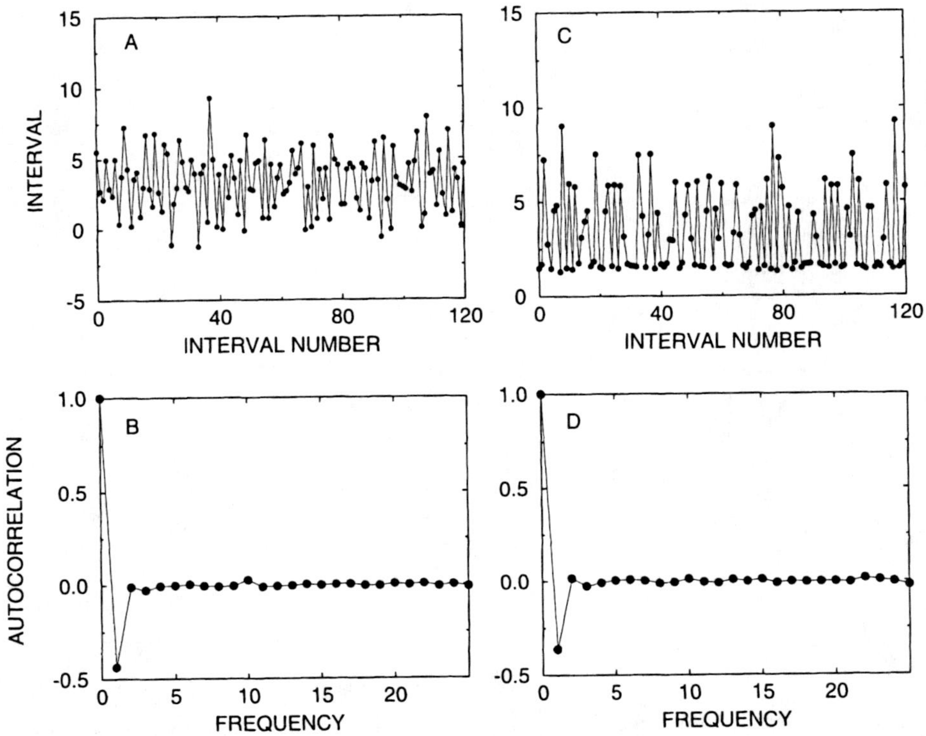

Figure 7 Sequences of interval values and their autocorrelation functions for surrogate data sets having the same linear correlations as the raw interval data. (a)-(b) FFT phase-randomized surrogates. (c)-(d) Amplitude-adjusted FFT phase-randomized surrogates. Note that the autocorrelation functions are similar to that for the raw intervals shown in Figure 4. The surrogate sequences have 9165 intervals. The autocorrelation was estimated using 50 lags.

method described in Theiler et al. [1991]. We then discuss the significance of these results in the context of point processes.

One realization of a surrogate sequence of ISI's is plotted in Figure 7(a) and Figure 7(c) for, respectively, the normal (Algorithm 1) and amplitude-adjusted (Algorithm 2) versions of the FFT-phase-randomized surrogates. The corresponding autocorrelation functions are plotted below them in Figure 7(b) and Figure 7(d). Both have autocorrelations similar to that of the raw ISI data shown in Figure 4. Figure 8(a) shows that the amplitude distribution of the normal surrogate has a Gaussian shape, while that for the amplitude-adjusted surrogate (Figure 8(c)) is practically identical to that for the raw data. The power spectra of intervals, shown in Figures 8(b) and 8(d), are similar to that of the raw intervals, plotted as the dotted line in each of those panels (note that this method of generating surrogates cannot reproduce exactly the same spectra, and there are even small discrepancies at low frequencies in Figure 8(d)). Thus, these surrogates adequately preserve the linear temporal properties of the ISI data.

The results for the normal and for the amplitude-adjusted surrogates are plotted with the results for the raw data in Figure 4. The NPE values for the amplitude-adjusted surrogates, which most closely mimick the raw data (see Figures 7-8), are higher than those for the normal surrogates, and are approximately 7% higher than

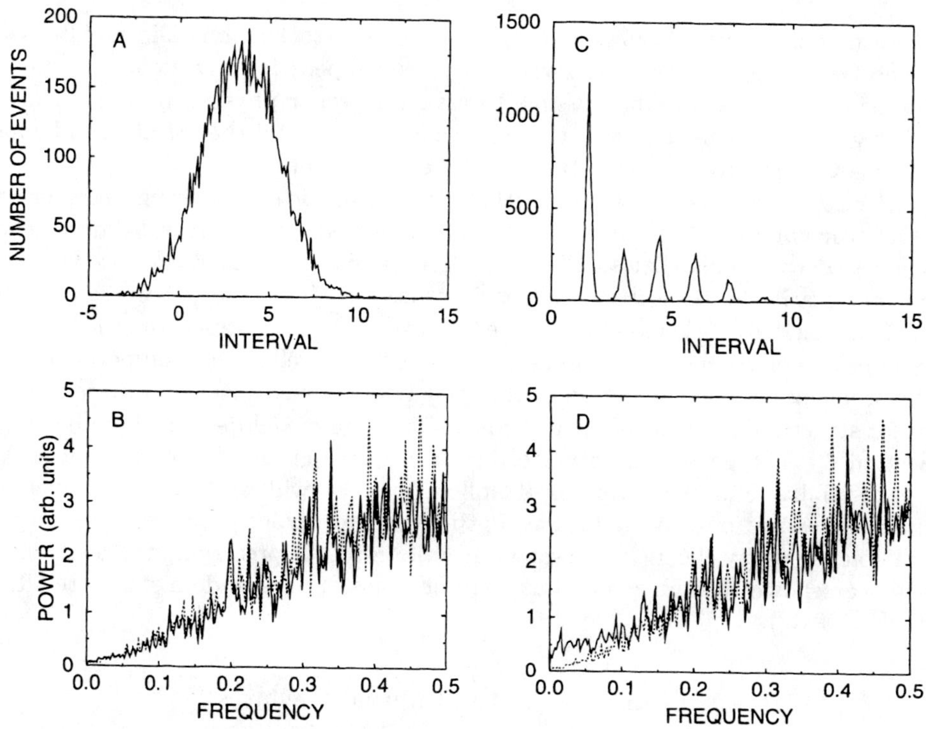

Figure 8 Amplitude distribution and power spectra of the surrogate interval sequences shown in Figure 7. (a) ISIH for the phase-randomized surrogate data. (b) Power spectrum of the first 8192 intervals, obtained by averaging 8 spectra of 1024 intervals. (c)-(d) Same as (a)-(b) except that the data are a realization of the amplitude-adjusted phase-randomized algorithm. The ISIH is practically identical to that in Figure 1(b). The dotted lines in (b) and (d) correspond to the spectrum of intervals of the raw data. While there are slight discrepancies between the raw and surrogate spectra, they both have the same upward trend. A Hanning window was used for the power spectra. The ISIH's contain 200 bins over the interval ranges shown in (a) and (c).

those for the raw data. It appears then that about half of the small forecastability in the raw data can be explained by the linear correlations in the interval sequences. Also, since the $\pm 2\sigma$-error bars on both FFT surrogates do not overlap the values for the raw data, we can reject the null hypothesis that the dynamics of ISI "generation" are linearly correlated noise (seen or not through some kind of "static nonlinearity" that operates on intervals – see below). Hence, we expect that there is forecastability due to nonlinear dynamical processes.

The interval distribution is Gaussianized as a result of the phase randomization used (in both algorithms) to generate the linearly correlated surrogates. This Gaussian distribution differs strongly from the original multimodal ISIH. Hence, it does not appear possible to reap the second benefit of surrogate data analysis, namely the rejection of a specific physically meaningful null hypothesis. While Figure 6 indicates clearly that all the nulls chosen can in principle be rejected, they do not, in our view, correspond to reasonable physical mechanisms. A problem of interpretation arises because we are dealing with variables that measure time between firing events in neurons. It is difficult to imagine that the ISI's are actually produced by linearly correlated Gaussian noise (with some values less than zero! – see Figure 8(a) seen, e.g., in the case of the amplitude-adjusted surrogates, through a static nonlinearity. At best, one could imagine a stochastic nonlinear filter that generates a random integer (number of cycles skipped) for any incoming value of Gaussian noise. It may be possible to assign a proper physical interpretation to other variables (such as phase) that can be extracted from the ISI data and which become Gaussianized through the use of these algorithms.

Unless we can imagine a process that alters or distorts these firing times, or acts as a static nonlinearity for firing phases, we are forced to reject specific mechanisms associated with nulls that significantly alter the distribution of events. This may not necessarily be the case for all point processes. Nevertheless, we can use the FFT surrogates to check whether the linear correlations present in the data are responsible for the measured forecastability. Alternatively, one could perhaps avoid the problems of Gaussianization by resorting to other surrogates that preserve some linear structure, such as the MA(1) moving average model described at the end of Section 3. The general question of how the distribution of intervals affects the forecastability and the rejection of null hypotheses will be further addressed in future investigations. We note that, in the context of point processes generated by radio-active decay events, phase-randomization without preserving the distribution has been shown to lead to spurious identifications of non-random structure (Rapp et al. [1994b]).

6 Analysis of the Interval Deviates

The sensory neurons of interest in this paper are driven nonlinear systems. Hence, it is expected that there may be a nonlinear map relating the phases, with respect to the external stimulus, at which firings occur. These dynamical relations may be obscured by stochastic processes. This possibility was discussed and analyzed in previous work (Longtin et al. [1993a]) by focussing on a variable similar to a phase. There, it was concluded that, since forecastability got better as the number of neighbors used for the prediction increased, these "phases" were probably linearly correlated. We now analyze the electroreceptor data for forecastability of this phase-like variable using the aforementioned forecasting algorithm rather than the more sophisticated local linear predictor used in Longtin et al. [1993a]. To distinguish this variable from the real phase at which the neuron fires with respect to the driving stimulus, we now give it the name "interval deviate" or "deviate" for short.

The extraction of these deviates from the ISI sequence consists in removing from the ISIH the multimodal peak structure. This is done by translating all the peaks to the left, and centering and superposing them at the origin. Thus this deviate reveals the fluctuations of intervals between the positive- and negative-sloped sides of the multimodal peaks. In other words, this variable measures the deviation of the ISI values from a multiple of the driving period (the multimodal peaks are almost perfectly aligned on the multiples of this period). Correlations of this variable underlie the elongation of the point clusters in the return map of Figure 1(d). However, this deviate does not carry the correlations between the absolute values of the ISI's which, as we know from Figure 6, tend to alternate between long and short skips.

The deviate is constructed specifically as follows. The mean period of the EOD stimulus is known to be $T = 1.51$ msec (it can be measured from zero crossings of the EOD, or estimated from the power spectrum of the spike train (not shown)). It fluctuates from cycle to cycle between 1.46 msec and 1.56 msec, with a significant

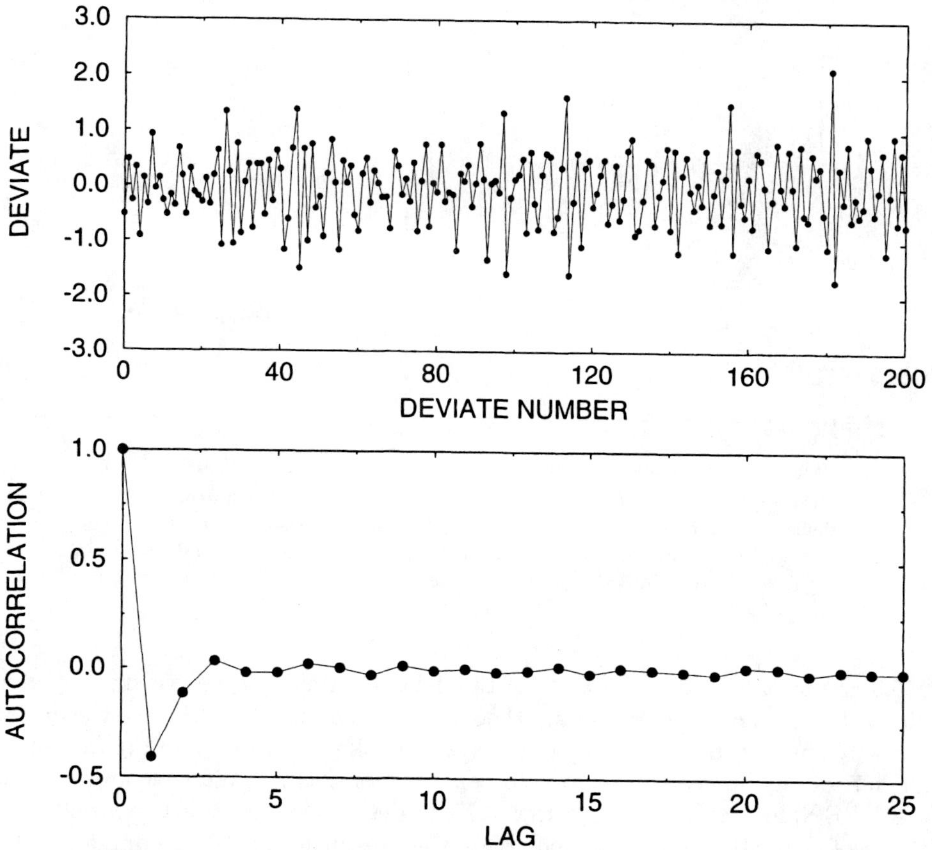

Figure 9 (a) Interval deviates extracted from the raw ISI data in Figure 1 (see text). 9165 phase values are plotted versus their position in the sequence of deviates (deviate number). (b) Autocorrelation function of the data in (a) using 50 lags.

alternance between a longer and a shorter value. The intervals in the range $(0, T_o/2)$ are left intact, those in the range $(T_o/2, T_o)$ are shifted to $(-T_o/2, 0)$, those between $(T_o, 3T_o/2)$ to $(0, T_o/2)$, and so on. The resulting sequence is divided by T_o such that all deviates belong to $(-0.5, 0.5)$. The resulting sequence has the same kind of time base as the original ISI sequence, i.e., time is counted in deviate number. A value of zero means that the interval I between two successive firings is a multiple of the stimulus period, i.e., $I = mT_o$, where m is an integer. A value belonging to $(-0.5, 0)$ implies that $(m - 0.5)T_o < I < mT_o$, and a value in $(0, 0.5)$ implies that $mT_o < I < (m + 0.5)T_o$.

The interval deviate values are plotted versus deviate number in Figure 9. A priori, there does not appear to be much forecastable structure in these data, nor in a return map of these data (not shown). There is however a noticeable elongation of the cluster of points in a direction of negative slope. This is expected, as it is present in the clusters of the return map in Figure 1(d). Thus, a positive deviate will be followed with higher probability by a null deviate or a negative deviate, and vice-versa. This is further substantiated by the autocorrelation function shown in Figure 9, which is similar to the autocorrelation of the ISI's themselves. Note that the autocorrelation in Figure 9 is nonzero up to the second lag, in distinction with the case of intervals where the decay to zero is complete after one lag.

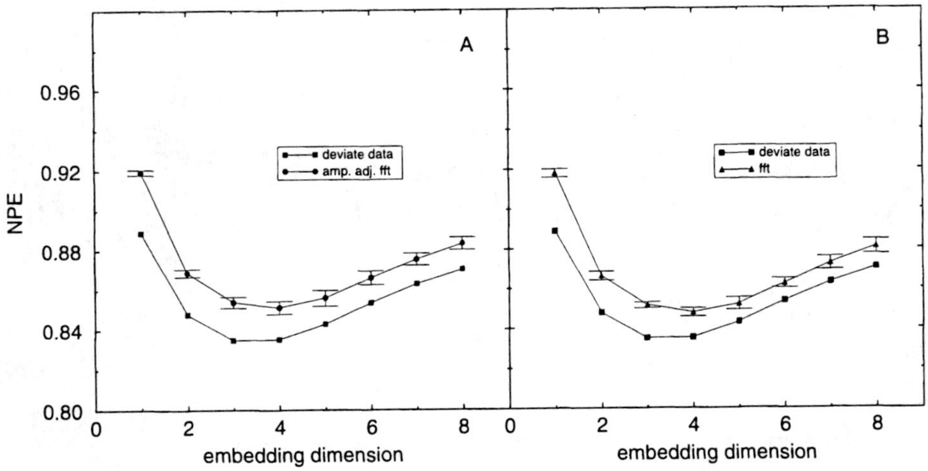

Figure 10 Normalized prediction error versus embedding dimension for the deviate data shown in Figure 8(a). (a) NPE for the phases and for amplitude-adjusted FFT surrogates of these data. (b) NPE for phase data and for normal FFT surrogates. Ten surrogates were used in each case. The error bars correspond to $\pm 2\sigma$.

The normalized forecast error of the deviate data is plotted versus embedding dimension in Figure 10. We obtain the striking result that the NPE for the deviates not only parallels but is almost equal to that for the raw ISI data in Figure 6, across all embedding dimensions used in our computations. One could have imagined that the NPE value of 0.85 for the raw ISI's is determined to a large extent by the shape of the ISIH, and in particular, by the prevalence of ISI's falling around T_o. Instead, the fact that the deviate variable has a similar NPE, but a very different distribution, suggests that the shape of the ISIH may not be as important as the ISI autocorrelation function, since this function is similar for both the raw ISI's and the deviates. Analysis using phase-randomized surrogates of these deviate data can again be used to see what effect the linear correlations have on the NPE. The curves for normal FFT-phase randomized surrogates and their amplitude-adjusted versions follow behaviors similar to that of the deviate data and, by extension, to that of the raw ISI data.

The significance with which the null hypothesis can be rejected is defined as the difference between the NPE for the raw data minus the average NPE for the surrogates, all divided by the standard deviation of the NPE for the surrogates (Theiler et al. [1991]). For both kinds of surrogates shown in Figure 10, the significance is always greater than six. Nevertheless, in terms of absolute NPE values, the surrogates seem to be capturing much of the predictability. Thus, as in Longtin [1993a] for other skipping data, we cannot conclude that the correlations detected are evidence for strongly nonlinear behavior. They may mostly result from the linear correlations in the deviates. Further work is needed to clarify this point.

The amplitude distribution of the deviate data appears almost Gaussian (not shown). This may explain why there is less difference between the NPE's for the amplitude-adjusted and normal surrogates, in comparison to the results for the ISI data. The distribution of deviates is not perfectly centered on the origin, because our procedure of deviate extraction is not accurate for long skipping events. The procedure is based on an estimation of the basic period of the EOD rhythm. Multiples of this period are used to determine left from right in all the peaks. The data is such that the alignment of the peaks with multiples of this period degrades with increasing multiples. This is likely a consequence of fluctuations in the EOD period. We expect that a small error in our assessment of the NPE arises from this misalignment. Preliminary results indicate that the NPE values and autocorrelation functions do not vary significantly for a 1-2% change in the mean of T_o.

It is not clear whether there is a more legitimate physical meaning to rejecting the null hypotheses on which surrogates for these deviate data are based. The null for the FFT-surrogate implies that the behavior of the deviates can be modelled as linearly correlated noise. Biophysical modeling studies do indeed indicate that noise is probably a key component of the firing dynamics (Longtin [1993b, 1995]). One can also think that the deviates are related by a linear map. This map could be a line with negative slope passing through the origin, with slope less than one in absolute value, i.e., with a stable fixed point at zero. If the deviate is perturbed away from this fixed point, these linear dynamics would reach equilibrium in an oscillatory fashion, alternating between positive and negative deviates, as seen in the data. This would be a reasonable null, given that the creation of the surrogate does not significantly alter the amplitude distribution (it is already nearly Gaussian). It is not clear however what amplitude-adjusted surrogates would mean. One would again have to have an idea of a biophysical mechanism that can act as a static nonlinearity for interval deviates. Again, as discussed in the previous section, this is not obvious for quantities relating to the timing of events.

Finally, we wish to point out that the real firing phases, measured as the time between a spike and the immediately preceding zero-crossing of the EOD stimulus, do not exhibit any correlations (not shown). Their autocorrelation is flat, and the NPE is around one. In other words, the firing probability as a function of stimulus phase is roughly constant from cycle to cycle. This supports the picture described in Section 3, where we argued that statistical fluctuations in the real firing phases induce a negative serial correlation in the deviates studied here. Refractoriness would then affect the overall probability to fire during a given cycle, without significantly affecting the phase-dependence of this probability.

7 Conclusion

We have presented analyses of correlations of time intervals between firing events in a sensory neuron of an electric fish. We have contrasted standard linear analyses with a recently developed nonlinear analysis, based on nonlinear forecasting of intervals derived from point processes. We have found significant correlations between intervals, despite their apparent randomness. It appears that much, but not all, of the correlations are attributable to linear correlations operating in the presence of noise.

This behavior appears to be characteristic of noisy phase-locked behavior in neurons (Longtin [1995]). There appear to be two mechanisms of correlation at work in the driven neuron we have studied. The first is an alternation between long skips and short skips. This anti-correlation is probably due to the decaying refractoriness to fire after a spike has occurred. The second is an alternation of ISI's between left and right sides of the multimodal peaks in the ISIH. This correlation can be ascribed to the phaselocking itself, whereby the firings have to keep in step with the stimulus; a firing at a more improbable phase of the stimulus will eventually result in a longer or shorter than normal ISI.

The predominant first peak in the ISIH could a priori endow the interval sequence with added predictability. In fact, its second mode is lower that the third mode, which is seldom the case in other skipping data sets and in models (Longtin [1993a,b, 1995] and Longtin and Racicot [1996]). However, we have found that the same level of forecastability, although low, is also present in the interval deviates which have a very different distribution. A similar NPE value was also found for the normal FFT-surrogates of the ISI data, which have a Gaussian distribution. It is perhaps a coincidence that the NPE's for both intervals and deviates are almost equal. It is more likely a consequence of the fact that these two data sets have similar autocorrelation functions. This is further evidenced by the fact that the NPE's for the phase-randomized surrogate data sets, which capture the linear properties of the data, are close to (yet distinct from) those for the interval (Section 4) and deviate (Section 6) data. This is true whether or not their distribution of amplitudes is the same as for the raw data.

Our analysis with surrogate data emphasizes the need for physically motivated surrogates, i.e., biophysical models. That analysis made it possible to verify what portion of the NPE was imputable to the linear properties of the correlations. However, it did not allow us to reject a physically significant mechanism associated with the null hypothesis on which the surrogate is based. Thus, one can turn to generic biophysical models for the skipping data (such as those studied in Longtin [1995]), and test whether such models behave similarly to the raw data with respect to nonlinear forecastability (this approach has been advocated in Longtin [1993a]). Promising results using this approach can be found in Longtin and Racicot [1996]. They support the hypothesis of noise-induced firing. Also, one could try to generate phase-randomized surrogates of the spike trains themselves. Such data have a degenerate distribution of amplitudes, since they are, in their simplest form, a succession of zeroes and ones. It would be interesting to study how to construct surrogate "spike trains" having e.g., the same power spectrum of spikes and/or intervals, yet be indistinguishable by eye from the raw spike train.

We remark that our NPE has a smooth behavior as a function of the embedding dimension; the same is true of the surrogates, with values of the NPE for the surrogates being always greater than those for the raw data. Also, we find only positive values of significance (not shown). This is in contrast to results in Longtin [1993a], where skipping data from another preparation sometimes yielded negative significance values. The difference may lie in the lack of stationarity of the data in Longtin [1993a]. It may also result from the use of a more complicated local linear forecasting algorithm, rather than the simpler algorithm used here; for skipping data, such an algorithm may generate systematic errors at certain m values. It is possible however, in the present study as in Longtin [1993a], that the dynamics are highly nonlinear, but too complicated (high-dimensional) to be easily extracted with nonlinear forecasting techniques. Future work should also determine how forecastability scales with prediction time. This could be done by plotting e.g., the coefficient of correlation between predicted and actual intervals versus prediction time. This would be an application of the time series results in Wales [1991], Tsonis and Elsner [1992] to point processes. This approach will be presented elsewhere. Also, it would be interesting to compare the behavior of the nonlinear predictor of intervals used here to that of a linear predictor, such as an ARMA-type model, as was done in Casdagli [1992], Farmer and Sidorowich [1987]. This may serve to better pinpoint the origin and nature of the correlations between firing times. Finally, it is worthwhile investigating the fluctuations and correlations in the EOD rhythm itself. Preliminary results indicate that the cycle-to-cycle EOD period fluctuations have significant negative correlation, which may be imposed on the ISI and deviate sequences studied here. This will be reported elsewhere.

Acknowledgements

This research was supported by NSERC (Canada) and by NIMH (USA) grant 1-R01-MH47184-01. The authors are grateful to Joe Bastian for providing the physiological data on which the analyses were performed. The authors would also like to thank Leonard Maler, Danny Kaplan, Martin Casdagli and James Theiler for useful comments on this work.

References

Abbott, L F [1994] *Decoding neuronal firing and modelling neural networks*, Quart Rev. Biophys **27**, 291-331.

Casdagli, M. C. [1992] *Chaos and deterministic versus stochastic nonlinear modeling*, J. Roy. Statist. Soc. **B54**, 303-328.

Cox, D. R. and Isham, V. [1980] *Point Processes*, Chapman and Hall, London.

Farmer, J. D. and Sidorowich, J. J. [1987] *Predicting chaotic time series*, Phys. Rev. Lett. **59**, 845-848.

Konishi, M. [1991] *Deciphering the brain's codes*, Neural Comput. **3**, 1-18.

Longtin, A. [1993a] *Nonlinear forecasting of spike trains from sensory neurons*, Int J. Bif Chaos **3**, 651-661.

Longtin, A. [1993] *Stochastic resonance in neuron models*, J Stat. Phys. **70**, 309-327.

Longtin, A. [1995] *Mechanisms of stochastic phase locking*, Chaos **5**, 209-215.

Longtin, A. and Racicot, D. [1996] *Spike train patterning and forecastability*, Biosystems (in press).

Moore, G. P., Perkel, D. H. and Segundo, J. P. [1966] *Statistical analysis and functional interpretation of neuronal spike data*, Ann. Rev. Physiol. **28**, 493-522.

Racicot, D. and Longtin, A. [1995] *Reconstructing dynamics from neural spike trains*, Proc. IEEE 17th Ann. Conf. Engin. Med. Biol. Soc., Montréal, Canada

Rapp, P. E., Zimmerman, I. D , Vining, E. P., Cohen, N., Albano, A. M. and Jiménez-Montaño, M. A. [1994a] *The algorithmic complexity of neural spike trains increases during focal seizures*, J. Neurosci. **14**, 4731-4739.

Rapp, P. E., Albano, A. M., Zimmerman, I. D. and Jiménez-Montaño, M. A. [1994b] *Phase-randomized surrogates can produce spurious identifications of non-random structure*, Phys Lett. A **192**, 27-33.

Sauer, T. [1994] *Reconstruction of dynamical systems from interspike intervals*, Phys. Rev. Lett. **72**, 3811-3814.

Sayers, B. McA. [1970] *Inferring significance from biological signals*, in Biomedical Engineering Systems (M. Clynes and J. Milsum, eds.), McGraw-Hill, New York, 84-164

Schiff, S. J., Jerger, K., Chang, T., Sauer, T. and Aitken, P. G. [1994] *Stochastic versus deterministic variability in simple neuronal circuits: II. Hippocampal slice*, Biophys. J. **67**, 684-691.

Sugihara, G. and May, R. M. [1990] *Nonlinear forecasting as a way of distinguishing chaos from measurement error in time series*, Nature **344**, 734-741.

Teich, M. C. [1992] *Fractal neuronal firing patterns;* In Single Neuron Computation (T. McKenna, J. Davis and S.F. Zornetzer, eds.), Academic Press, 589-625.

Theiler, J., Galdrikian, B., Longtin, A , Eubank, S. and Farmer, J D. [1991] *Using surrogate data to detect nonlinearity in time series*; in Nonlinear Prediction and Modeling (M. Casdagli and S. Eubank, eds.), Addison-Wesley, Redwood City, Ca, 163-188.

Tsonis, A. A. and Elsner, J. B. [1992] *Nonlinear prediction as a way of distinguishing chaos from random fractal sequences*, Nature **358**, 217-220.

Wales, D. J. [1991] *Calculating the rate of loss of information from chaotic time series by forecasting*, Nature **350**, 485-488.

Fields Institute Communications
Volume **11**, 1997

Markov Chain Methods in the Analysis of Heart Rate Variability

Stephen J. Merrill
Department of Mathematics, Statistics and Computer Science
Marquette University
Milwaukee, WI, USA 53201-1881

John R. Cochran
San Antonio Pediatric Pulmonary and Critical Care Associates
Methodist Plaza
4499 Medical Drive, Suite 255
San Antonio, TX, USA 78229

Abstract. Heart rate variability as seen through the sequence of RR intervals is analyzed via a Markov chain model in three pediatric data sets. The technique used for this analysis involves construction of a transition matrix and the utilization of information from this matrix and the limit distribution of the chain to address aspects of complexity of the RR sequence.

Due to the ergodic properties of stationary Markov chains, information dimension of the RR sequence is easily approximated. The nature of the eigenvalues of the transition matrix provides another measure of complexity of the signal. The stability of the nature and level of complexity is indicated by the position of the eigenvalues of the transition matrix with respect to the unit circle. Ability to predict imminent changes in the qualitative properties of the signal is one of the applications that suggests clinical usefulness. The technique does not require intensive computation, making possible the real-time use of the method in cardiac monitors.

1 Introduction

Heart rate variability (HRV) describes the variation in the heart rate due to the nature of the factors which influence and control the behavior of the heart. To quantify and describe HRV, a family of indices computed from a standard electrocardiogram (ECG) have been proposed and studied. The heart rate and how it changes in time reflect an interplay between several factors. These include respiration, hormones, the autonomic nervous system, and autonomic receptors in the periphery (baroreceptors reporting pressure from several important sites). A change in "normal" variability indicates a change in one of these systems or the heart's response to the controlling signals.

1991 *Mathematics Subject Classification.* Primary 60J20, 62P10; Secondary 92C50.
Partially supported by the Elaine Kohler Fund for Healing.

The importance of variability of the heart rate was first recognized in 1965 in the context of heart rate patterns in a fetus (Hon and Lee [1965]). At that time, Hon and Lee reported that a decrease in the variability of the fetal heart rate was correlated with decreased viability of the fetus. Since that time, many other investigators have used HRV as a window to the functioning of aspects of the central nervous system and defined characteristics of HRV in individuals having various pathologies or diseases. It should be noted again that HRV is measured by a family of indexes, each capturing some important aspect of the way in which the heart rate changes. A general introduction to the measurement of HRV can be found in Porges and Byrne [1992]. The literature is extensive and description of the indices and how they relate to one another can be found in medical and mathematical reviews of the subject (e.g., Spiers et al. [1993] and Kaplan and Talajic [1991]). Here we are motivated by applications in pediatric populations, and are primarily interested in developing techniques to track changes in HRV during and after surgery. This places constraints on the complexity of the techniques and the amount of data (time) required before useful information is obtained.

A common first step for most methods which measure some aspect of HRV is to create a sequence of times between adjacent identifiable features of the QRS complex of a single heart beat recorded by an ECG. The most prominent feature, the R wave, is also relatively unaffected by noise. As a result, many techniques utilize the times between R wave initiations, the RR intervals, to create indices describing HRV. These methods can be categorized as either *time domain methods* or as *frequency domain methods.* Generally, frequency domain methods involve a Fourier transform applied either to the original data or to resampled data after an interpolation (Berger [1986]). All other approaches are considered time domain methods. In this paper an approach not previously described, involving modeling the RR interval sequence as a Markov chain, is developed. This method has features of both time and frequency domain methods. Although a "time domain" method, frequency information usually obtained from the power spectrum is available (although in a different form). This method is applied to three pediatric data sets which serve to illustrate important aspects of the approach. One primary benefit of the Markov chain method is that it incorporates both a complexity measure of the RR interval data and a measure of "stability" of that complexity. The stability of the complexity seems not to have been identified in previous indices and relates directly to the clinical "stability" of the patient. A second benefit is that the approach requires little computational effort, and promises to be usable in real-time.

2 Clinical Motivation

There have been several studies indicating that HRV in pediatric patients could be used to investigate clinical conditions which affect the autonomic nervous system. These studies have related changes in HRV to illness severity in neonates (Pincus et al. [1991]), diabetic neuropathy (Akinci et al. [1993]), prediction of cardiac arrest in children post operative for cardiac surgery (Gordon et al. [1988]), and the prediction of sudden infant death (Gordon et al. [1986]). There have been many more studies in adult populations, especially describing decreased HRV as predictive of negative outcomes after heart attacks and congestive heart failure. However, much of this work may not apply directly to children. The primary reason is that HRV changes naturally with age (Kaplan et al. [1991]). In addition, frcm birth to two years, multiple prenatal and antenatal factors influence a nervous system under rapid development. Techniques designed and tested for the study of intact mature systems may not be well-suited for these patients. It is hoped to develop methods which could be designed for and tested on children, to monitor the development of the central nervous system (CNS) under normal and pathological conditions.

A related motivation is to develop measures of HRV which could be used to assess the effects of anesthesia. Current modern anesthesia is a carefully controlled balance to produce a state of amnesia, analgesia, and paralysis. To obtain this state, multiple medications are administered to the patient. Several problems faced by the anesthesiologist are:

- complex cardiac and respiratory hemodynamic reflexes which are both stimulated and blocked by anesthetics;
- uncertainty about anesthetics' effects on the CNS and a lack of information regarding CNS function during surgery;
- determining the depth of anesthesia to avoid premature arousal or prolonged anesthesia;
- varying amounts of surgical stress at different times during the procedure which require varying the amount of anesthesia, and
- individual sensitivity to different anesthetics due partially to age.

There are limits as to what is actually measurable in a timely manner during surgery. Gelb [1994] put it that anesthesiologists "monitor what we can measure, rather than measure what we should monitor." Using HRV to assess the depth of anesthesia was first investigated in the dog (McCrady et al. [1965]) and more recently, Rosenbaum et al. [1992] and Latson [1992] have proposed that aspects of HRV (especially the power spectrum) would be a potentially valuable aid to the anesthesiologist.

3 Mathematical Motivation

Samples of an RR interval sequence do not appear to be governed by a simple deterministic process. In fact, it has been proposed by Goldberger et al. [1990] that the heart rate of a healthy individual is a (deterministic) chaotic system. While that may be true, it is also the case that many deterministic and stochastic systems can be approximated by the use of Markov chains having relatively simple structure. This approximation utilizes "Markovian coarse-graining" (Nicolis and Nicolis [1990]) or cell-to-cell mappings (Hsu [1987]). The resulting chains have the advantage of capturing the important dynamic properties of the system while making available a rich theory for the computation and interpretation of important quantities. These quantities include limit distributions and the rate of approach to those distributions. It is through the ergodic nature of these chains (assuming that stationarity of the sequence holds for that finite length series) that *every sample path will result in an approximation of the same transition matrix.* Properties of the transitions matrix can then be used to describe the nature of all sample paths (one of which is the path that created it). It is in this sense that this method has promise for this application, even with fairly short original series.

4 Methods and Data

4.1 Mathematical Procedures and Analysis. The mathematical procedure to create an approximate transition matrix utilizes the following steps:

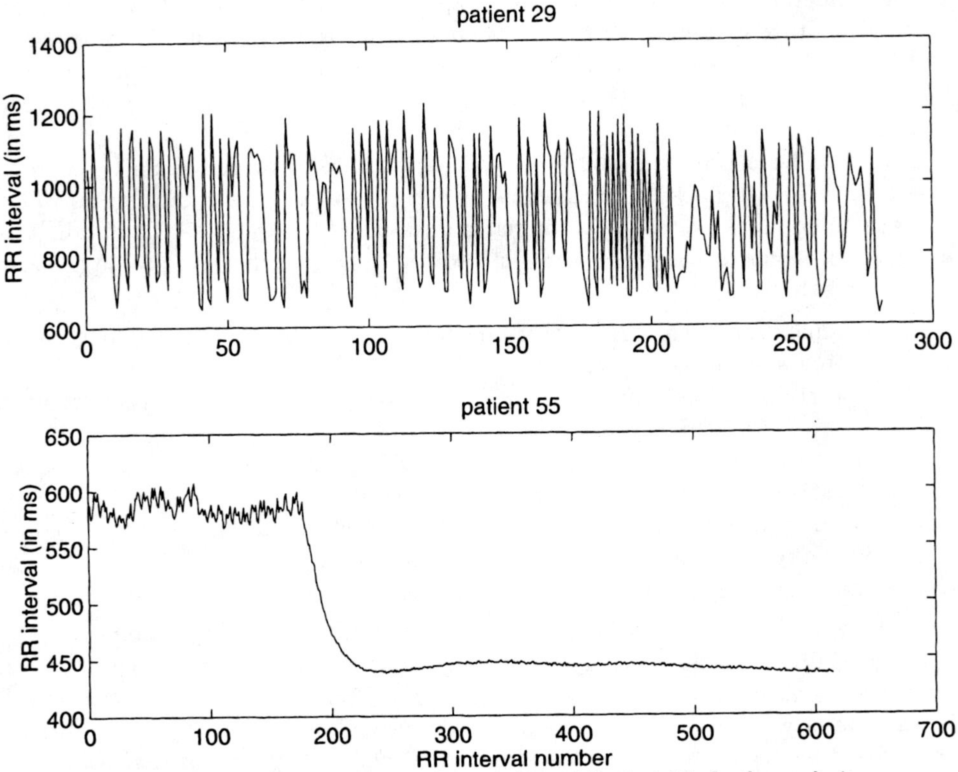

Figure 1 RR interval data from Patient 29 and Patient 55. In the analysis, the second set will be divided into an early and late phase.

1. Starting with the sequence $\{t_i\}_{i=1}^{n}$, divide the interval $[\min(\{t_i\}), \max(\{t_i\})]$ into m equal subintervals I_k, $k = 1, 2, ..., m$.
2. Create a new sequence $\{s_i\}_{i=1}^{n}$ by $s_i = k$ if t_i lies in interval I_k.
3. Create an m by m transition matrix P by estimating

$$p_{jk} = \Pr\{\ s_{i+1} = j | s_i = k\}\,.$$

Interval boundaries were chosen so that there were no data points lying on a boundary. Otherwise, boundary points must be assigned to one interval or the other.

All calculations and simulations employed MATLAB 4.2 (The MathWorks, Natick, MA).

4.2 The Data. Data from two patients, part of a prospective cohort study approved by the Human Resources Review Committee of the Children's Hospital of Wisconsin and the Medical College of Wisconsin, are presented here. The cohort consisted of healthy children having elective minor surgery at Children's Hospital. These patients showed no severe systemic disturbance or disease (as determined by the American Society of Anesthesia Classification). There was also no evidence of autonomic nervous system dysfunction. The two patient data sets are #29, a 7.5 year old boy, breathing spontaneously at the rate of about 10/minute and #55, a 2.5 year old girl whose breathing was controlled at 15/minute. The two data sets are displayed in Figure 1.

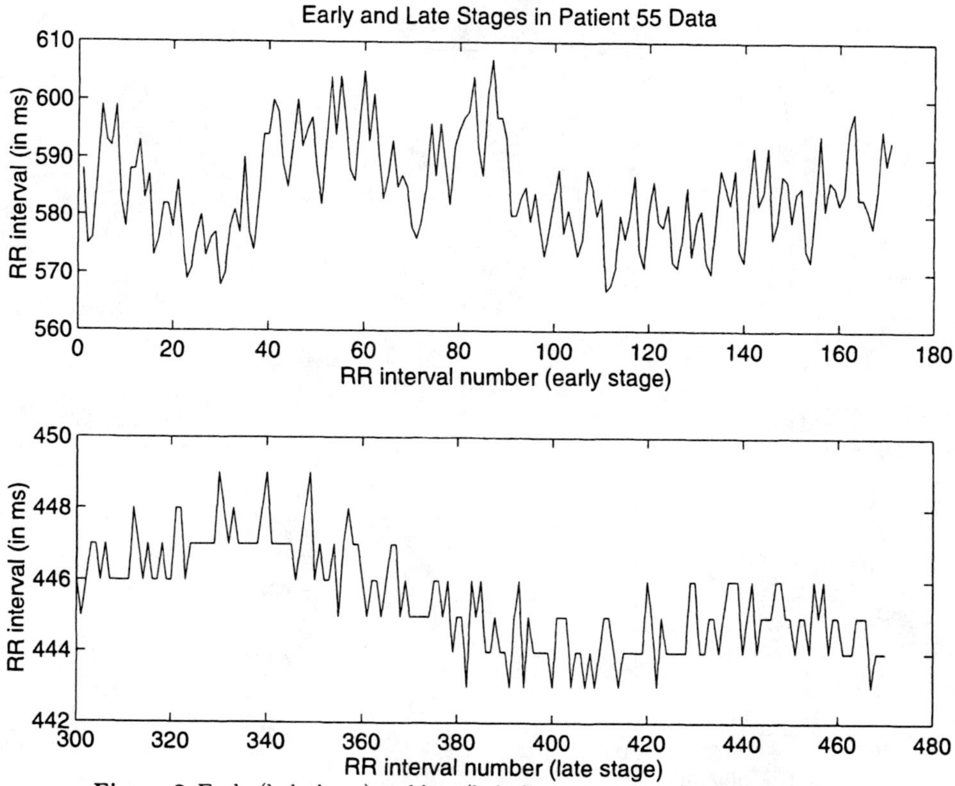

Figure 2 Early (halothane) and late (halothane + atropine) stages in Patient 55 data.

For the analysis, the second set, from Patient 55, is divided into early phase (under the effects of halothane only) and late phase (after atropine was administered in addition to halothane). The two data sets are shown separately in Figure 2.

ECG data was stored on a Vetter PCM FM recorder and later sampled at 1 KHz (DAP 2400 A to D board, Microstar corp., Bellview, WA). An R wave detection scheme utilizing software provided on the DAP board was used to create the RR interval sequence. The interval series was screened for artifacts using the method of Schectman et al. [1988].

5 Data Sets and Preliminary Analysis

For each of the three resulting data sets, lag 1 maps were constructed. These are displayed in Figure 3. The map from Patient 29 shows a complex "butterfly wing" structure (281 points are plotted). It illustrates that there is in fact a sequence of "long" and "short" interval lengths – a random number of "longs" followed by a random number of "shorts" and so on. Patient 55 early (170 points) is a fairly typical pattern, some of the movement up and down the "torpedo" probably due to the period of the controlled breathing. With the addition of atropine, Patient 55 late (170 points) shows a possible higher order periodic orbit. Again the controlled breathing frequency is probably responsible for the lag map's move up and down the "torpedo."

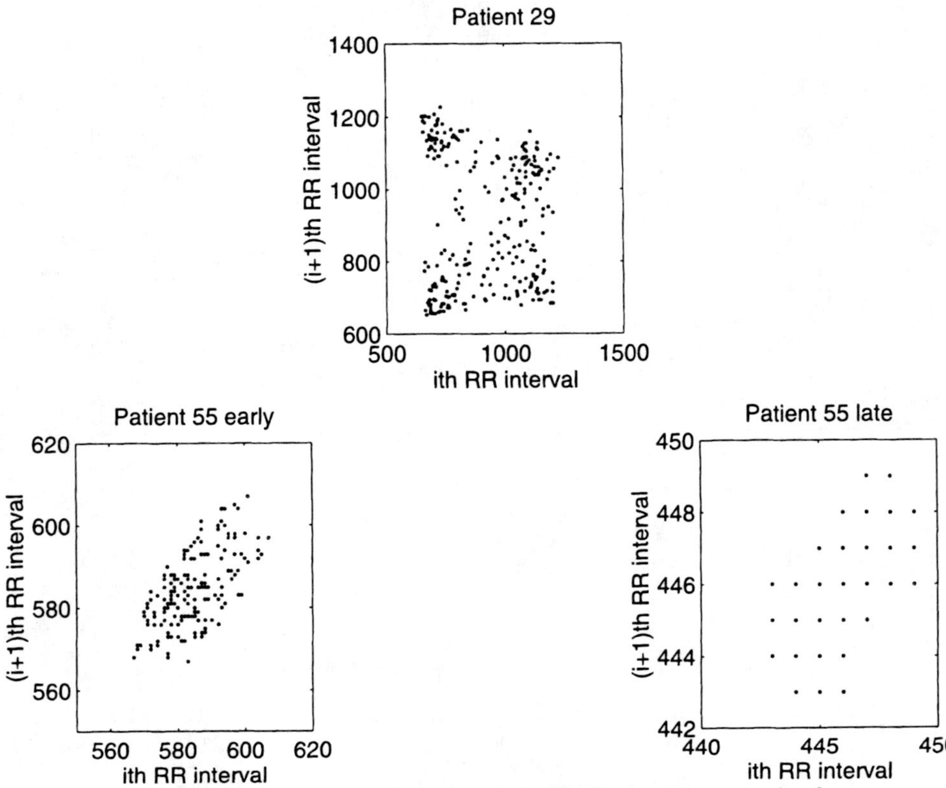

Figure 3 Lag 1 plots for the three data sets. The Patient 29 pattern has been described previously as a "complex pattern" (Woo et al. [1992]). Patient 55 early and late data plots would be classified as "torpedo patterns" by Woo et al. [1992].

6 Markov Chain Models

As outlined in the Methods section above, several transition matrices were created for each of the three data sets. The parameter m, the number of subintervals used, was varied, always lying between 5 and 30. The number of useful data pairs (no transition is recorded for the last interval) varied from 170 to 281 in the three sets. Due to the definition of the transition matrix, P, described above, the structure of each matrix is approximately that of the lag 1 plot for that data set rotated clockwise 90 degrees and reflected. As a result, it should not be surprising that the transition matrices contain important variability information. Here, we will be primarily interested in the structure of these matrices, and what they tell us of the underlying RR sequences.

6.1 Appropriateness of a Markov Model. Our approach will be to study the Markov chain whose transition matrix was created by an RR interval data set. Although there is no difficulty in creating such a transition matrix, that does not ensure that the Markov assumption is satisfied by the process that determines RR intervals. The Markov assumption is related directly to the definition of the transition probabilities, and specifically requires that

$$\Pr\{X_{n+1} = j \mid X_n, X_{n-1}, \dots, X_0\} = \Pr\{X_{n+1} = j \mid X_n\}. \tag{6.1}$$

In this particular application, the Markov assumption requires that there be no memory of previous RR interval lengths (beyond the current one) that is used to determine the next interval length. However, factors affecting rate, such as respiration, occur on different time scales, and previous interval lengths may contain, for example, information as to the time since the last respiration. Due to recognition of this difficulty, we view our approach primarily as a tool for data analysis.

A second problem is one which is common to most analytic methods, that of an assumption of stationarity. It is clear that, taken as a whole, the Patient 55 data set does not represent a stationary process. This is due, in that patient, to the administration of the drug atropine. To ensure at least a chance of stationarity, the Patient 55 data set needed to be divided into an early and late phase. To test stationarity on each of the two resulting data sets, transition matrices (with small m) were constructed from subsets of the data sets. The approximate transition matrices computed from early, middle, and late periods had properties that were the "same" as those from the entire set. The most straightforward test of "sameness" is based on a goodness-of-fit test utilizing the sum of chi-squared statistics computed from the rows of the matrices. (The implementation of this goodness-of-fit test of stationarity will be presented elsewhere.) Although this test statistic is easily computed and evaluated, other tests, based on essential properties of the matrices, are needed. Property-based tests of stationarity would allow one to determine if the nature of the RR interval signal has changed over time. This has clinical meaning in several contexts including post surgical monitoring.

It will be assumed here that the RR interval data sets considered are stationary in the sense that properties under consideration are not changing over time. Related to the above is the fact that the relative frequency histogram of interval lengths always approximates a stationary vector for the empirical transition matrix so constructed. If the interval sequence has reached the ergodic states (transient states are not present in the sequence) then this is expected.

6.2 Results. Transition matrices with $m = 10$ were constructed for the three data sets with the results shown graphically in Figure 4. The lighter gray corresponds to relatively larger probabilities. This format was selected to suggest that if m were chosen to be larger, a similar image would be the result (with appropriate scaling).

Interpretation of the characteristics of the chains whose transition matrices are now in hand requires several theoretical results on finite Markov chains. The theory of Markov chains can be found in any one of a number of monographs including Chung [1967] and Karlin and Taylor [1975]. For this application, we will be primarily interested in the "Fundamental Theorem of Markov Chains" and descriptions of the ergodic properties of chains. The Fundamental Theorem describes the nature of the eigenvalues of the transition matrix, gives conditions guaranteeing the existence of a limit distribution, and describes the rate at which the n step transition probabilities approach the limiting transition matrix.

Theorem 6.1 (Fundamental Theorem) *Let P be a transition matrix of a finite state Markov chain, with m the number of states. Then*

1. $\lambda_1 = 1$ *is always an eigenvalue of* P.
2. $|\lambda_i| \leq 1$ *for all eigenvalues* λ_i *of* P.
3. *if* $|\lambda_i| < 1$ *for* $i = 2$ *to* m, *then*

$$P^{\infty} = \lim_{n \to \infty} P^n \tag{6.2}$$

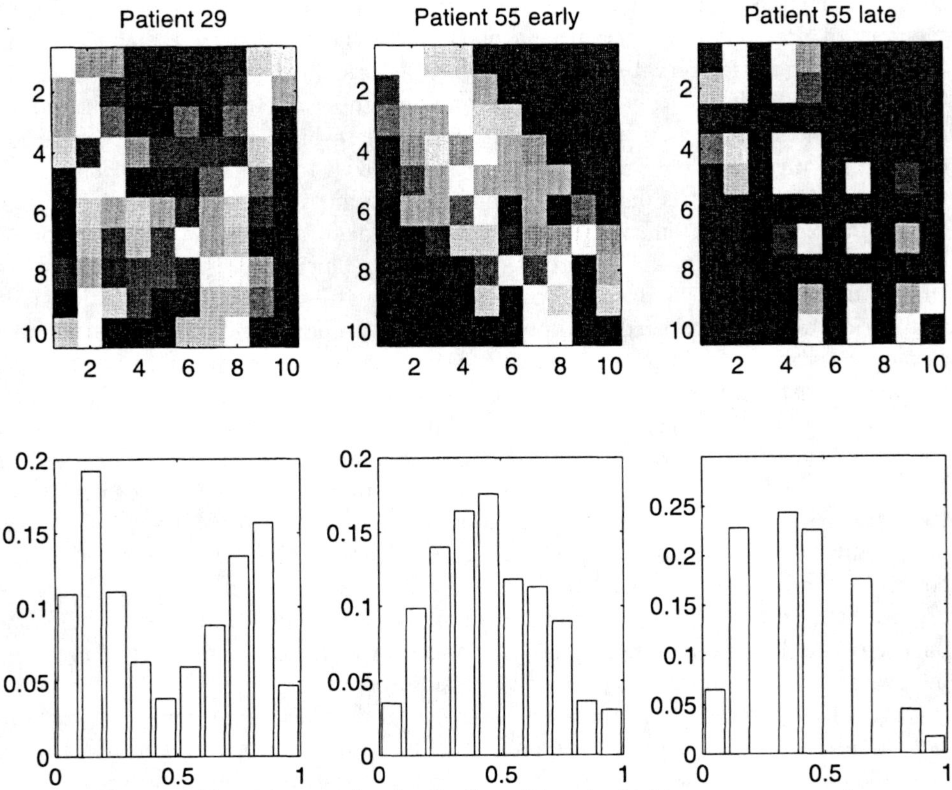

Figure 4 Transition matrices for the three data sets. Lighter gray corresponds to higher probabilities. Below the image of each matrix is the limiting vector for the matrix as described in the Fundamental Theorem of Markov chains.

exists. Each row of P^∞ *is an identical copy of the limit distribution of the Markov chain.*

4. *if* P^∞ *exists,*

$$|P^n - P^\infty| \leq c\hat{\lambda}^n \tag{6.3}$$

where c is a constant, and $\hat{\lambda}$ *is the* $\max_{2 \leq i \leq m}\{|\lambda_i|\}$

We are interested in the following properties of the Markov chains determined by these transition matrices: the number of complex conjugate pairs in the eigenvalues of the transition matrix, the largest "non-1" eigenvalue in modulus, and the nature of the limiting distribution (to compute an approximate information dimension). In Figure 5, eigenvalues for each of the transition matrices are displayed.

The number of complex conjugate pairs of eigenvalues is related to the complexity of the signal. This can be seen either through an approximate power spectrum computed from these eigenvalues or through the Markov chain theory. These eigenvalues indicate an oscillatory approach to the limit distribution. From a deterministic dynamics sense, the complex eigenvalues indicate the existence of an unstable periodic orbit. If one starts near an unstable periodic orbit, the orbit will shadow this unstable orbit until attracted to the stable behavior (distribution of visits to states) indicated by the limit distribution. The approximate power spectrum is constructed by examining the angles θ_i corresponding to eigenvalues in the upper half plane. For each, a frequency is computed

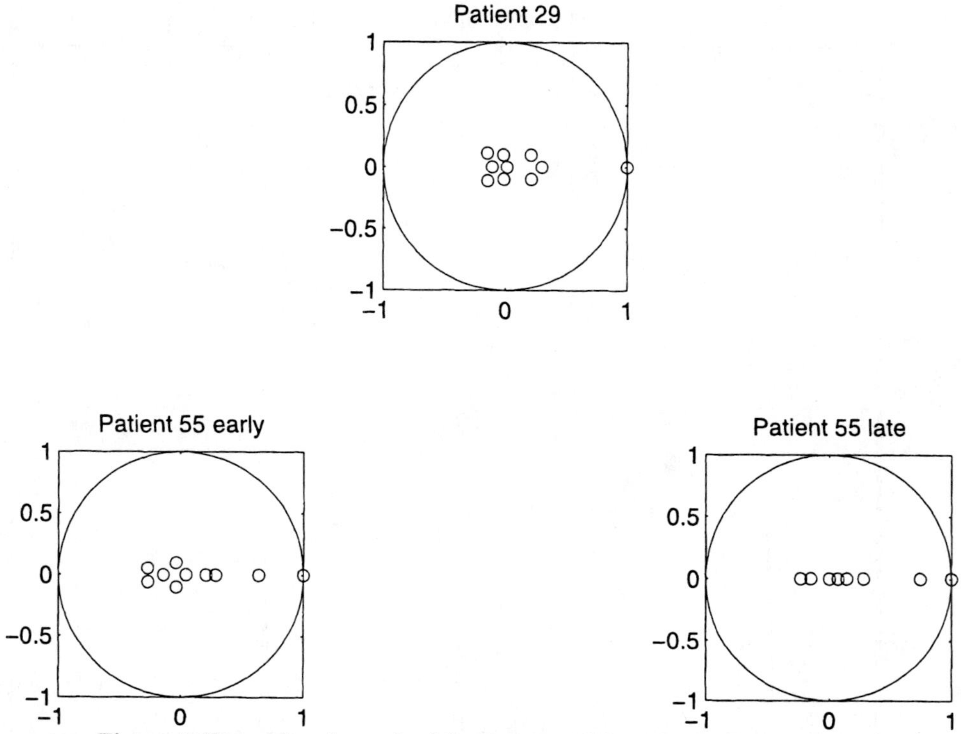

Figure 5 Eigenvalues for each of the three transition matrices are shown in relation to the unit circle. Besides the nearness of the "non-1" eigenvalues to the unit circle, notice the differences in the number of complex eigenvalues.

$$f_i = 2\pi/\theta_i. \tag{6.4}$$

For Patient 29, this approximate power spectrum is plotted in Figure 6. The height that is plotted is twice the modulus of the eigenvalue (due to the complex conjugate). Real eigenvalues correspond to an angle of zero or π and the sum of their lengths is plotted. Either approach indicates that a more complex signal corresponds to a larger number of complex conjugate eigenvalues. For $m = 10$, with a possible 8 complex eigenvalues, in Patient 29 data, we have 6 being complex. For this data set, with $m = 30$, 26 are complex. In Patient 55 early, for $m = 10$, 4 are complex. For this data set with $m = 30$, 20 are complex. For Patient 55 late, for all possible m values, there are no complex eigenvalues. It seems reasonable to take the number of complex eigenvalues at a given degree of resolution, m, to be a measure of complexity in sample paths.

The rate at which a sample path will approach the frequencies of the limit distribution is indicated by the modulus largest "non-1" eigenvalue. For $m = 10$, in Patient 29 it is .32, in Patient 55 early it is .59, and in Patient 55 late it is .75. The nearer to 1, the slower the approach to the stable patterns of RR intervals indicated by the limit distribution. In a dynamics sense, near a bifurcation point, eigenvalues of a corresponding Markov chain approach the unit circle. The nearness of an impending bifurcation (and resulting change in the limit distribution) is directly indicated by this modulus. It is in this way that we are measuring the stability of this limiting behavior.

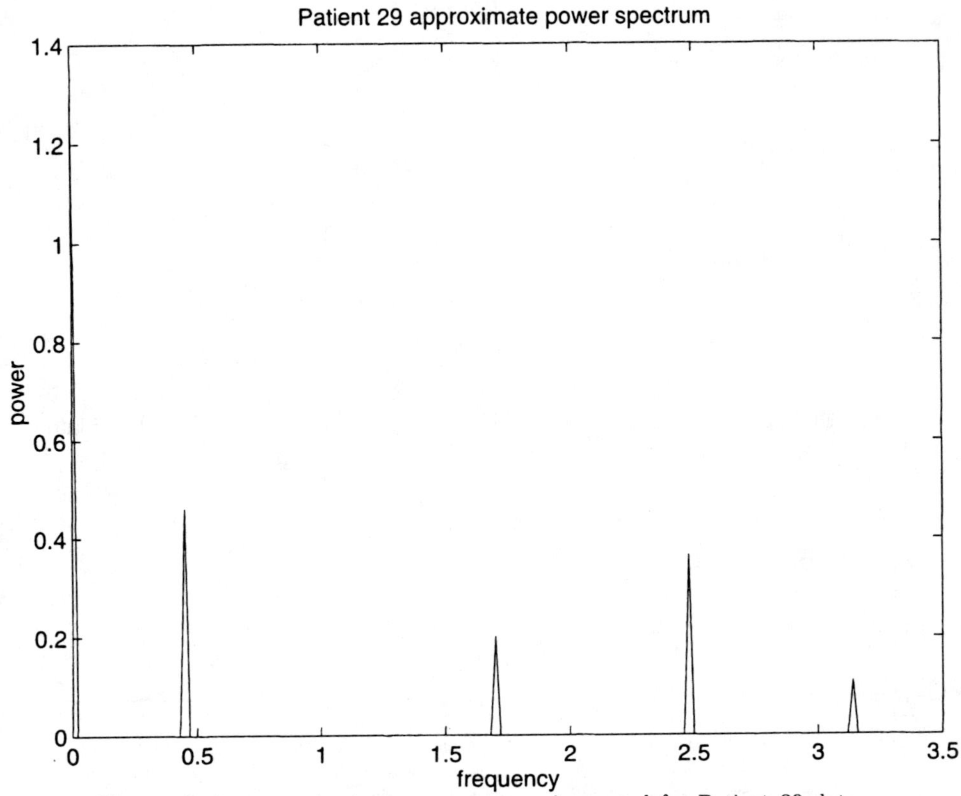

Figure 6 An approximate power spectrum is created for Patient 29 data. Each complex eigenvalue gives rise to a frequency peak. More subintervals expose more of the detail seen in the lower power peaks.

The complexity of the dynamics indicated by the limit distribution and the Markov chain can be measured by approximation of the information dimension of the limit distribution. This is possible due to the ergodic properties of a subset of the states of the chain as determined by the limit distribution. States corresponding to positive probabilities in the limit distribution are said to be *positive* or *ergodic*, and these ergodic states form their own Markov chain, an *ergodic chain.* The properties of this chain are summarized in the second theorem which describes the relationship between sample paths and the limit distribution in an ergodic chain.

Theorem 6.2 (Ergodic Properties) *In an ergodic Markov chain with limiting distribution* $\boldsymbol{p}$*, let* $N_{ij}(n)$ *be the number of times a given sample path is in state* j *up to and including transition* n*, given that the initial state was* i*. Then with probability one,*

$$\lim_{n\to\infty} \frac{N_{ij}}{n} = p_j \tag{6.5}$$

This theorem says that (essentially) every path spends p_j of its time in state j. As a result, histograms of relative frequencies of visits to states from different sample paths are approximately the same.

Due to the ergodic theorem, we can directly compute the information (Shannon and Weaver [1949]) $I(\epsilon)$ for $\epsilon =$ the length of the subintervals with the number of intervals m.

$$I(\epsilon) = \sum_{i=1}^{N(\epsilon)} p_i \ln(\frac{1}{p_i}). \tag{6.6}$$

The p_i in this formula is fraction of time that a path is in the ith state. Here, p_i is the ith component of the limit distribution. The actual information dimension d_I is computed by

$$d_I = \lim_{\epsilon \to 0} \frac{I(\epsilon)}{\ln(\frac{1}{\epsilon})}. \tag{6.7}$$

As infinite resolution is not available with data, we will compute $I(\epsilon)$ only. At $m = 30$, we have .951, .92, and .72 respectively for the three sets. This number can be interpreted by comparing with Markov chains constructed from known systems such as the discrete logistic equation at a similar degree of refinement.

Through these three pediatric data sets, we have shown that the Markov chain created through the empirical transition matrix contains dynamic information concerning the complexity of the signal and its stability. These quantities require little computation and can be estimated after only a few minutes of RR interval data. Continued evaluation and development of this technique will require careful attention to defining conditions under which the Markov model is appropriate and the development of several new statistical tools.

New statistical tools are needed to determine the optimal value of the number of intervals used, m, given a fixed amount of data, and determination of the amount and accuracy of data needed to detect the presence or absence of certain dynamic properties of the underlying sequence. Related to this is the need for tests of stationarity which are based on important (dynamic) properties of a transition matrix (such as the nature of its eigenvalues).

References

Akinici, A., Celiker, A., Baykal, E. and Težic, T. [1993] *Heart rate variability in diabetic children: sensitivity of the time- and frequency-domain methods*, Pediatr. Cardiol. **14**, 140–146.

Berger, R. D., Axelrod, S., Gordon, D. and Cohen, R. J. [1986] *An efficient algorithm for spectral analysis of heart rate variability*, IEEE Trans. Biomed. Eng. **33**, 900–904.

Chung, K. L. [1967] *Markov Chains with Stationary Transition Probabilities, 2nd edition*, Springer-Verlag, New York.

Gelb, A. W. [1994] Discussion of *Control of anesthesia during neurosurgery*, lead discussant I.P. Osborn, Neurological Times: Current Issues in Neuroanesthesiology, Advanstar Comm. Inc., Sept., p. 9.

Goldberger, D. R., Rigney, D. R. and West, B. J. [1990] *Chaos and fractals in human physiology*, Sci. Amer. **262**, 42–49.

Gordon, D., Herrera, V. L., McAlpine, L., Cohen, R. J., Akselrod, S., Lang, P. and Norwood, W. I. [1988] *Heart-rate spectral analysis: a noninvasive probe of cardiovascular regulation in critically ill children with heart disease*, Pediatr. Cardiol. **9**, 69–77.

Gordon, D., Southall, D. P., Kelly, D. H., Wilson, A., Akselrod, S., Richards, J., Kenet, B., Cohen, R. J. and Shannon, D. [1986] *Analysis of heart rate and respiratory patterns in sudden infant death syndrome victims and control infants*, Pediatr. Res. **20**, 680–684.

Hon, E. H. and Lee, S. T. [1965] *Electronic evaluation of the foetal heart rate patterns preceding foetal death, further observations*, Am. J. Obstet. Gynecol. **87**, 814–826.

Hsu, C. S. [1987] *Cell-to-Cell Mapping*, Springer-Verlag, New York.

Kaplan, D. T., Furman, M. I., Pincus, S. M., Ryan, S , Lipsitz, L. and Goldberger, A. L. [1991] *Aging and the complexity of cardiovascular dynamics*, Biophys. **59**, 945–949.

Kaplan, D. T. and Talajic, M. [1991] *Dynamics of heart rate*, Chaos **1**, 251–256.

Karlin, S. and Taylor, H. W. [1975] *A First Course in Stochastic Processes*, Academic Press, New York.

Kleiger, R. E., Stein, P. K., Bosner, M. S. and Rottman, J. N. [1992] *Time domain measurements of heart rate variability*, Cardiology Clinics **10**, 487–498.

Latson, T. W. [1992] *Editorial: Heart rate variability and anesthesiology: reasons for cautious optimism*, J. Cardiothorac. Vasc. Anesth. **6**, 647–650.

McCrady, J. D., Vallbona, C. V. and Hoff, H. E. [1965] *The effect of preanesthetic and anesthetic agents on the respiration-heart rate response of dogs*, Am. J. Vet. Res. **26**, 710–715.

Nicolis, G. and Nicolis, C. [1990] *Chaotic dynamics, Markovian coarse-graining and information*, Physica A **163**, 215–231.

Pincus, S. M., Gladstone, I. M. and Ehrenkranz, R. A. [1991] *A regularity statistic for medical data analysis*, J. Clin. Monit. **7**, 335–345.

Porges, S. W. and Byrne, E. A. [1992] *Research methods for measurement of heart rate and respiration*, Biological Psychology **34**, 93–130.

Rosenbaum, S. H. and Fleisher, L. A. [1992] *Editorial: Was Einthoven a 21st century visionary?*, J. Clin. Anesth. **4**, 263–264.

Schectman, V. L., Kluge, K. A. and Harper, R. M. [1988] *Time-domain system for assessing variation in heart rate*, Med. & Biol. Eng. & Comput. **26**, 367–373.

Shannon, C. E. and Weaver, W. [1949] *The Mathematical Theory of Information*, University of Illinois Press, Urbana, IL.

Spiers, J. P., Silke, B., McDermott, U., Shanks, R. G., and Harron, D. W. G. [1993] *Time and frequency domain assessment of heart rate variability: a theoretical and clinical approach*, Clinical Autonomic Research **3**, 145–189.

Woo, M. A., Stevenson, W. G., Moser, D. K., Trelease R. B. and Harper, M. H. [1992] *Patterns of beat-to-beat heart rate variability in advanced heart failure*, Am. Heart J. **123**, 704–710.